Climate Change, Coasts and Coastal Risk

Climate Change, Coasts and Coastal Risk

Special Issue Editors

Roshanka Ranasinghe
Ruben Jongejan

MDPI • Basel • Beijing • Wuhan • Barcelona • Belgrade

MDPI

Special Issue Editors

Roshanka Ranasinghe
IHE Delft Institute for Water Education/University of Twente
The Netherlands

Ruben Jongejan
Jongejan Risk Management Consulting/Delft University of Technology
The Netherlands

Editorial Office
MDPI
St. Alban-Anlage 66
4052 Basel, Switzerland

This is a reprint of articles from the Special Issue published online in the open access journal *Journal of Marine Science and Engineering* (ISSN 2077-1312) in 2018 (available at: https://www.mdpi.com/journal/jmse/special_issues/clim_chang_coast_risk)

For citation purposes, cite each article independently as indicated on the article page online and as indicated below:

LastName, A.A.; LastName, B.B.; LastName, C.C. Article Title. *Journal Name* **Year**, *Article Number*, Page Range.

ISBN 978-3-03897-481-9 (Pbk)
ISBN 978-3-03897-482-6 (PDF)

Cover image courtesy of pixabay.com.

Contents

About the Special Issue Editors

Roshanka Ranasinghe (RR) holds the AXA Chair in Climate Change Impacts and Coastal Risk at IHE Delft and at the Department of Water Engineering and Management, University of Twente. He is also the Head of the Department of Water Science and Engineering at IHE Delft and holds a Visiting Professor appointment at the University of Queensland, Australia. RR obtained his Bachelor of Science Engineering (1st Class Honours) from the University of Peradeniya, Sri Lanka in 1993 and his PhD in Coastal Engineering and Oceanography from the Centre for Water Research, University of Western Australia in 1998. Since then he has worked in the academic and government sectors in Australia, the USA, and The Netherlands. RR has undertaken cutting-edge research on various nearshore coastal processes, spanning the event scale (individual storms) to climate change impacts (100-year time scales). His current focus is on the development and application of novel reduced-complexity physical-based models to assess the various climate change impacts on coasts (erosion and inundation), within probabilistic frameworks that will enable quantitative risk assessments, thereby enabling risk-informed decision-making. To date, RR has published over 100 peer-reviewed articles (including full length articles in high impact journals such as *Nature Climate Change, Earth Science Reviews, Scientific Reports (Nature)*, and *Climatic Change*) and delivered over 25 invited presentations around the world. He is regularly invited to provide expert advice on adaptation to climate change and coastal zone management by national and local governments and international agencies. RR has supervised/co-supervised 10 postdoctoral fellows, 20+ PhD candidates, and numerous MSc students to date and has lead several advisory projects in Sri Lanka, India, VietNam, Thailand, and Sao Tome and Principe funded by national governments, the Asian Development Bank, and the World Bank. RR was recently appointed the co-ordinating lead author of the IPCC AR6 Working Group 1 Chapter on *Climate change information for regional impact and for risk assessment (Chapter 12)* and as a national expert (from The Netherlands) to the *United Nations World Ocean Assessment (second edition)*.

Ruben Jongejan (RJ) is the statutory director and principal consultant of Jongejan Risk Management Consulting BV. RJ studied Civil Engineering at Delft University of Technology and Political Science at Leiden University in the Netherlands. He obtained his PhD from Delft University of Technology in 2008. During his PhD studies, he served on committees of The Netherlands Hazardous Materials Council, worked as a visiting academic at the London School of Economics, and performed advisory work. RJ then started as an independent consultant. He is still affiliated with the Department of Hydraulic Engineering of Delft University and publishes regularly in scientific journals. RJ focuses on probabilistic design, quantitative risk analysis, and risk evaluation in the field of flood risk management. He is actively involved in the development and application of new methods for levee safety risk analysis and the design of flood defenses, in The Netherlands and abroad. Over the years, RJ has worked with numerous governments, research institutes, regional water authorities, and engineering firms. Amongst others, he is an advisor to The Netherlands Commission for Environmental Assessment (NCEA) and a member of the Safety and Flood Risk Approach Workgroup of the Dutch Expertise Network for Flood Protection (ENW).

Preface to "Climate Change, Coasts and Coastal Risk"

The coastal zone is the most heavily populated and developed land zone in the world, with rapid expansions in settlements, urbanization, infrastructure, economic activities, and tourism. Sea level rise and projected climate change-driven variations in wave conditions, storm surge, and river flow will affect the coastal zone in many ways. This Special Issue contains 13 scientific papers related to the development of much-needed models for facilitating a risk-informed approach to coastal management. Such an approach will be essential for addressing the challenges posed by climate change and population pressure. The editors wish to thank the contributors and the support of the *JMSE* editorial staff, whose professionalism and dedication have made this issue possible.

Roshanka Ranasinghe, Ruben Jongejan
Special Issue Editors

Journal of
Marine Science and Engineering

MDPI

Editorial

Climate Change, Coasts and Coastal Risk

Roshanka Ranasinghe [1,2,3,]* and **Ruben Jongejan [4,5,]**

1   Department of Water Science and Engineering, UNESCO-IHE, PO Box 3015, 2601 DA Delft, The Netherlands
2   Water Engineering and Management, University of Twente, PO Box 217, 7500 AE Enschede, The Netherlands
3   Harbour, Coastal and Offshore Engineering, Deltares, PO Box 177, 2600 MH Delft, The Netherlands
4   Department of Hydraulic Engineering, Faculty of Civil Engineering and Geosciences, Delft University of Technology, PO Box 5048, 2600 GA, Delft, The Netherlands
5   Jongejan Risk Management Consulting B.V., Schoolstraat 4, 2611 HS Delft, The Netherlands
*   Correspondence: r.ranasinghe@un-ihe.org (R.R.); ruben.jongejan@jongejanrmc.com (R.J.)

Received: 8 November 2018; Accepted: 15 November 2018; Published: 19 November 2018

1. Introduction

Projected climate change driven variations in mean sea level (i.e., sea level rise), wave conditions, storm surge, and river flow will affect the coastal zone in many ways. The coastal zone is the most heavily populated and developed land zone in the world with rapid expansions in settlements, urbanization, infrastructure, economic activities and tourism.

The combination of coastal climate change impacts and their effects on the ever-increasing human utilization of the coastal zone will invariably result in increasing coastal risk in the coming decades. But, while the economic damage (potential consequence) that can be caused by Climate change driven coastal inundation and erosion (potential hazard) is likely to increase, foregoing land-use opportunities in coastal regions is also costly (opportunity cost). Thus, a 'zero risk'-policy could have severe economic consequences, while high risk policies could lead to risks that are unacceptable to society and individuals.

To avoid unacceptable future risks, due to coastal hazards and/or sub-optimal land use, it is imperative that risk informed and sustainable coastal planning/management strategies are implemented sooner rather than later. This requires comprehensive coastal risk assessments which combine state-of-the-art consequence (or damage) modelling and coastal hazard modelling. Apart from being of crucial importance to coastal managers/planners, this type of risk quantification will also be invaluable to the insurance and re-insurance industries for insurance pricing, which may have a follow-on effect on coastal property values. However, the present level of knowledge on generally applicable coastal hazard and risk assessment approaches, especially at local scale (~10 km) is rather limited. This Special Issue contains 13 papers that aim to address this knowledge gap.

2. Thematic Contributions

The 13 contributions have been organized around the following themes:

1.   Climate change impacts on forcing and coastline change,
2.   Socio-economic and environmental impacts of coastline change,
3.   Probabilistic methods for analysing system reliability, and
4.   Risk-informed decision making and probability-based economic optimization.

2.1. Climate Change Impacts on Forcing and Coastline Change

The papers of Watson [1], Helman & Tomlinson [2], Bigalbal et al. [3], Bamunawala et al. [4] and Kumbier et al. [5] aim to improve our understanding of the links between climate change, coastal forcing (SLR, storm surge, waves) and coastline change. Watson [1] and Helman & Tomlinson [2] look backward and discuss observed climate change impacts. Bigalbal et al. [3], Bamunawala et al. [4] and Kumbier et al. [5] look forward and discuss methods for modelling the impacts of climate change, presenting applications from around the world.

Watson [1] compares observed rates of sea-level rise from observational data records (tide gauges) against the ensemble mean of the model-projection products used in IPCC AR5 at 19 sites around the world over the period 2007–2016. Helman & Tomlinson [2] examine two centuries of observed climate and coastline response along the central east coast of Australia, between Fraser Island and Coffs Harbour. Looking backwards and describing historic trends and contrasting model predictions to actual behavior are essential for improving our understanding of relevant physical processes, improving models and quantifying model uncertainties.

The contributions of Bigalbal et al. [3], Bamunawala et al. [4] and Kumbier et al. [5] show that the impacts of climate change on coastal forcing and coastal change depend strongly on local circumstances, and that estimating these impacts requires physics-based extrapolations into an uncertain future. Bigalbal et al. [3] examine the potential impacts of sea level rise and marsh migration on the hydrodynamics and wave conditions within four natural protected areas of the Chesapeake Bay during storm surge events. The impacts of climate change are shown to be site-specific and suggest the presence of a critical limit for conservation. Kumbier et al. [5] present numerical simulations of coastal flooding under different sea-level scenarios for two Australian estuaries that are at different geomorphological evolutionary stages of infilling. The two estuaries display very different responses to sea level rise illustrating the limitations of applying simple rules of thumb. Bamunawala et al. [4] present reduced complexity modelling results for several small tidal inlets in Australia and Vietnam which show that human induced changes in fluvial sediment supply to the coast from e.g., urbanization or deforestation, should not be ignored when assessing the long-term behavior of such coastal systems in a changing climate.

2.2. The Socio-Economic and Environmental Impacts of Coastline Change

O'Neill et al. [6] and Erikson et al. [7] present the USGS's innovative Coastal Storm Modeling System (CoSMoS) for coastal zones affected by sea-level rise and changing coastal storms. CoSMoS allows coastal managers to not only quantify hazards but also to examine the associated socio-economic impacts on coastal communities. The contribution of O'Neill et al. [6] addresses the hazard quantification aspect, while the contribution by Erikson et al. [7] addresses the link between hazards and socio-economic consequences. The example applications in Southern California illustrate the inner workings of CosMoS and show that climate change may have considerable effects on the built environment. Moving from climate change impacts on the coastal zone built environment to the natural coastal environment, Mehvar et al. [8] examine environmental losses and present a methodology for valuing ecosystem services in monetary terms, allowing them to be included in economic assessments.

2.3. Probabilistic Methods for Analysing System Reliability

Because of the inherent and epistemic uncertainties related to coastal forcing, probabilistic methods for analyzing system reliability are increasingly gaining traction in coastal zone management circles. Oosterlo et al. [9] present a modular modelling framework for propagating the uncertainties related to forcing in wave overtopping assessments. The framework allows the computation of overtopping failure probabilities for sea dikes with shallow foreshores, as illustrated by a case study in southern Netherlands. Aguilar-López et al. [10] present a probabilistic method for assessing the resistance of grass covered dikes to overtopping and specifically examine the effect of roads on the crests of such dikes. Finally, on this theme, Ngo et al. [11] present a modelling approach to support probabilistic assessment of flood hazards at Cantho city in in the Mekong Delta, Vietnam.

2.4. Risk-Informed Decision Making and Probability-Based Economic Optimization

Probabilistic estimates of coastal change and the reliability of coastal defenses are indispensable for risk-informed decision making. Lowering the probability of socio-economic and/or environmental losses typically comes at a cost, turning coastal zone management into a balancing act. Economic optimization methods can be used for establishing economically optimal failure probabilities, to support risk-informed decision making. Along these lines, Galiatsatou et al. [12] present the economic optimization of measures to improve the reliability of a rubble mound breakwaters. In the Final paper of this issue, Dastgheib et al. [13] examine potential shoreline changes for sites in Sri Lanka using a probabilistic coastline recession model, and balance risk and reward to determine economically optimal setback lines.

3. Conclusions

This Special Issue provides an overview of the different types of research that are essential for responsibly managing the risks to coastal zones posed by climate change. The 13 contributions cover the physics of climate change driven shoreline change, associated impacts on communities and the environment, and methods for balancing risk and reward. Climate change poses a global challenge. Yet its impacts are felt locally. Risk management actions will therefore have to be tailored to local circumstances. As illustrated by studies from across the globe, the uniqueness of coastal zones does not permit convenient generalizations.

Author Contributions: R.R. and R.J. jointly developed the concept and co-wrote this editorial.

Funding: R.R. is supported by the AXA Research fund and the Deltares Strategic Research Programme 'Coastal and Offshore Engineering'.

Acknowledgments: The authors wish to thank all contributors to this Special Issue. The authors also wish to thank the very professional and efficient JMSE editorial staff without whose excellent assistance, this issue would not have been possible

Conflicts of Interest: The authors declare no conflict of interest.

References

1. Watson, P. How Well Do AR5 Sea Surface-Height Model Projections Match Observational Rates of Sea-Level Rise at the Regional Scale? *J. Mar. Sci. Eng.* **2018**, *6*, 11. [CrossRef]
2. Helman, P.; Tomlinson, R. Two Centuries of Climate Change and Climate Variability, East Coast Australia. *J. Mar. Sci. Eng.* **2018**, *6*, 3. [CrossRef]
3. Bigalbal, A.; Rezaie, A.; Garzon, J.; Ferreira, C. Potential Impacts of Sea Level Rise and Coarse Scale Marsh Migration on Storm Surge Hydrodynamics and Waves on Coastal Protected Areas in the Chesapeake Bay. *J. Mar. Sci. Eng.* **2018**, *6*, 86. [CrossRef]
4. Bamunawala, J.; Maskey, S.; Duong, T.; van der Spek, A. Significance of Fluvial Sediment Supply in Coastline Modelling at Tidal Inlets. *J. Mar. Sci. Eng.* **2018**, *6*, 79. [CrossRef]
5. Kumbier, K.; Carvalho, R.; Woodroffe, C. Modelling Hydrodynamic Impacts of Sea-Level Rise on Wave-Dominated Australian Estuaries with Differing Geomorphology. *J. Mar. Sci. Eng.* **2018**, *6*, 66. [CrossRef]
6. O'Neill, A.; Erikson, L.; Barnard, P.; Limber, P.; Vitousek, S.; Warrick, J.; Foxgrover, A.; Lovering, J. Projected 21st Century Coastal Flooding in the Southern California Bight. Part 1: Development of the Third Generation CoSMoS Model. *J. Mar. Sci. Eng.* **2018**, *6*, 59. [CrossRef]
7. Erikson, L.; Barnard, P.; O'Neill, A.; Wood, N.; Jones, J.; Finzi Hart, J.; Vitousek, S.; Limber, P.; Hayden, M.; Fitzgibbon, M.; et al. Projected 21st Century Coastal Flooding in the Southern California Bight. Part 2: Tools for Assessing Climate Change-Driven Coastal Hazards and Socio-Economic Impacts. *J. Mar. Sci. Eng.* **2018**, *6*, 76. [CrossRef]
8. Mehvar, S.; Filatova, T.; Dastgheib, A.; de Ruyter van Steveninck, E.; Ranasinghe, R. Quantifying Economic Value of Coastal Ecosystem Services: A Review. *J. Mar. Sci. Eng.* **2018**, *6*, 5. [CrossRef]

9.   Oosterlo, P.; McCall, R.; Vuik, V.; Hofland, B.; van der Meer, J.; Jonkman, S. Probabilistic Assessment of Overtopping of Sea Dikes with Foreshores including Infragravity Waves and Morphological Changes: Westkapelle Case Study. *J. Mar. Sci. Eng.* **2018**, *6*, 48. [CrossRef]

10.   Aguilar-López, J.; Warmink, J.; Bomers, A.; Schielen, R.; Hulscher, S. Failure of Grass Covered Flood Defences with Roads on Top Due to Wave Overtopping: A Probabilistic Assessment Method. *J. Mar. Sci. Eng.* **2018**, *6*, 74. [CrossRef]

11.   Ngo, H.; Pathirana, A.; Zevenbergen, C.; Ranasinghe, R. An Effective Modelling Approach to Support Probabilistic Flood Forecasting in Coastal Cities—Case Study: Can Tho, Mekong Delta, Vietnam. *J. Mar. Sci. Eng.* **2018**, *6*, 55. [CrossRef]

12.   Galiatsatou, P.; Makris, C.; Prinos, P. Optimized Reliability Based Upgrading of Rubble Mound Breakwaters in a Changing Climate. *J. Mar. Sci. Eng.* **2018**, *6*, 92. [CrossRef]

13.   Dastgheib, A.; Jongejan, R.; Wickramanayake, M.; Ranasinghe, R. Regional Scale Risk-Informed Land-Use Planning Using Probabilistic Coastline Recession Modelling and Economical Optimisation: East Coast of Sri Lanka. *J. Mar. Sci. Eng.* **2018**, *6*, 120. [CrossRef]

Journal of
*Marine Science
and Engineering*

MDPI

Article

How Well Do AR5 Sea Surface-Height Model Projections Match Observational Rates of Sea-Level Rise at the Regional Scale?

Phil J. Watson

School of Civil and Environmental Engineering, The University of New South Wales, Sydney, NSW 2052, Australia; philwatson.slr@gmail.com

Received: 26 November 2017; Accepted: 24 January 2018; Published: 1 February 2018

Abstract: The reliance upon and importance of climate models continues to grow in line with strengthening evidence of a changing climate system and the necessity to provide credible projections for risk assessment to guide policy development, mitigation and adaptation responses. The utility of the models to project regional rates of sea-level rise over the course of the 21st century is reliant on evaluating model outputs against global observational data (principally altimetry products). This study compares rates of sea-level rise from observational data records (tide gauges) against the ensemble mean of the model-projection products used in AR5 at 19 sites around the world over the decade of common data coverage (2007–2016) using enhanced time-series analysis techniques. Although it could be concluded that the observational and model-projected average velocity agree (95% confidence level (CL)), error margins are comparatively wide, masking the fact that the mean velocity for the model-projection products exceed observational records for nearly all stations and Representative Concentration Pathway (RCP) experiments, and are likely in the range of 1.6–2.5 mm/year. The analysis might provide an early warning sign that the evaluation of ocean model components with respect to projected mean sea level could be relevantly improved.

Keywords: mean sea level; velocity; AR5 regional projection modelling

---

1. Introduction

Climate models are key tools in assisting in understanding and planning for the predicted impacts of a changing climate system. These models form central elements of the Assessment Reports of the Intergovernmental Panel on Climate Change (IPCC) [1,2] with great reliance placed on projection outputs to facilitate appropriate policy, adaptation and mitigation responses. By coordinating the design and distribution of global climate model simulations of the past, present, and future climate, the Coupled Model Intercomparison Project (CMIP) has become one of the foundational elements of climate science [3].

CMIP—Phase 5 (or CMIP5), provided an ensemble of climate models to support the IPCC's Assessment Report 5 (AR5) [2], each of which are required to meet established protocols including adherence to fundamental laws of nature and large-scale observational constraints across a range of physical parameters [4]. The relative sea surface-height (SSH) projection products used in AR5 [5] include the following 10 geophysical sources that drive long-term changes in relative SSH [6]:

- 5 ice components (Greenland dynamic ice and surface-mass balance, Antarctic dynamic ice and surface-mass balance, and glaciers);
- 3 ocean-related components, all of which are derived from CMIP5 models (dynamic SSH, global thermosteric SSH anomaly, and the inverse barometer effect from the atmosphere);
- land water storage (also called terrestrial water); and

- glacial isostatic adjustment (GIA, as a change in sea level "relative" to land).

The next-generation models forming part of CMIP6 [3] have been under development since about 2012 to support the IPCC's Assessment Report 6 (AR6) [7], seeking amongst other things to capitalize on improving knowledge of complex integrated components of the climate system, enhanced computing power and outputs at increasingly finer resolution (or more localized scale).

In addition to advances in understanding and integrating key physical processes at an increasingly finer scale, improved projection outcomes from climate models are also critically reliant on better evaluation of modelling components against observational data. Recent literature has focused upon improving the regional resolution of SSH from CMIP5 by accounting for additional elements of the sea-level budget using improved offline modelled components and observation-based contributions [8]. Slangen et al. (2017) [9] provided an extensive evaluation of model simulations of 20th century sea level rise at the global scale compared to observational data from tide gauges and satellite altimetry spanning 1900–2015. This work made provision for sea-level contributions associated with groundwater depletion, reservoir storage, and dynamic ice-sheet mass changes that are not simulated by climate models.

A companion paper [10], provided a detailed evaluation of model performance at the regional scale by comparing these enhanced ensemble climate model outputs of SSH to tide gauge data (spanning the same period) at 27 sites around the world. The analysis concluded that for most of the tide gauge records, climate models tend to underestimate the observed twentieth-century trends. The analysis concluded the average difference between observed and modeled sea-level trends of the order of 0.27 ± 0.77 mm/year (90% confidence level (CL)) with discrepancies potentially explained by an underestimation in the uncertainty in Glacial Isostatic Adjustment (GIA).

These results are encouraging, suggesting agreement in sea-level trends between climate models and observational records at a regional level on a quasi-centennial timescale. This current analysis augments these previous works, taking a step further by comparing the AR5 SSH ensemble model projection products against a global network of tide gauges (refer Figure 1) for the period of overlapping coverage (2007–2016).

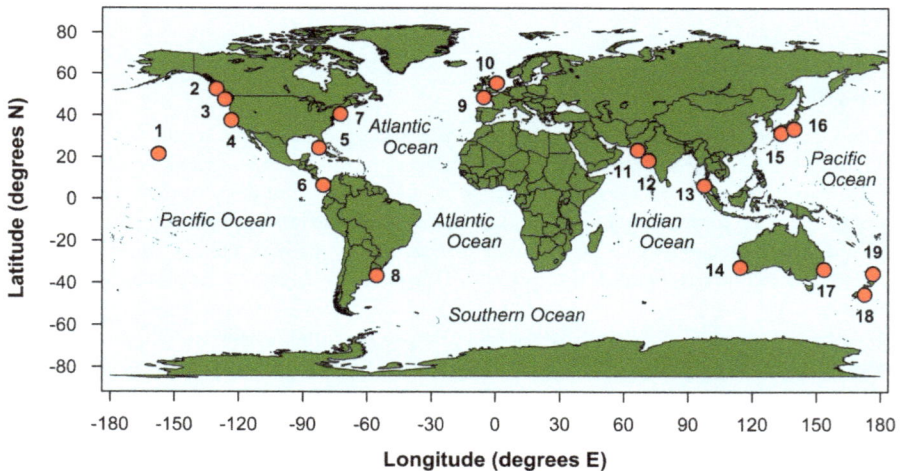

Figure 1. Location of analysis sites used in the study. Sites have been delineated with a station "ID" commencing with 1 (Honolulu, HI, USA) moving easterly to 19 (Auckland, New Zealand). Full details of data records are summarized in Table 1.

The analysis is also enhanced by several key features that attempt to overcome ubiquitous issues associated with sea-level trend analysis. Firstly, the application of improved time-series analysis techniques, using singular spectrum analysis (SSA) enables separation of the inter-annual to multi-decadal variability from the external (or greenhouse-gas induced) forcing for both the SSH projection model ensembles and tide gauge records. The scale of internal climate-mode signals can substantially bias or contaminate a mean sea-level trend estimate, particularly when the quantum of externally forced sea-level rise (of primary interest) is only around a couple of millimetres per year over both the historical and near-term forecasts.

Secondly, isolating or accommodating the encumbrance of vertical land motion (VLM) embedded within tide gauge records has been addressed as a sensitivity analysis by looking at both the "relative" and "geocentric" rates of sea-level rise at each of the sites considered.

Having removed the influence of internal climate modes (and other contaminating signals), the results would suggest that although the rate of rise in both the AR5 projection modelling and observational SSH data tend to agree (95% CL), the mean rates from the ensemble modelling products over the period of common coverage would appear biased on the high side at this point in time.

2. Data and Methods

The selection of sites used in this study were based on the existence of tide-gauge records which maximized length and quality (limited gaps and absence of known datum problems) whilst providing regional representativeness and global spatial coverage. Nineteen sites were selected that met these criteria (refer to Figure 1 and Table 1 for more details). Each site has notionally been assigned a station ID commencing with Honolulu (1) in the Central Pacific, progressing west to east to Auckland (19). By graphically representing characteristics associated with records based on the station ID, spatially dominant patterns are more readily apparent.

Annual average time-series data from the public archives of the Permanent Service for Mean Sea Level (PSMSL) [11,12] have been used in the analysis. Extended datasets from the extensive composite time series work by Hogarth (2014) [13] using near-neighbour tide gauge records have been used to augment the PSMSL records. Where possible, every effort has been made to source data up to and including 2016 (refer to Acknowledgements).

AR5 data [5] have been used to extract model ensemble outputs for total projected relative sea-level rise at each site. These data are publicly available in netCDF format from the Integrated Climate Data Center (ICDC) [6] with yearly outputs spanning the period 2007 to 2100 on a spatial resolution grid of $1° \times 1°$ for the Representative Concentration Pathway (RCP) 2.6, 4.5 and 8.5 experiments. The CMIP5 multi-model ensemble contains only 16 models for the RCP2.6 experiment; however, the RCP4.5 and 8.5 experiments are based on all 21 models (Dr. Mark Carson, Institute of Oceanography, ICDC, University of Hamburg, 2017, pers.comm., 19 June). Time series data for each of the respective RCP experiments have been extracted at the nearest grid point to the respective tide-gauge record, for which there is complete ensemble model coverage (refer Table 1 for details).

Satellite altimeter products by Ssalto/Duacs distributed by Archiving, Validation and Interpretation of Satellite Oceanographic (AVISO), with support from the Centre National d'Etudes Spatiales (CNES) [14] have been used to extract time series of SSH. These data have been made available for this research in netCDF format from the Integrated Climate Data Center (ICDC) [15] with daily outputs spanning the period 1 January 1997 to 6 January 2017 on a spatial resolution grid of $0.25° \times 0.25°$ (Cartesian). Daily outputs have been converted into annual time series at the nearest grid point to the respective tide gauge record (refer to Table 1 for details) and compared with the tide-gauge record to provide one of the estimates of VLM used in the study.

However, the key initial step in the process is to estimate the externally forced (or climate-change) component by removing the internal climate mode and other higher frequency signals from both the tide-gauge and ensemble projection-model time series. SSA has proven an optimal analytic for this task in sea-level studies [16] as a powerful data adaptive technique capable of decomposing a time series into the sum of interpretable components with no a priori information about the time-series structure [17,18]. Specifically, SSA can efficiently decompose an original record into a series of components of slowly varying trend, oscillatory components with variable amplitude, and a structureless noise [19].

Following sensitivity testing of parameterisation and visual inspection of individual components from the SSA decomposition across a range of station records and projection-model outputs, the internal climate mode and other higher frequency signals are effectively removed using 1-dimensional SSA, a default-embedding dimension of half the time series length and frequency-thresholding techniques [20] aggregating only components in which the relative contribution in the lowest frequency bin (0 to 0.02 cycles per year) is set above a threshold of ≈0.80. Figure 2 highlights the efficiency of these parameter settings in isolating the externally forced component at Fremantle, Australia which was one of the sensitivity-testing sites used for this purpose.

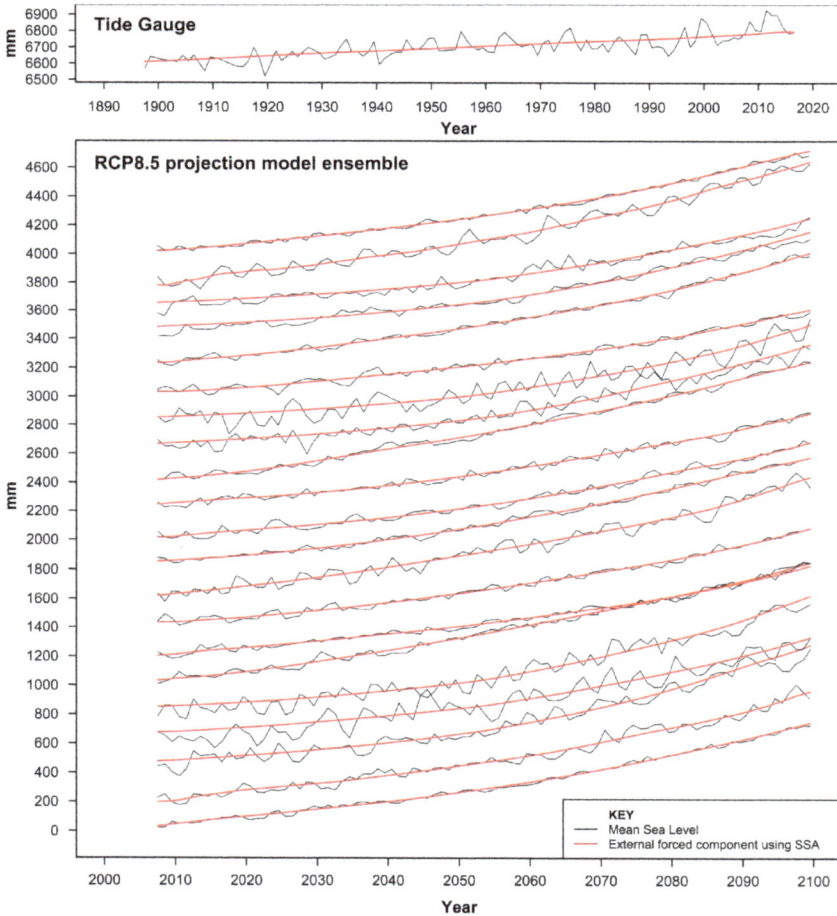

Figure 2. Optimization of singular spectrum analysis (SSA) to isolate the external forced component of both the tide-gauge records and projection-model ensemble. The example from Fremantle, Australia, shows the decomposition of the tide-gauge record in the top panel and all 21 representative concentration pathway (RCP) 8.5 projection-model outputs in the bottom panel. Each time series has been notionally offset vertically in the bottom panel by 200 mm for clarity.

Table 1. Summary of tide-gauge, AR5 model-projection and altimetry data used in this study.

Station ID [1]	Tide Gauge Record [2]	Start (year) [2]	End (year) [2]	AR5 Lat (° N) [5]	AR5 Long (° E) [5]	Altimetry Lat (° N) [6]	Altimetry Lat (° E) [6]
1	Honolulu [3]	1891	2016	21.5	202.5	21.125	202.125
2	Prince Rupert [3]	1903	2016	52.5	229.5	54.125	228.875
3	Seattle	1899	2016	47.5	233.5	47.375	235.125
4	San Francisco [3]	1855	2016	37.5	236.5	37.625	237.375
5	Key West	1913	2016	24.5	277.5	24.375	278.125
6	Balboa	1908	2016	6.5	279.5	8.875	280.625
7	New York [3]	1853	2016	39.5	287.5	40.375	286.125
8	Buenos Aires [3,4]	1905	2016	−36.5	304.5	−34.625	301.625
9	Brest	1807	2016	48.5	354.5	48.125	355.375
10	North Shields [3]	1895	2016	55.5	0.5	54.875	358.875
11	Karachi	1868	2014	23.5	66.5	24.625	66.875
12	Mumbai	1878	2012	18.5	71.5	18.875	72.675
13	Ko Taphao Noi	1940	2016	7.5	96.5	7.625	98.375
14	Fremantle	1897	2016	−32.5	114.5	−32.125	115.625
15	Hosojima	1930	2016	31.5	133.5	32.375	131.875
16	Aburatsubo	1930	2016	33.5	139.5	31.625	131.625
17	Sydney	1886	2016	−33.5	153.5	−33.875	151.375
18	Dunedin [3]	1900	2016	−45.5	172.5	−46.125	170.375
19	Auckland	1899	2016	−35.5	176.5	−36.675	174.875

[1] Station ID ordered moving west to east commencing at Honolulu, HI, USA (refer Figure 1). [2] Permanent Service for Mean Sea Level (PSMSL) annual average data used in this study [11,12]. [3] PSMSL data has been further extended by Hogarth (2014) [13]. [4] 2015 and 2016 data added to Hogarth (2014) [13] by filling from comparison with nearby Palermo station record which dates back to 1960. [5] Nearest projection-model output to tide-gauge record with complete ensemble coverage. [6] Nearest satellite altimetry grid point to tide-gauge record with complete coverage. Altimeter products by Ssalto/Duacs distributed by Aviso, with support from Centre National d'Etudes Spatiales (CNES) [14].

All analysis and graphical outputs have been developed by the author from customized scripting code within the framework of the R Project for Statistical Computing [21] and are available upon request. The applied methodology can be appropriately partitioned into analysis of the historical tide-gauge record and that of the AR5 ensemble projection-model outputs. The selection and application of SSA and other key steps in the analysis are broadly underpinned by unprecedented time-series research, development, and analysis of ocean water-level data [16,22–26].

2.1. Historical Tide-Gauge Analysis

The methodology applied in analysing the observational tide gauge records can be broadly summarised in the following 5 steps:

Step 1: Gap-filling of time series. This is a necessity in order to decompose the time series using SSA. Whilst the longest and most complete tide-gauge records have been used for analysis, missing data persists in several records. Where required, records have been filled using an iterative SSA procedure [27] in the first instance, which has an (assumed) advantage in preserving the principal spectral structures of the complete portions of the original data set in filling the gaps. Station records filled using this procedure included Prince Rupert, Key West, Balboa, North Shields, Mumbai, Ko Taphao Noi, Hosojima, Aburatsubo, Sydney and Auckland. Visual inspection confirmed the suitability of the gap-filled sections. Where this method provided visually unreliable results, gaps were filled using the comparatively simpler Stineman's interpolation [28]. This procedure was suitably applied to the Brest, Fremantle and Dunedin records.

Step 2: Estimation of "relative" mean sea level. Having necessarily filled time series in Step 1, the record is decomposed using 1 dimensional SSA to isolate components of slowly varying trend (i.e., mean sea level due to external climate forcing) from oscillatory components with variable amplitude, and noise. As advised earlier, frequency thresholding has been used to facilitate this step with mean sea level estimated by summing the components of the SSA decomposition in which the relative contribution in the lowest frequency bin [0–0.02] exceeds 0.80.

Step 3: Estimation of average "relative" mean sea-level velocity over period of common coverage (post 2007). For all but 2 of the records considered (Karachi and Mumbai), the period of common coverage between the tide-gauge records and the projection-model outputs extends from 2007 to 2016. The average "relative" velocity over this period has been determined by simple linear least squares regression of the respective portion of the time series determined in Step 2.

Step 4: Estimation of errors. This process initially involves fitting an autoregressive time-series model to remove the serial correlation in the residuals between the SSA derived trend (Step 2) and the gap-filled time series (Step 1). The estimation of error in the average "relative" mean sea-level velocity is then based on bootstrapping techniques where the uncorrelated residuals are randomly recycled and the process in steps 2 and 3 is repeated 1000 times. From the extensive pool of outputted relative velocities, standard deviations are readily calculated to derive robust confidence intervals.

Step 5: Correction to estimate "geocentric" velocity. The correction from "relative" to "geocentric" velocity has been undertaken using 3 separate estimates for VLM as a form of sensitivity analysis given the general uncertainty with this parameter. The first method (referred to hereafter as "Type 1") uses VLM estimates provided by Systeme d'Observation Du Niveau Des Eaux Littorales (SONEL) based on continuous global-positioning system (GPS) measurements of the land mass in the vicinity of the tide gauge [29,30]. However, "Type 1" VLM rates are only available for 12 of the 19 tide gauge records. The second method estimates VLM considering the difference between the tide-gauge record and that of the nearest gridded data time series from satellite altimetry using the approach of Ostanciaux et al. (2012) [31]. This second method (referred to hereafter as "Type 2") estimates VLM from the linear regression of the difference in the annualised time series of both data sets providing a VLM estimate for every station record considered. A third method (referred to hereafter as "Type 3") proposed by Pfeffer and Allemand (2016) [32], estimates VLM via a similar approach to Ostanciaux et al. (2012) [31] with some further advancements to the time-series analysis procedure, providing a VLM estimate for 15 of the 19 station record considered. VLM rates applied in this study are summarised in Table 2. Error margins in quantifying "geocentric" velocity are calculated in quadrature by combining the error in the VLM rates with those determined in Step 4.

2.2. AR5 Sea Surface-Height (SSH) Projection-Model Output Analysis

The following methodology has been applied to the AR5 projection modelling SSH data products for RCP2.6, 4.5 and 8.5 experiments, to assimilate and compare them to the observational tide gauge data:

Step 1: Estimation of "relative" mean sea level. The same SSA procedure advised for the tide gauge records is then applied to the projection-model ensemble products to remove the internal variability and other dynamic signals in order to isolate the slowly varying trend (i.e., mean sea level due to external climate forcing). As advised previously, the SSH projection-modelling outputs used in AR5 [5], made available by the ICDC [6], are already provided as "relative" sea-level products. This was done by inverting an allowance for GIA and applying this "correction" to the original "geocentric" model outputs. The GIA allowance used in AR5 for this purpose was an average of both the Peltier (2004) [33] and Lambeck et al. (1998) [34] estimates (see Table 2 for details). With the internal variability (and other high-frequency signals) removed, the ensemble mean has been used to normalise projection-model outputs to each of the respective tide-gauge datums. The ensemble model SSH output products are based on a 20-year moving average with the modelling start point set at 1986–2005 (i.e., centred around 1995). The annual time series output products from the ICDC for AR5 start at 2007 and have therefore been normalized to each of the respective tide-gauge records by using the mean sea-level estimate in 1995.

Step 2: Estimation of average "relative" mean sea-level velocity over period of common coverage (post 2007). The same process advised for the tide-gauge analysis has been applied to each of the outputted time series from Step 1 (above), providing a pool of 16 outputted velocities over the period of common coverage for the RCP2.6 experiment, and 21 outputted velocities for the RCP 4.5 and

8.5 experiments at each location. The average "relative" mean sea-level velocity at each location, for each of the respective RCP experiments, is then determined simply from the mean of each of the outputted pool of velocities. The associated errors are also readily determined from the standard deviation of the pool of outputted velocities.

Table 2. Summary of vertical land motion (VLM) and glacial isostatic adjustment (GIA) rates used in this study.

Station ID [1]	Site	Peltier GIA [3] (mm/Year)	AR5 GIA [2,4] (mm/Year)	Type 1 VLM [2,5] (mm/Year)	Type 2 VLM [2,6] (mm/Year)	Type 3 VLM [2,7] (mm/Year)
1	Honolulu	−0.17	0.23 ± 0.07	−0.23 ± 0.18	−0.84 ± 0.52	0.30 ± 0.31
2	Prince Rupert	0.33	0.64 ± 0.51	NA	0.56 ± 1.17	−2.10 ± 0.93
3	Seattle	−1.10	−0.99 ± 0.59	−0.99 ± 0.22	−0.27 ± 0.75	−0.99 ± 0.65
4	San Francisco	−1.00	−0.73 ± 0.61	−0.04 ± 0.15	0.37 ± 0.64	0.27 ± 0.86
5	Key West	−0.75	−0.65 ± 0.52	−1.76 ± 0.43	−1.23 ± 0.92	−1.18 ± 0.66
6	Balboa	−0.26	0.04 ± 0.08	NA	7.55 ± 1.93	1.89 ± 1.85
7	New York	−1.80	−1.39 ± 0.43	−2.12 ± 0.62	0.96 ± 1.37	−1.95 ± 0.89
8	Buenos Aires	0.70	0.29 ± 0.16	1.03 ± 0.24	0.84 ± 1.80	NA
9	Brest	−0.60	−0.48 ± 0.12	0.01 ± 0.11	−0.64 ± 0.85	−1.60 ± 0.59
10	North Shields	0.11	0.10 ± 0.27	1.39 ± 0.67	0.46 ± 0.75	−0.72 ± 0.61
11	Karachi	0.33	0.06 ± 0.17	NA	−2.53 ± 0.44	NA
12	Mumbai	0.31	0.14 ± 0.18	NA	0.61 ± 2.17	NA
13	Ko Taphao Noi	0.16	0.13 ± 0.10	NA	−9.51 ± 2.40	−3.12 ± 1.83
14	Fremantle	−0.33	0.07 ± 0.01	NA	−0.30 ± 0.66	−0.76 ± 0.71
15	Hosojima	0.43	0.26 ± 0.11	NA	1.33 ± 0.59	0.59 ± 0.79
16	Aburatsubo	0.47	0.31 ± 0.10	NA	−1.48 ± 1.29	−2.58 ± 0.67
17	Sydney	−0.23	0.02 ± 0.06	−0.33 ± 0.25	−1.27 ± 0.98	2.12 ± 0.75
18	Dunedin	−0.02	0.20 ± 0.04	−1.02 ± 0.12	−1.53 ± 0.61	0.46 ± 0.47
19	Auckland	0.03	0.17 ± 0.01	−0.62 ± 0.23	−1.24 ± 0.80	NA

[1] Refer to Figure 1 and Table 1 for further details. [2] Error margins advised are standard errors (i.e., 1σ). [3] Peltier GIA estimates [33] at tide-gauge sites available from PSMSL [12]. Aligned with VLM prediction for consistency (i.e., positive is up, negative down). [4] AR5 GIA estimates based on average of Peltier [33] and Lambeck [34] provided by Integrated Climate Data Center (ICDC) [6] and aligned with VLM prediction for consistency (i.e., positive is up, negative down). Gridded error margins for the GIA estimates are also provided by ICDC. [5] Type 1 VLM estimates are based on ULR6 global-positioning system (GPS) solutions [30] provided by Systeme d'Observation Du Niveau Des Eaux Littorales (SONEL) [29]. [6] Type 2 VLM estimates are based on the difference between the annual time series from satellite altimetry [14] and the tide gauge using the procedure advised in Ostanciaux et al. (2012) [31]. [7] Type 3 VLM estimates are based on the procedure advised in Pfeffer and Allemand (2016) [32].

Step 3: Correction to estimate "geocentric" velocity. The correction from "relative" to "geocentric" velocity is straightforward, requiring merely the addition of the respective gridded GIA allowance discussed in Step 2 above to the average "relative" mean sea-level velocity determined in Step 2 (refer Table 2 for AR5 GIA estimates). Error margins in quantifying "geocentric" velocity are calculated in quadrature by combining the error in the AR5 GIA rates with those determined in the "relative" mean sea-level velocity in Step 2.

3. Results

The results of the analysis are graphically represented in Figures 3–6. Figure 3 summarises the average "relative" velocity determined over the period of common coverage at each location, directly comparing estimates derived from the observational tide-gauge record with those derived from the model ensemble of SSH projections used in AR5 for each of the respective RCP experiments. From this analysis, at the 95% confidence level depicted, the observational and model-projected average velocity agree for 18 of the 19 records for all RCP experiments (excluding only Balboa, Panama). However, owing to the comparatively large error margins, particularly for the model-ensemble products, the mean velocity for the model-projection products are higher than the mean of the observational records for 16 of 19 stations across all RCP experiments. For these 16 records, the average gap is in the range of 1.6–1.8 mm/year. When all station records are considered across all RCP experiments the average gap is slightly lower, in the range of 1.2–1.4 mm/year.

Figure 4 summarises the average "geocentric" velocity analysis using the "type 1" VLM correction for the tide gauge records. However, it should be noted that there are only 11 tide gauge records where direct GPS VLM measurements are available for this analysis. Of these records, at the 95% confidence level depicted, the observational and model-projected average velocity agree across all RCP experiments. However, similar to the "relative" velocity analysis, the mean "geocentric" velocity for the model-projection products are higher than the mean of the observational records for all 11 records with Type 1 VLM corrections available, across all RCP experiments. For these 11 records, the average gap is similarly in the range of 1.6–1.8 mm/year.

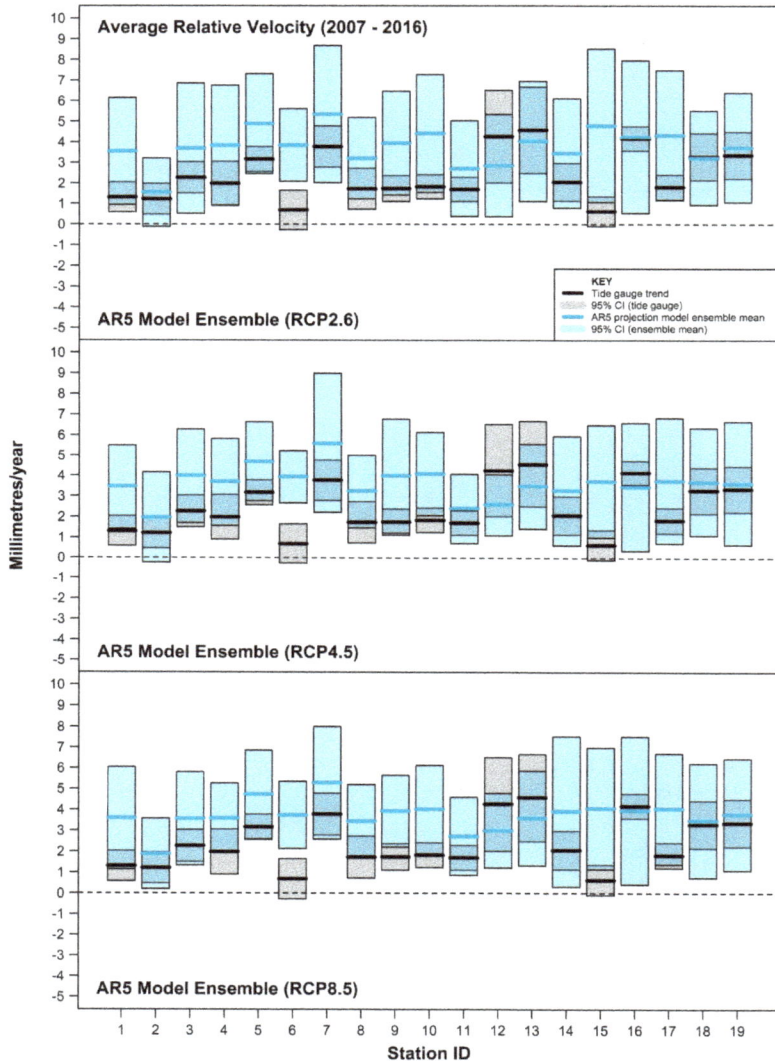

Figure 3. Average "relative" velocity determined over the period of common coverage (2007–2016). These charts directly compare the observational record (tide-gauge analysis) and corresponding AR5 ensemble-model projections at the grid point of interest. Refer to Figure 1 and Table 1 for station ID details.

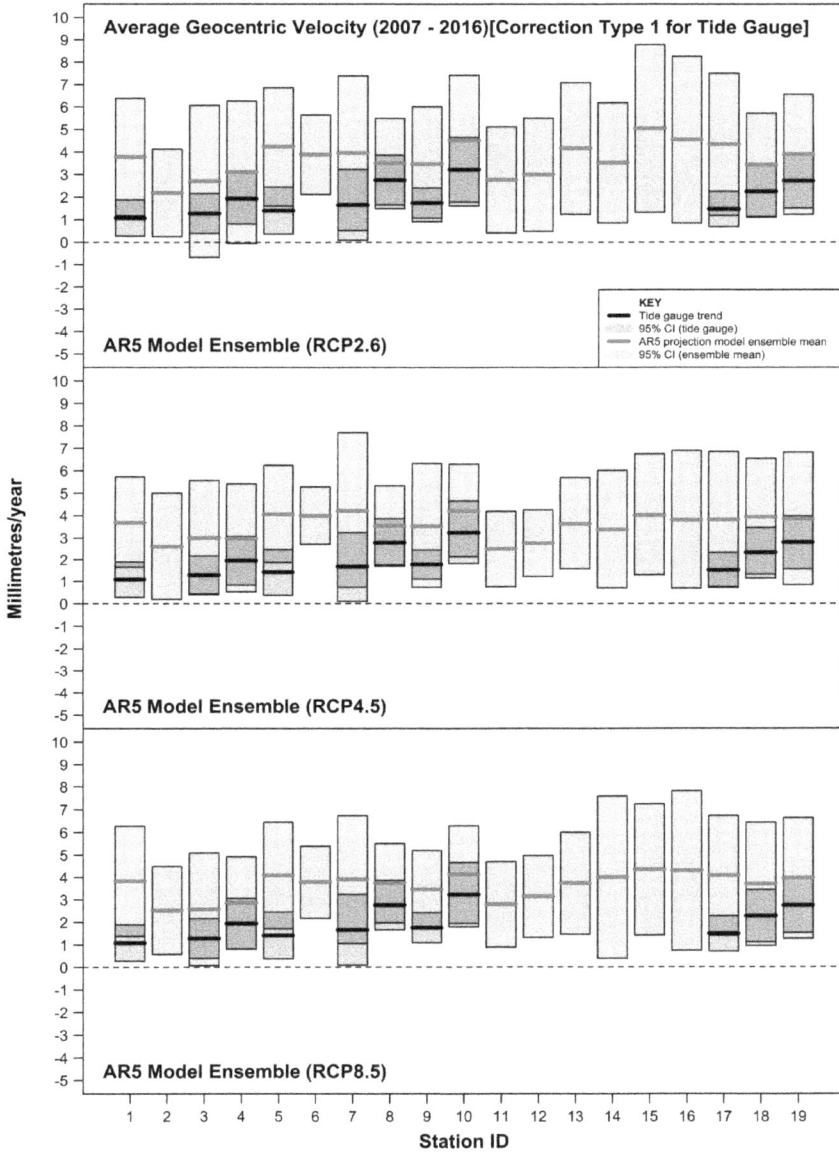

Figure 4. Average "geocentric" velocity determined over the period of common coverage (2007–2016) using Type 1 correction for VLM. Type 1 corrections are based on direct GPS measurements at or near the tide gauge. These charts directly compare the observational record (tide-gauge analysis) and corresponding AR5 ensemble-model projections at the grid point of interest. Refer to Figure 1 and Table 1 for station ID details.

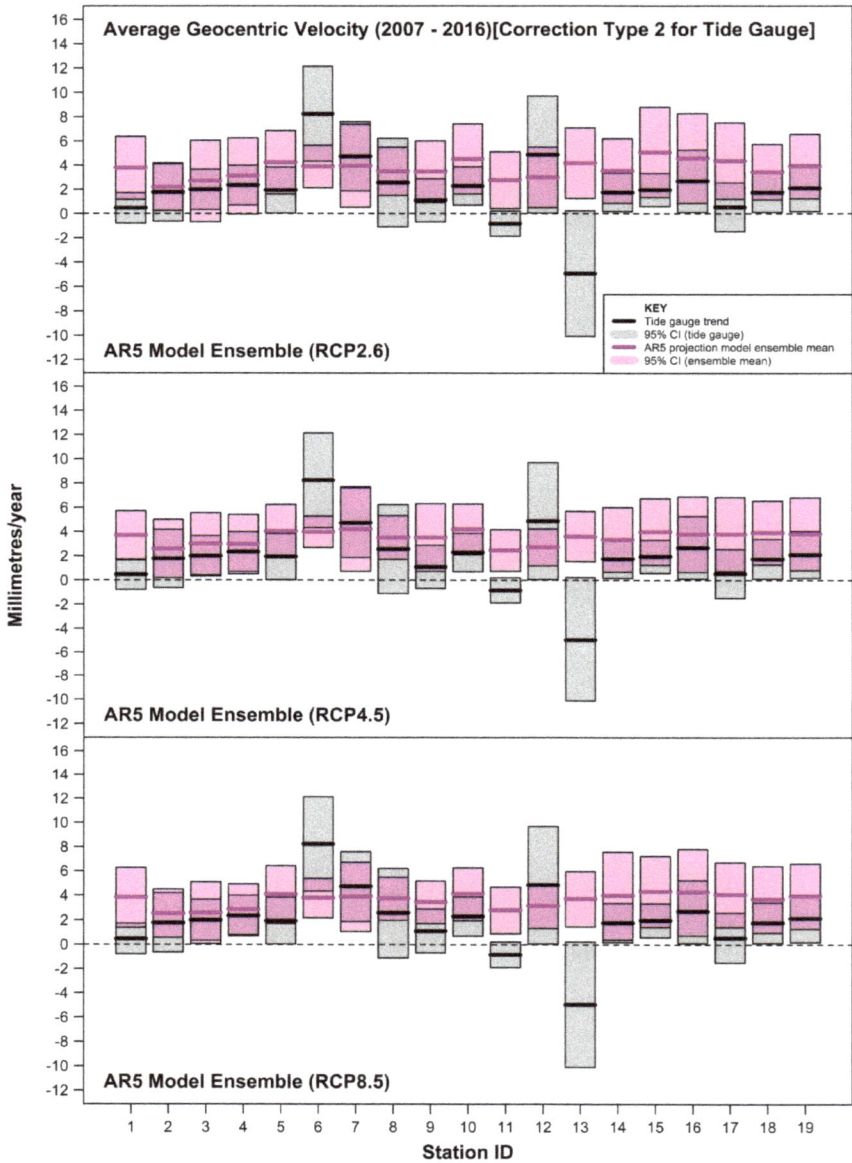

Figure 5. Average "geocentric" velocity determined over the period of common coverage (2007–2016) using Type 2 correction for VLM. Type 2 corrections are based on differences between altimetry and the tide-gauge records via the procedure advised in Ostanciaux et al. (2012) [31]. These charts directly compare the observational record (tide-gauge analysis) and corresponding AR5 ensemble-model projections at the grid point of interest.

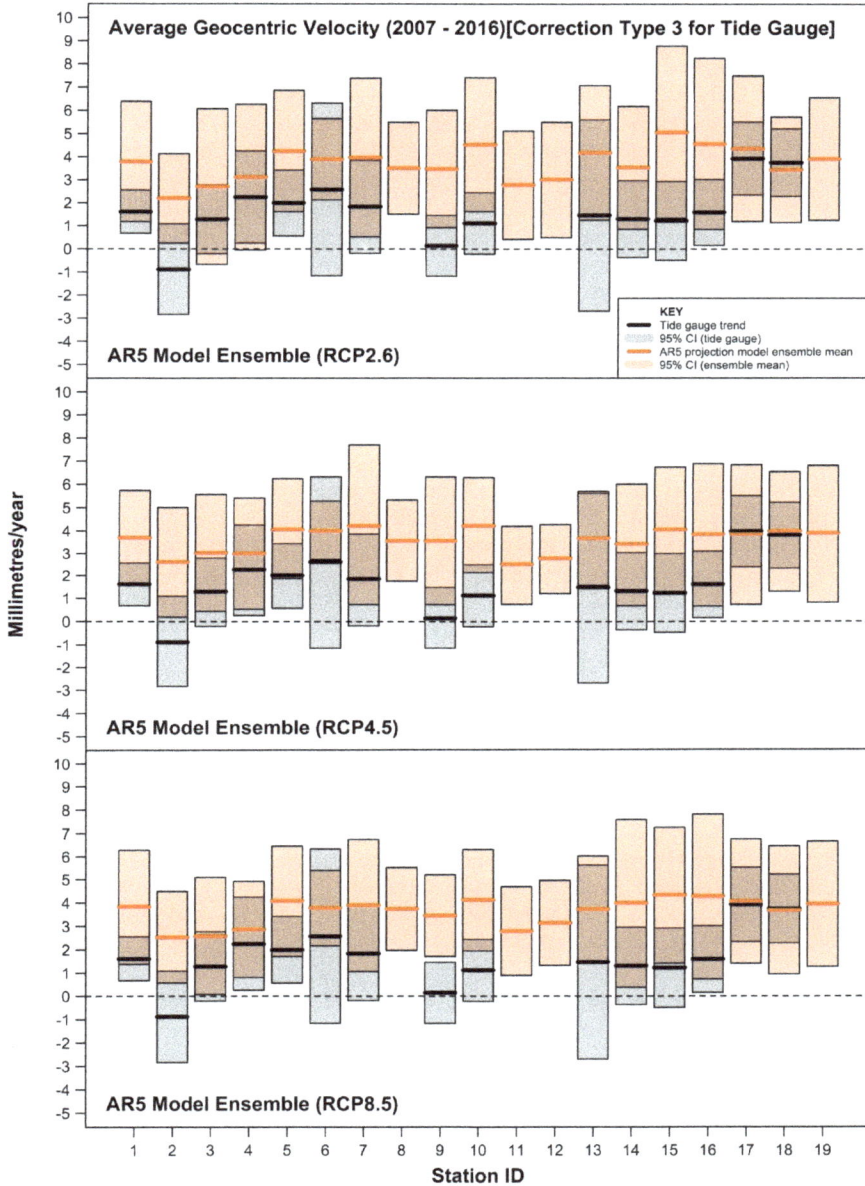

Figure 6. Average "geocentric" velocity determined over the period of common coverage (2007–2016) using Type 3 correction for VLM. Type 3 corrections are based on differences between altimetry and the tide-gauge record advised by Pfeffer and Allemand 2016 [32]. These charts directly compare the observational record (tide-gauge analysis) and corresponding AR5 ensemble-model projections at the grid point of interest. Refer to Figure 1 and Table 1 for station ID details.

Figure 5 summarises the average "geocentric" velocity analysis using the "type 2" VLM correction for the tide-gauge records, which is estimated from the difference between the satellite altimetry and tide-gauge record. Unlike, "type 1" VLM estimates, the "type 2" method provides consistent estimates for all sites based on nearly 20 years of altimetry records. From this analysis, at the 95% confidence level depicted, the observational and model-projected average velocity agree for 18 of the 19 records for all RCP experiments (excluding only Ko Taphao Noi, Thailand where ground subsidence is prevalent). As with the aforementioned analyses, the mean "geocentric" velocity for the model-projection products are higher than the mean of the observational records for 16 of 19 stations across all RCP experiments. For these 16 records, the average gap is in the range of 2.3–2.5 mm/year. When all station records are considered across all RCP experiments the average gap reduces to around 1.5–1.8 mm/year.

Figure 6 summarises the average "geocentric" velocity analysis using the "type 3" VLM correction for the tide-gauge records, which is estimated via the process espoused in Pfeffer and Allemand 2016 [32], although it should be noted that there are only 15 sites where "type 3" corrections are available for analysis. From this analysis, at the 95% confidence level depicted, the observational and model-projected average velocity agree for all stations across all RCP experiments with the exception of only Brest, France, for the RCP 8.5 experiment whereby the observational record was lower than the projection-model average. Similar to the other VLM sensitivity analyses, the mean "geocentric" velocity for the model-projection products are higher than the mean of the observational records for 14 of 15 stations with "type 3" VLM corrections available, across all RCP experiments. For these 14 records, the average gap is in the range of 2.0–2.3 mm/year. When all station records are considered across all RCP experiments the average gap is slightly lower, in the range of 1.9–2.2 mm/year.

4. Discussion

There will always be difficulties associated with comparing rates of change in SSH from observational and projection-modelling sources for the comparatively short timeframe (2007–2016) available. For this reason, different "relative" and "geocentric" velocity comparisons have been undertaken to provide a sensitivity test for the results of the analysis. However, it is worthwhile noting that the error margins associated with the respective data analyses, particularly those involving the ensemble projection-modelling products, are comparatively large. Given that subtle emerging trends might risk being obscured by the width of the confidence margins depicted (95%), further analysis has been undertaken using 80% confidence intervals as a further sensitivity check. At the narrower confidence levels, a clearer trend emerges whereby up to 6 of the 19 station records for the "relative" velocity assessment highlight observational records that are statistically lower than the projection-modelled products across all RCP experiments. For the "geocentric" velocity assessment using the 11 station records with "Type 1" VLM corrections available, some 4 records indicate observational velocities statistically lower than the projection-modelled products (80% CL).

Similarly for the "geocentric" velocity assessment using the full coverage "Type 2" VLM corrections, 6 of the observational records are statistically different to the projection-modelling products, with 5 being lower (80% CI). When "type 3" VLM corrections are applied, 4 observational records exhibit a "geocentric" velocity statistically different to the projection-modelling products, with all being lower. Across the majority of analysis types and RCP experiments at the 80% confidence level, 4 sites consistently emerge for which the projection-modelling products are statistically higher than the observational data record over the period of common coverage: Honolulu (USA), Key West (USA), Brest (France) and Sydney (Australia).

From the analysis undertaken, evidence suggests the AR5 projection-model outputs for SSH appear to be rising at a faster rate than the observational (tide gauge) records over the decade of common coverage. AR5 alluded to this possibility noting that for the global mean sea-level analysis, the rate of rise at the start of the RCP projections was about 3.7 mm/year, slightly above the observational (altimetry) range of 3.2 ± 0.4 mm/year for 1993–2010, surmising the simulated rate of climate warming being greater than that observed [5]. The current analysis, benefitting from improved time-series analysis techniques across 19 global sites with good global spatial coverage, observes a wider difference at site-specific (or regional) scales.

Although the overlapping period (2007–2016) of coverage of both types of data permit a decade over which relevant comparisons can be made between observational and projection-modelling products, one should appreciate the conclusions of the work at this point in time have particular caveats and limitations which might include (but not be limited to):

- AR5 projection-model outputs were not designed to be necessarily rigorous at the decadal time scale;
- an overlapping decade is a relatively small window over which to compare the characteristics of long-term phenomena such as mean sea-level rise. Furthermore, time-series analysis techniques are ubiquitously affected by end effects and the analysis herein is attempting to investigate and compare the characteristics of the end of one time series (tide-gauge data to present) with the start of another (AR5 projections from 2007);
- estimating and accommodating the ubiquity of VLM resulting from multiple origins (including tectonics, subsidence, etc., as well as GIA) embedded within tide-gauge records. As evident in this study, co-located GPS measurements are not available for many tide-gauge sites or are at best quite short records (often <10 years) [29,30]. The more complete and longer VLM estimates based on the method of Ostanciaux et al. 2012 [31] are reliant on satellite altimetry data [14] on a spatial resolution grid of 0.25° × 0.25° and could be a maximum of ≈20 km from the tide-gauge site. Similarly, the AR5 approach of correcting projection outputs for GIA to better approximate mean sea-level rise "relative" to the land ignores the many other VLM origins present at tide-gauge sites. This is highlighted clearly in Figure 7, where the GIA estimates at Balboa [ID = 6] and Ko Taphao Noi [ID = 13] differ markedly from the actual VLM observed via differing techniques. At both sites, the estimated GIA is <0.2 mm/year, whereas the VLM associated with the tectonic uplift of the Panama Arc at Balboa [35] and subsidence due to groundwater and aquifer mining at Ko Taphao Noi [36] are of the order of several mm/year;
- the highest density of long tide-gauge records are predominantly clustered around Europe and North America in the northern hemisphere [25]. There is only a relatively small pool of records available outside these domains with sufficient length and apparent robustness (i.e., absence of warning flags in the PSMSL database) to consider for meaningful global coverage; and
- the postulated theory that the Mt Pinatubo eruption in the Philippines in 1991 has had the effect of masking (or delaying) the rate of global sea-level rise [37] which would also be reflected in the tide-gauge records.

Notwithstanding the aforementioned issues, the analysis undertaken of tide-gauge records and projection-modelling outputs permits one to take advantage of their considerable length, enabling the application of improved time-series techniques to isolate the trend (associated with external climate forcing) from the contaminating dynamic influences that persist on decadal to multi-decadal and longer cycles [38–43]. By comparison, the altimetry products used to evaluate the CMIP5 SSH projection-model outputs [4], though having the advantage of broad ocean coverage to a fixed reference datum, have limited utility in separating out the aforementioned type of contaminating influences on longer timescales owing to shortness of the records at this point in time (post late 1992).

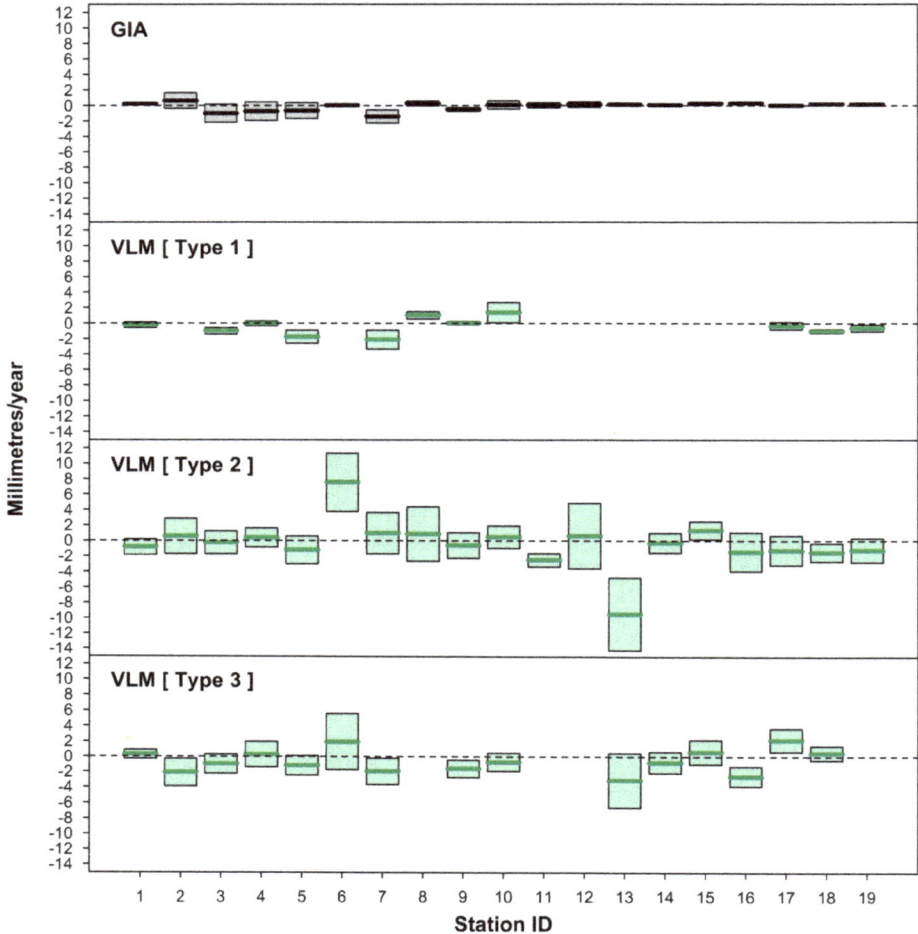

Figure 7. Comparison of VLM and GIA rates for each station record. GIA estimates are those used in AR5 [5] based on an average of Peltier [33] and Lambeck [34] provided by ICDC [6]. "Type 1" VLM are based on ULR6 GPS solutions [30] provided by SONEL [29]. "Type 2" VLM are based on the difference between the annual time series from satellite altimetry [14] and the tide gauge using the procedure advised in Ostanciaux et al. (2012) [31]. "Type 3" VLM are based on the procedure advised in Pfeffer and Allemand (2016) [32]. Error margins depicted in each panel are 95% CL. Refer to Figure 1 and Table 1 for station ID details.

It is worthwhile appreciating differences between the current paper and the contemporary analysis undertaken in Meyssignac et al. (2017) [10]. This recently published work compared 20th-century regional sea-level trends between an ensemble of 12 climate-model simulations using enhanced input on a range of sea-level contributions and that of 29 tide-gauge observations. This work concluded a general agreement between simulated sea-level and tide-gauge records in terms of interannual to multidecadal variability over the period 1900–2015. It is worth appreciating that this work acknowledges that the trends determined from the tide-gauge records are potentially dominated by decadal to multidecadal internal variability, and that the trend analysis of both types of data relies on simple linear regression techniques applied over the period corresponding to the available tide-gauge record (Dr. Benoit Meyssignac, Laboratoire d'Etudes en Géophysique et Océanographie Spatiales, 2017, pers.comm., 12 October). By comparison, this current study takes advantage of more advanced techniques to first remove the influence associated with decadal to multidecadal internal variability from both the tide-gauge data and AR5 projection-model ensembles of sea surface heights before comparing the results for the decade of common coverage (2007–2016). Both studies provide broad agreement between observational and simulated sea levels when error margins are taken into consideration and within the limits of the respective analyses undertaken. A logical next step might be to reapply the techniques espoused in this paper to the gridded outputs of the enhanced climate-model simulations developed in Meyssignac et al. (2017) [10] when they are available.

5. Conclusions

The utility and importance of climate models continues to grow in line with strengthening evidence of a changing climate system [2,44–47], in order to provide credible projections for risk assessment (e.g., [48–50]) to guide necessary policy development, mitigation and adaptation responses. The climate models themselves are an integral tool in understanding and interpreting the complex science and interrelationships of Earth systems and climate.

Despite the increasing complexity and resolution of these models, their utility for future projections will always be conditional on their ability to replicate historical and recent observational global and regional data trends of importance (such as temperature, sea level, CO_2 trends, etc.).

This paper provides a snapshot of how closely current rates of sea-level rise from observational data records (tide gauges) are represented by the ensemble mean of the AR5 model-projection products at the regional scale, considering 19 sites across the global ocean over the period of common coverage (2007–2016). The application of SSA (described in Section 2) provides the means by which to efficiently isolate the externally (or climate-change) forced signal from all other contaminating dynamic influences (including internal climate modes) for both types of data. This represents a significant improvement in overcoming the bias associated with trend comparisons in the presence of internal climate mode influences.

Corrections for VLM based on the difference between the tide-gauge record and satellite altimetry permit a more consistent approach for correcting the "relative" velocity from the tide-gauge record to "geocentric" rates for direct comparison to the projection-model outputs. With such corrections, it could be concluded that the observational and model-projected average velocity agree across all RCP experiments at the 95% confidence level. However, the error margins are quite wide, masking the fact that the mean "geocentric" velocity for the model-projection products are higher than the mean of the observational records for nearly all stations across all RCP experiments, and are likely in the range of 1.6–2.5 mm/year. At the 95% confidence level across all RCPs, there is no clear spatial pattern to the larger differences between the mean velocities of projection model outputs and the observational records.

Different VLM estimates have been considered as a sensitivity analysis in this paper with all respective analyses confirming similar results and providing more robustness to the general findings herein. However, significant discrepancies have been observed between various VLM estimates at the sites under consideration in this paper. Notwithstanding this, every effort should be made to continue investigating methodologies to make better use of the lengthening satellite altimetry records in this regard, with benchmarking against direct GPS measurements, such as espoused in Pfeffer and Allemand (2016) [32], which will continue to augment and improve the utility of the long tide-gauge records available.

The analysis within this study might provide an early warning sign that the evaluation of ocean model components with respect to projected mean sea level could be relevantly improved. In addition to integrating more advanced oceanographic phenomena at increasingly finer resolution into the ocean model components (e.g., [8–10]), the techniques espoused might also be considered part of the evolutionary process by which to improve the robustness and veracity of these critical projection-modelling tools at increasingly finer resolution over the course of CMIP6 [3] and AR6 [7]. With the rate of sea-level rise (or velocity) a key parameter upon which to evaluate model outputs, the lengthy tide-gauge records provide an additional level of enhanced capacity in this regard at increasingly finer resolution.

This situation could be further improved by augmenting the available records in the global data repository of the PSMSL through:

- searching for additional tide-gauge data records of sufficient length (at least 75–80 years) to improve the global spatial coverage of the station records used in this paper, in particular around important margins including the southern coasts of the African continent and along the coastlines of China;
- investigating methods to fill long records in other key areas such as Aden, Yemen (dating back to 1879) and Takoradi, Ghana (dating back to 1930), supplementing the critical work already undertaken to extend PSMSL records (e.g., [13]); and
- contributing data custodians ensuring up-to-date data is supplied in a timely manner.

Supplementary Materials: The Supplementary Materials are available online at http://www.mdpi.com/2077-1312/6/1/11/s1.

Acknowledgments: This research work has not benefitted from any grants or financial assistance. The author would like to thank the following (in alphabetical order):

- David Bate and Port of Auckland Limited (Auckland, New Zealand) for permitting access to recent data for Auckland;
- Mark Carson (Institute of Oceanography, University of Hamburg, Germany) who provided advice on specifics of the SSH projection-model outputs used in AR5, made available via the Integrated Climate Data Center, University of Hamburg;
- James Chittleborough (Tidal Unit, Bureau of Meteorology, Australia) for providing recent data at Fremantle and Sydney;
- Stefan Kern (Institute of Oceanography, University of Hamburg, Germany) who provided access to the Ssalto/Duacs satellite altimetry products. The altimeter products were produced by Ssalto/Duacs in collaboration with LOcean and CTOH and distributed by Aviso, with support from Cnes (http://www.aviso.altimetry.fr/);
- Glen Rowe, LINZ (Land Information, New Zealand) for providing recent data at Dunedin and Auckland;
- Thiago Dos Santos (Land and Atmospheric Science, University of Minnesota, MN, USA) who provided advice on R scripting code to facilitate extraction of data from netCDF format files;
- Aimée Slangen (Royal Netherlands Institute for Sea Research, The Netherlands) for wide-ranging advice and discussions on climate-modelling products that improved the study; and
- PSMSL, ICDC and SONEL for their publicly accessible data repositories.

Conflicts of Interest: The authors declare no conflict of interest.

References

1. Solomon, S.; Qin, D.; Manning, M.; Chen, Z.; Marquis, M.; Averyt, K.B.; Tignor, M.; Miller, H.L. (Eds.) *Climate Change 2007: The Physical Science Basis*; Contribution of Working Group I to the Fourth Assessment Report of the Intergovernmental Panel on Climate Change; Cambridge University Press: Cambridge, UK; New York, NY, USA, 2007; p. 333.
2. Stocker, T.F.; Qin, D.; Plattner, G.-K.; Tignor, M.; Allen, S.K.; Boschung, J.; Nauels, A.; Xia, Y.; Bex, V.; Midgley, P.M. (Eds.) *Climate Change 2013: The Physical Science Basis*; Contribution of Working Group I to the Fifth Assessment Report of the Intergovernmental Panel on Climate Change; Cambridge University Press: Cambridge, UK; New York, NY, USA, 2013. [CrossRef]
3. Eyring, V.; Bony, S.; Meehl, G.A.; Senior, C.A.; Stevens, B.; Stouffer, R.J.; Taylor, K.E. Overview of the Coupled Model Intercomparison Project Phase 6 (CMIP6) experimental design and organization. *Geosci. Model Dev.* **2016**, *9*, 1937–1958. [CrossRef]

4. Flato, G.; Marotzke, J.; Abiodun, B.; Braconnot, P.; Chou, S.C.; Collins, W.; Cox, P.; Driouech, F.; Emori, S.; Eyring, V.; et al. Evaluation of Climate Models. In *Climate Change 2013: The Physical Science Basis*; Stocker, T.F., Qin, D., Plattner, G.-K., Tignor, M., Allen, S.K., Boschung, J., Nauels, A., Xia, Y., Bex, V., Midgley, P.M., Eds.; Contribution of Working Group I to the Fifth Assessment Report of the Intergovernmental Panel on Climate Change; Cambridge University Press: Cambridge, UK; New York, NY, USA, 2013; pp. 741–866. [CrossRef]
5. Church, J.A.; Clark, P.U.; Cazenave, A.; Gregory, J.M.; Jevrejeva, S.; Levermann, A.; Merrifield, M.A.; Milne, G.A.; Nerem, R.S.; Nunn, P.D.; et al. Sea Level Change. In *Climate Change 2013: The Physical Science Basis*; Stocker, T.F., Qin, D., Plattner, G.-K., Tignor, M., Allen, S.K., Boschung, J., Nauels, A., Xia, Y., Bex, V., Midgley, P.M., Eds.; Contribution of Working Group I to the Fifth Assessment Report of the Intergovernmental Panel on Climate Change; Cambridge University Press: Cambridge, UK; New York, NY, USA, 2013; pp. 1137–1216. [CrossRef]
6. Integrated Climate Data Center (ICDC) Website for AR5 Sea Level Data. Available online: http://icdc.cen.uni-hamburg.de/1/daten/ocean/ar5-slr.html (accessed on 8 April 2017).
7. IPCC. Press Release—IPCC Agrees Special Reports. AR6 workplan, 2016/03/PR, 14 April 2016. Available online: http://www.ipcc.ch/news_and_events/pdf/press/160414_pr_p43.pdf (accessed on 15 April 2017).
8. Slangen, A.B.A.; Carson, M.; Katsman, C.A.; van de Wal, R.S.W.; Köhl, A.; Vermeersen, L.L.A.; Stammer, D. Projecting twenty-first century regional sea-level changes. *Clim. Chang.* **2014**, *124*, 317–332. [CrossRef]
9. Slangen, A.A.; Meyssignac, B.; Agosta, C.; Champollion, N.; Church, J.A.; Fettweis, X.; Ligtenberg, S.R.M.; Marzeion, B.; Melet, A.; Palmer, M.D.; et al. Evaluating Model Simulations of Twentieth-Century Sea-Level Rise. Part I: Global Mean Sea-Level Change. *J. Clim.* **2017**, *30*, 8539–8563. [CrossRef]
10. Meyssignac, B.; Slangen, A.A.; Melet, A.; Church, J.A.; Fettweis, X.; Marzeion, B.; Agosta, C.; Ligtenberg, S.R.M.; Spada, G.; Richter, K.; et al. Evaluating Model Simulations of Twentieth-Century Sea-Level Rise. Part II: Regional Sea-Level Changes. *J. Clim.* **2017**, *30*, 8565–8593. [CrossRef]
11. Holgate, S.J.; Matthews, A.; Woodworth, P.L.; Rickards, L.J.; Tamisiea, M.E.; Bradshaw, E.; Foden, P.R.; Gordon, K.M.; Jevrejeva, S.; Pugh, J. New data systems and products at the permanent service for mean sea level. *J. Coast. Res.* **2012**, *29*, 493–504. [CrossRef]
12. Permanent Service for Mean Sea Level (PSMSL) Website. Available online: http://www.psmsl.org (accessed on 8 April 2017).
13. Hogarth, P. Preliminary analysis of acceleration of sea level rise through the twentieth century using extended tide gauge data sets (August 2014). *J. Geophys. Res. Oceans* **2014**, *119*, 7645–7659. [CrossRef]
14. AVISO Website. Available online: http://www.aviso.altimetry.fr/duacs/ (accessed on 10 June 2017).
15. Integrated Climate Data Center (ICDC) Website for AVISO SSH Data. Available online: http://icdc.cen.uni-hamburg.de/1/daten/ocean/ssh-aviso.html (accessed on 10 June 2017).
16. Watson, P.J. Identifying the best performing time series analytics for sea-level research. In *Time Series Analysis and Forecasting: Contributions to Statistics*; Rojas, I., Pomares, H., Eds.; Springer International Publishing: Berlin, Switzerland, 2016; pp. 261–278. [CrossRef]
17. Alexandrov, T.; Bianconcini, S.; Dagum, E.B.; Maass, P.; McElroy, T.S. A review of some modern approaches to the problem of trend extraction. *Econ. Rev.* **2012**, *31*, 593–624. [CrossRef]
18. Golyandina, N.; Zhigljavsky, A. *Singular Spectrum Analysis for Time Series*; Springer Science & Business Media: Berlin, Germany, 2013.
19. Golyandina, N.; Nekrutkin, V.; Zhigljavsky, A.A. *Analysis of Time Series Structure: SSA and Related Techniques*; Chapman and Hall/CRC Press: London, UK, 2001.
20. Alexandrov, T.; Golyandina, N. Automatic extraction and forecast of time series cyclic components within the framework of SSA. In Proceedings of the 5th St. Petersburg Workshop on Simulation, St. Petersburg, Russia, 26 June–2 July 2005; pp. 45–50.
21. Project for Statistical Computing Website. Available online: https://www.r-project.org/ (accessed on 8 April 2017).
22. Watson, P.J. Development of a unique synthetic data set to improve sea-level research and understanding. *J. Coast. Res.* **2015**, *31*, 758–770. [CrossRef]
23. Watson, P.J. How to improve estimates of real-time acceleration in the mean sea level signal. *J. Coast. Res.* **2016**, *75*, 780–784. [CrossRef]
24. Watson, P.J. Acceleration in US Mean Sea Level? A New Insight using Improved Tools. *J. Coast. Res.* **2016**, *32*, 1247–1261. [CrossRef]

25. Watson, P.J. Acceleration in European Mean Sea Level? A New Insight Using Improved Tools. *J. Coast. Res.* **2017**, *33*, 23–38. [CrossRef]

26. Watson, P.J. Improved Techniques to Estimate Mean Sea Level, Velocity and Acceleration from Long Ocean Water Level Time Series to Augment Sea Level (and Climate Change) Research. Ph.D. Thesis, University of New South Wales, Sydney, Australia. under review, submitted for examination October 2017.

27. Kondrashov, D.; Ghil, M. Spatio-temporal filling of missing points in geophysical data sets. *Nonlinear Process. Geophys.* **2006**, *13*, 151–159. [CrossRef]

28. Stineman, R.W. A consistently well-behaved method of interpolation. *Creat. Comput.* **1980**, *6*, 54–57.

29. SONEL Website. Available online: http://www.sonel.org/spip.php?page=gps&idStation=2722 (accessed on 8 April 2017).

30. Santamaría-Gómez, A.; Gravelle, M.; Collilieux, X.; Guichard, M.; Míguez, B.M.; Tiphaneau, P.; Wöppelmann, G. Mitigating the effects of vertical land motion in tide gauge records using a state-of-the-art GPS velocity field. *Glob. Planet. Chang.* **2012**, *98*, 6–17. [CrossRef]

31. Ostanciaux, É.; Husson, L.; Choblet, G.; Robin, C.; Pedoja, K. Present-day trends of vertical ground motion along the coast lines. *Earth-Sci. Rev.* **2012**, *110*, 74–92. [CrossRef]

32. Pfeffer, J.; Allemand, P. The key role of vertical land motions in coastal sea level variations: A global synthesis of multisatellite altimetry, tide gauge data and GPS measurements. *Earth Planet. Sci. Lett.* **2016**, *439*, 39–47. [CrossRef]

33. Peltier, W.R. Global glacial isostasy and the surface of the ice-age Earth: The ICE-5G (VM2) model and GRACE. *Annu. Rev. Earth Planet. Sci.* **2004**, *32*, 111–149. [CrossRef]

34. Lambeck, K.; Smither, C.; Johnston, P. Sea-level change, glacial rebound and mantle viscosity for northern Europe. *Geophys. J. Int.* **1998**, *134*, 102–144. [CrossRef]

35. O'Dea, A.; Lessios, H.A.; Coates, A.G.; Eytan, R.I.; Restrepo-Moreno, S.A.; Cione, A.L.; Collins, L.S.; de Queiroz, A.; Farris, D.W.; Norris, R.D.; et al. Formation of the Isthmus of Panama. *Sci. Adv.* **2016**, *2*, e1600883. [CrossRef] [PubMed]

36. Phien-Wej, N.; Giao, P.H.; Nutalaya, P. Land subsidence in Bangkok, Thailand. *Eng. Geol.* **2006**, *82*, 187–201. [CrossRef]

37. Fasullo, J.T.; Nerem, R.S.; Hamlington, B. Is the detection of accelerated sea level rise imminent? *Sci. Rep.* **2016**, *6*. [CrossRef]

38. Minobe, S. Resonance in bidecadal and pentadecadal climate oscillations over the North Pacific: Role in climatic regime shifts. *Geophys. Res. Lett.* **1999**, *26*, 855–858. [CrossRef]

39. Sturges, W.; Douglas, B.C. Wind effects on estimates of sea level rise. *J. Geophys. Res. Oceans* **2011**, *116*. [CrossRef]

40. Qiu, B.; Chen, S. Multidecadal sea level and gyre circulation variability in the northwestern tropical Pacific Ocean. *J. Phys. Oceanogr.* **2012**, *42*, 193–206. [CrossRef]

41. Chambers, D.P.; Merrifield, M.A.; Nerem, R.S. Is there a 60-year oscillation in global mean sea level? *Geophys. Res. Oceans* **2012**, *39*, 39. [CrossRef]

42. Calafat, F.; Chambers, D. Quantifying recent acceleration in sea level unrelated to internal climate variability. *Geophys. Res. Lett.* **2013**, *40*, 3661–3666. [CrossRef]

43. Houston, J.R.; Dean, R.G. Effects of sea-level decadal variability on acceleration and trend difference. *J. Coast. Res.* **2013**, *29*, 1062–1072. [CrossRef]

44. Richter-Menge, J.; Overland, J.E.; Mathis, J.T. (Eds.) Arctic Report Card. 2016. Available online: http://www.arctic.noaa.gov/Report-Card (accessed on 25 May 2017).

45. Provisional World Meteorological Organization Statement on the Status of the Global Climate in 2016. Press Release No. 15, Published 14 November 2016. Available online: https://public.wmo.int/en/media/press-release/provisional-wmo-statement-status-of-global-climate-2016 (accessed on 6 May 2017).

46. IOC-UNESCO and UNEP. *Open Ocean: Status and Trends, Summary for Policy Makers*; United Nations Environment Programme (UNEP): Nairobi, Kenya, 2016.

47. Bolch, T.; Kulkarni, A.; Kääb, A.; Huggel, C.; Paul, F.; Cogley, J.G.; Frey, H.; Kargel, J.S.; Fujita, K.; Scheel, M.; et al. The state and fate of Himalayan glaciers. *Science* **2012**, *336*, 310–314. [CrossRef] [PubMed]

48. Wuebbles, D.J.; Fahey, D.W.; Hibbard, K.A.; Dokken, D.J.; Stewart, B.C.; Maycock, T.K. (Eds.) *Climate Science Special Report: Fourth National Climate Assessment*; U.S. Global Change Research Program: Washington, DC, USA, 2017.

J. Mar. Sci. Eng. **2018**, *6*, 11

49. Neumann, B.; Vafeidis, A.T.; Zimmermann, J.; Nicholls, R.J. Future coastal population growth and exposure to sea-level rise and coastal flooding-a global assessment. *PLoS ONE* **2015**, *10*, e0118571. [CrossRef] [PubMed]

50. Watkiss, P. *The ClimateCost Project. Final Report, Volume 1: Europe*; Stockholm Environmental Institute: Stockholm, Sweden, 2011; ISBN 978-91-86125-35-6.

Journal of
*Marine Science
and Engineering*

MDPI

Article

Two Centuries of Climate Change and Climate Variability, East Coast Australia

Peter Helman and Rodger Tomlinson *

Griffith Centre for Coastal Management, Griffith University, Gold Coast 4222, Australia; big.gannet@gmail.com
* Correspondence: r.tomlinson@griffith.edu.au; Tel.: +61-7-5552-8499

Received: 11 December 2017; Accepted: 19 December 2017; Published: 3 January 2018

Abstract: On the east Australian coast, climate change is expressed as a slowly rising sea level. Analysis of records, dating back over two centuries, also shows oscillating multidecadal 'storm' and 'drought' dominated climate periods that are distinct from long-term climate change. Climate variability, as expressed by these distinct multidecadal periods, is generally associated with phases of the Interdecadal Pacific Oscillation Index (IPO). Two centuries of climate and coastline response are examined for the central east coast of Australia, between Fraser Island and Coffs Harbour. The long record has been compiled by analysing a wide range of indicators and observations, including: historical accounts, storm records, sea level trends, assessment of storm erosion faces, and coastal movement in relation to fixed monuments, surveys, and maps. Periods of suppressed sea level, beach accretion, and drought were found to be associated with strongly positive IPO. Periods of higher sea level, increased storminess, and beach erosion were associated with strongly negative IPO. Understanding the behaviour of climate variability over different timescales has the potential to improve the understanding of, and responses to, climate change. This will be important in the sustainable management of geomorphic and ecological systems.

Keywords: IPO; climate change; climate variability; sea level rise; coastal erosion; Australia

1. Introduction

Climatic and geomorphic evidence indicates that the east Australian coastline observed by the first Europeans settlers 200 years ago was experiencing a relatively stable climatic period, when the sea level was slightly below that in the present. Consistent findings of slowly receding shorelines and erosion of Holocene dunes suggest that the coast was influenced by slowly rising sea levels from the early 1800s to present day [1]. Simulation from HadCM3 [2] shows a trend of global sea level rise from 1820. The east coast sea level rise follows, but has been below the global mean, due to the non-uniform distribution of thermal expansion resulting in different regional rates of sea level rise [3]. Pacific Ocean sea surface temperature (SST) has been recorded in ships' logs from the mid-1800s and has been modelled to show an Interdecadal Pacific Oscillation (IPO). The IPO index oscillates irregularly as positive and negative phases over several decades, resulting in major climate shifts [4–7]. References [8,9] relate oceanic ecosystem production and salmon catches in the North Pacific Ocean to decadal phases of IPO. The authors of Reference [10] observed that sea level at Auckland rises at a slower rate during positive IPO phases than in negative phases. IPO interacts with, and modifies, the inter-annual variation described by El Niño-Southern Oscillation (ENSO). The Southern Oscillation Index (SOI) gives an indication of the development and intensity of El Niño or La Niña events in the Pacific Ocean and is calculated using the pressure differences between Tahiti and Darwin. The variability in SOI is largely due to the influence of the widely understood and discussed two- to seven-year ENSO cycles.

Inter-annual variation of the annual mean sea level from Sydney shows a far greater magnitude of change than long-term trends (over 100 years). The IPO also appears to modulate the frequency of El Niño events, whereby the occurrence of El Niño events is shown to be significantly elevated during the positive phase, with the last 350 years showing that the frequency of La Niña is significantly higher during negative IPO phases [6]. Reference [11] reported that for recent decades (1979–2015), coastal vulnerability across the Pacific correlates with ENSO. The combination of IPO and ENSO provides the broad framework for understanding Australian climate episodic behaviour.

2. Methods

Projecting trends from studies such as Reference [11], which analyse data from periods of years to decades, is questioned. Analysis over decadal or centennial scales provides a method for separating the influence of climate change from climatic variation in coastal dynamics. Two centuries of climate and coastline response are examined for the central east coast of Australia, between Fraser Island and Coffs Harbour. The long record has been compiled by analysing a wide range of indicators and observations, including historical accounts, storm records, sea level trends, assessment of storm erosion faces, coastal movement in relation to fixed monuments, surveys, and maps [1].

To identify underlying climate change trends, recorded changes to the coastline were compared to estimates of sea level, observed records of annual mean sea level, and projected sea level trends. Climate variability was analysed by identifying storm, drought, and flood events on the coast.

3. Findings

3.1. Sea Level and Climate Change

On the east coast of Australia, a slowly rising sea level (Figure 1) has resulted in permanent coastal changes: breaching of coastal dunes, barrier island formation, loss of spits, and growth of flood tide deltas [1,12–14]. The coastline has been moving inland throughout the last two centuries, largely unperceived as the rate of sea level rise has been low and movement well within the range of climatic variability. Little concern has been raised as most of the eroded land is public reserve which has acted as an effective coastline buffer.

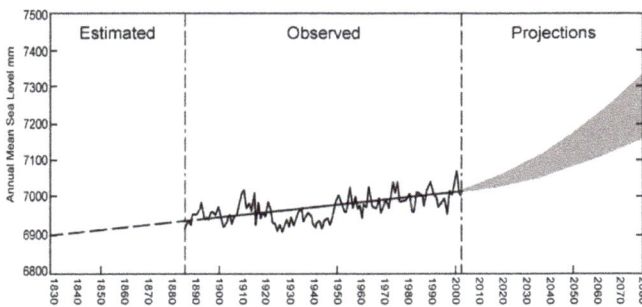

Figure 1. Trend of sea level rise, east coast Australia, 1830–2080. Estimated from 1840 [15]; observed (Sydney Permanent Service for Mean Sea Level (PSMSL) data), projections—Intergovernmental Panel on Climate Change (IPCC) 2007.

3.2. IPO and Sea Level

Quarterly IPO indices from the Hadley Centre, UK[1] were converted to annual means from 1871 and compared to detrended Sydney annual mean sea level (Figure 2). When IPO is plotted inversely,

---

[1] IPO (plotted inverse) 1856 to 1870 after Reference [4]. 1871–2013 version HADISST IPO JFM1871-AMJ2004, low pass 11-year Chebyshev filtered from Hadley Centre, Meteorology Office, UK. Dataset supplied by Scott Power, Research Centre, Bureau of Meteorology, Melbourne.

the relationship with the annual mean sea level is clear and can be statistically analyzed. On the basis of 91 paired comparisons of annual mean detrended Sydney sea level and IPO, the time series correlation is $p < 0.0001$, R = -0.3842, demonstrating a significant inverse relationship between the two time series. An apparent change in correlation occurred around 1976 at the end of the previous negative phase, with mean sea level being elevated during positive IPO. However, this trend did not continue into the 21st century, and the record is too short for definitive statistical analysis to assess whether this may indicate any significant climate change influence.

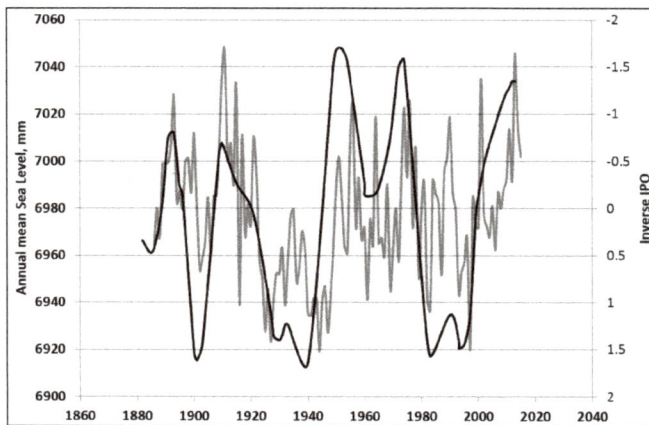

Figure 2. Sydney detrended annual mean sea level from 1886 compared to episodic oscillation of the Interdecadal Pacific Oscillation Index.

3.3. IPO and Climate Variability

The relationship between IPO and climatic factors affecting the Australian east coast have been analysed with past records being used to explore the relationship between drought, floods, storminess, and IPO.

3.3.1. Drought Periods

Due to the impact on agricultural production, droughts are reasonably well recorded in regional and local histories (Figure 3). From the late 1880s droughts have been well documented, especially since the Federation Drought in 1901–1902. When the central eastern Pacific SST is warm and IPO is positive, eastern Australia tends to be warm and dry [4]. Phases of positive IPO are associated with east coast drought conditions and the most severe droughts occur when IPO is positive and SOI is negative [16].

During drought on the coast, river flow is low with little scouring of entrances, especially flood tide deltas, and there is limited hydraulic pressure on spits and barriers. Drought periods as shown in Figure 4 are associated with low numbers of severe storms, lower than trend sea level and, in the past, inshore sand accretion of beaches and barriers along the coast.

3.3.2. Storm Periods

The variation in storminess throughout the 200-year record has several distinct features, especially where it is related to inter-annual variation and IPO phases. Inter-annual oscillations related to ENSO are usually associated with rainfall; however, wet years may not always be associated with severe storms on the coast. Severe coastal storms are those which have caused severe impacts such as beach erosion, inundation, or damage to property or natural assets. The annual number of severe storms per calendar year was derived from past records [17]. Between 1860 and 2014, the number of severe storms per year shows considerable annual variation. 1967 was the most extreme storm year, occurring when both ENSO and IPO were neutral. By contrast, the period between the 1860s and 1890s was the stormiest recorded during a mostly negative IPO.

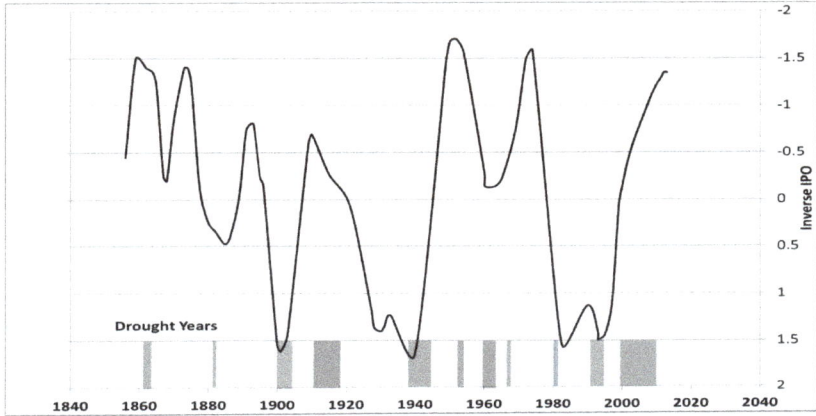

Figure 3. Major drought periods correlate with phases of positive IPO after Reference [16]. (IPO plotted inversely).

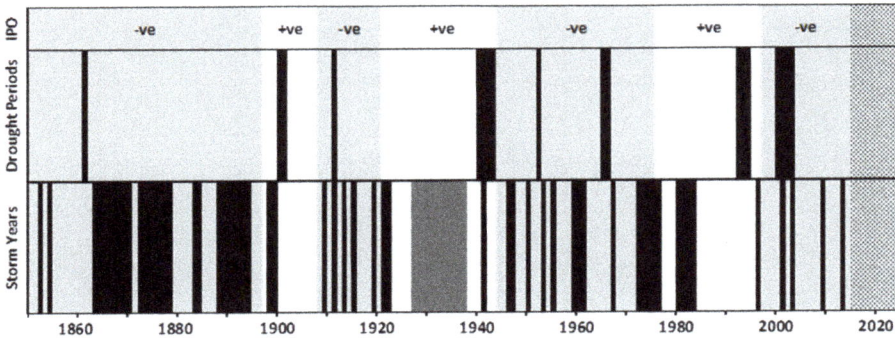

Figure 4. Relationship between IPO phase and severe storm and drought years along the east coast of Australia (storm years in the 1930s are shown but are not considered to be related to IPO). The current negative IPO phase is anticipated to continue into the foreseeable future.

Records have been used to classify erosion years, as shown against IPO phase in Figure 5, and reveal erosion years during negative IPO phase. Erosion years between 1926 and 1938 were from severe storms not related to IPO (as above).

Examination of the storm erosion record indicates significant erosion events are most likely related to individual very severe storms or a sequence of severe storms (storm periods) [12]. Many problems assessing individual, sequential, or annual storminess from incomplete and changing records can be overcome by considering coastline change outcomes of erosion and accretion [17].

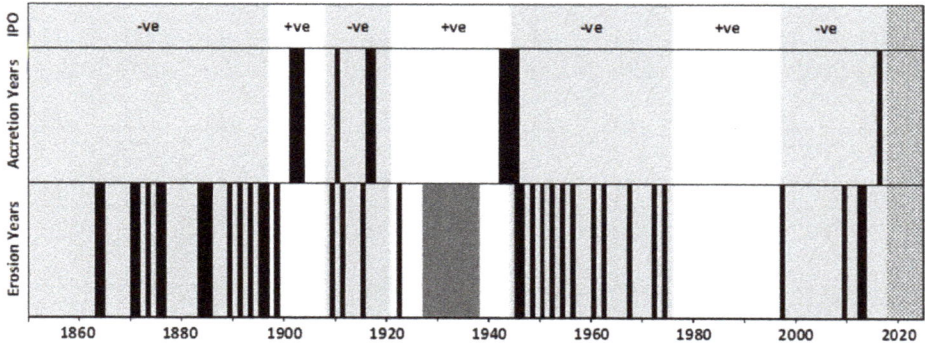

Figure 5. Erosion and accretion years shown against IPO phases. Erosion years in the 1930s are shown but are not considered to be related to IPO.

Long-term erosion response is plotted against inverse IPO (Figure 6). Erosion period 'A' from 1857 to 1898 is related to the long period of negative IPO and is the most severe storm period in the record. During this period, there was significant coastal change:

- Exposure of heavy mineral lenses on beaches (mined by hand for gold).
- Serious problems with construction of breakwaters; at Trial Bay, New South Wales they were continually destroyed during construction and abandoned; at the Clarence River, New South Wales they were continually damaged due to breakwaters during construction.
- Permanent breach of the barrier dune and the permanent division of Stradbroke Island, southern Queensland.
- Storm erosion restricted the use of the beach for travel along the coast's most eastern beaches (Seven Mile and Tallow beaches, northern New South Wales).

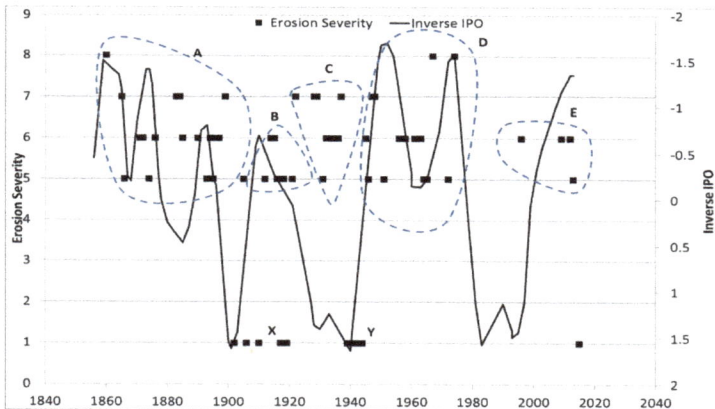

Figure 6. Long-term erosion response related to IPO. Erosion Severity values of 1 indicate an accretion year.

There appears to be a significant sequencing of erosion events, for example the long storm period from 1860 to 1890 culminating in 1893, with record Brisbane River floods and severe erosion on the coast. The following years, 1894 to 1897, were low storm years, yet erosion and barrier dune wash-overs continued. The impacts of extreme individual storms and 'storm periods' or 'sequences of storms' have a cumulative impact, for example the breaches of barrier dunes at Southport and Jumpinpin, Stradbroke Island in southern Queensland occurred from cumulative impacts of several events over a number of years [1].

Erosion period 'B' is related to negative IPO and is short-lived. The significant erosion period 'C', in the mid-1930s, is not related to IPO. The reason for this inconsistency is uncertain, although a similar anomaly has been recognised in ENSO [18]. Erosion period 'D', 1946 to 1974, is well documented [19,20] and is related to negative IPO, but includes 1967, an unusual year that is discussed later in this paper. The erosion face cut at the end of this period, in 1974, remains the most inland face on the majority of the coast. Recent significant erosion events 'E' appear to correlate to the current negative IPO, which is anticipated to continue into the foreseeable future.

Prior to beach photographs in late 1800s, accretion years are difficult to determine from past records. Later surveys and photographs have provided evidence of foredune accretion with two accretion periods identified and referred to as 'X' and 'Y' in Figure 6.

Accretion period 'X' was initiated during a period of positive IPO and severe Federation Drought. Accretion years continue between 1911 and 1918 during a period of below average rainfall and years of generally low storminess. Accretion period 'Y' was initiated during a positive IPO phase during World War II drought, continuing as IPO changed phase before storm period 'D'.

3.3.3. IPO Influence on Climate and Coastal Change

IPO is characterized by positive and negative phases that are linked to distinct periods of Australian climate and coastal behaviour:

- Negative IPO phase—sea level is above long-term trend, with storm periods and major coastal erosion. For example, the 1860s to 1890s and from the early 1950s to mid-1970s.
- Positive IPO phase—sea level is below long-term trend with drought periods and beach accretion. For example, early 1900s, 1940s.
- Periods of transition between these phases—an example is from 1946 to 1952, when IPO was changing from positive to negative. During this transition, the sea level rose rapidly with increasing storminess and erosion.

With positive IPO over the last few decades, accretion would have been expected on the coast, but has not occurred. Instead there has been erosion, especially in the southern end of compartments and western shores of barrier islands. The absence of accretion is attributed to more rapidly rising sea levels.

A long-term study of beach change on the Californian coast [21] showed IPO-related phase behaviour but an inverse response; a positive IPO phase related to storm energy and coastal erosion, and a negative IPO phase with low energy and beach accretion.

3.3.4. Anomalous Periods

Reference [18] considers that the importance of SOI was not recognised earlier due to the anomalous relationship between SOI and climate in the late 1920s and 1930s. It is suggested that a similar anomalous relationship appears between IPO and climate during this period, particularly the mid-1930s. For example, the storm period between 1933 and 1936 resulted in the severe erosion of the southern Queensland coast, and similarly in New Zealand the most severe storms of the 20th century occurred during this period. The Great Cyclone in February 1936 caused widespread damage to the north island of New Zealand and was followed by a second severe storm in March 1936 [22].

The most severe storm year recorded for southeast Queensland was 1967, with severe erosion experienced along the coast. The year was dominated by severe storms from January to August, including four tropical cyclones (including Tropical Cyclone Dinah with a central pressure of 945 hPa) and a record storm surge on the Gold Coast of 1.5 m. Of six east coast lows, the one between 25 and

28 June was very severe [19]. The year corresponded with a reversal in IPO; however, it is considered to have been in an overall negative IPO phase.

4. Discussion

Past positive IPO phases have been associated with lower than trend sea levels and periods of beach and dune accretion with abundant inshore sand supply. Periods of higher sea level, increased storminess, and beach erosion have been shown to be associated with strongly negative IPO. However, in the most recent positive IPO phase, with characteristically low storm activity, erosion is observed in the southern sections of ocean compartments and on the western shorelines of barrier islands. This minor erosion and lack of dune accretion provides evidence for the influence of sea level rise and raises the possibility that sections of the present coastal alignment are under-adjusted to sea level change due to the unusually long period of low storm energy from 1974 to 2009. However, Reference [10] suggested that climate change-driven sea level rise on the New Zealand coast was masked by the positive phase of IPO experienced over the past 30 years, and that for the decades following the year 2000, sea level rise from climate change would likely to be accompanied by a return of a negative IPO phase, associated with higher than trend sea levels, storm energy, and severe erosion.

An analysis of the current trend in IPO (Figure 6), annual mean sea level, and the evidence of more storms in recent years suggest that on the east coast of Australia, there has been a return to climate conditions and responses associated with a negative IPO phase. In the Gold Coast region, recent storminess has been experienced in 2012 and 2013, with more widespread storms along the east coast in 2009 and 2016.

The coincidence of the underlying climate trend of sea level rise and the return of the storm energy associated with negative IPO phases raises a concern that the impacts of future erosion periods are likely to be far more severe than any experienced during the last 200 years.

Coastal management decisions are commonly based on analysis from aerial photography and even shorter periods of measured data from wave rider buoys. Projection of trends from these short period studies is questioned. When compared to the variability shown over the last 200 years, coastal responses in recent decades, during which rapid infrastructure development has occurred, appear unrepresentative of long-term patterns. Analysis over long time periods provides a method for separating the influence of climate change from climatic variation in coastal dynamics.

Although projections of future sea level rise [23] are considered in coastline planning, the impact of climate variability resulting in shorter decadal periods with higher than long-term mean sea levels, accompanied by more storminess, should not be ignored. As has been seen on the east coast of Australia, these periods result in major coastal responses in terms of beach erosion and inundation. In future coastline planning, an allowance for sea level rise from climate change and decadal climate variability will need to be made.

Acknowledgments: The authors acknowledge the valuable contributions to this work by Danielle Verdon and Scott Power.

Author Contributions: Peter Helman conceived and designed the historical data analysis methods and undertook the analysis; Peter Helman and Rodger Tomlinson wrote the paper.

Conflicts of Interest: The authors declare no conflict of interest.

References

1. Helman, P. Two Hundred Years of Coastline Change and Future Change, Fraser Island to Coffs Harbour, East Coast Australia. Ph.D. Thesis, Southern Cross University, Lismore, Australia, 2007.
2. Hadley Centre MetOffice. Available online: http://www.cru.uea.ac.uk/cru/projects/soap/pw/data/model/hadcm3/hadcm3_sealevel.htm (accessed on 3 March 2017).
3. Cazenave, A.; Nerem, R.S. Present-day sea level change: Observations and causes. *Rev. Geophys.* **2004**, *42*. [CrossRef]

4. Power, S.; Casey, T.; Folland, A.; Colman, A.; Mehta, V. Inter-decadal modulation of the impact of ENSO on Australia. *Clim. Dyn.* **1999**, *15*, 319–324. [CrossRef]
5. Salinger, M.J.; Renwick, J.A.; Mullan, A.B. Interdecadal Pacific Oscillation and South Pacific climate. *Int. J. Climatol.* **2001**, *21*, 1705–1721. [CrossRef]
6. Verdon, D.C.; Franks, S.W. Long term behaviour of ENSO—Interactions with PDO over the last 400 years inferred from paleoclimate records. *Geophys. Res. Lett.* **2006**, *33*. [CrossRef]
7. Verdon, D.C. Pacific and Indian Ocean Climate Variability-Implications for Water Resource Management in Eastern Australia. Ph.D. Thesis, University of Newcastle, Callaghan, Australia, 2007.
8. Francis, R.C.; Hare, S.R. Decadal scale regime shifts in the large marine ecosystems of the north-east Pacific: A case of historical science. *Fish. Oceanogr.* **1994**, *3*, 279–291. [CrossRef]
9. Francis, R.C.; Hare, S.R.; Mantua, N.J.; Wallace, J.M.; Zhang, Y. A Pacific Interdecadal climate oscillation with impacts on salmon production. *Bull. Am. Meteorol. Soc.* **1997**, *78*, 1069–1079.
10. Goring, D.; Bell, R. Sea level on the move? *Water Atmos.* **2001**, *9*, 20–21.
11. Barnard, P.L.; Short, A.D.; Harley, M.D.; Splinter, K.D.; Vitousek, S.; Turner, I.L.; Allen, J.; Banno, M.; Bryan, K.R.; Doria, A.; et al. Coastal vulnerability across the Pacific dominated by El Nino/Southern Oscillation. *Nat. Geosci.* **2015**, *8*, 801–807. [CrossRef]
12. Helman, P.; Tomlinson, R.B. Coastal Storms and Climate Change Over the Last Two Centuries, East Coast Australia. In *Solutions to Coastal Disasters 2008*; Wallendorf, L., Ewing, L., Jone, C., Jaffe, B., Eds.; ASCE: Reston, VA, USA, 2008; pp. 139–146.
13. Goodwin, I.D.; Stables, M.A.; Olley, J.M. Wave climate, sediment budget and shoreline alignment Eeolution of the Iluka—Woody Bay sand barrier, northern New South Wales, Australia since 3000 yr BP. *Mar. Geol.* **2006**, *226*, 127–144. [CrossRef]
14. Sloss, C.R.; Jones, B.G.; Murray-Wallace, C.V.; McClennan, C.E. Holocene Sea Level Fluctuations and the Sedimentary Evolution of a Barrier Estuary: Lake Illawarra, New South Wales, Australia. *J. Coast. Res.* **2005**, *21*, 943–959. [CrossRef]
15. Pugh, R.S. Briefing: Some observations on the influence of recent climate change on the subsidence of shallow foundations. *Proc. Inst. Civ. Eng. Struct. Build.* **2002**, *155*, 23–25. [CrossRef]
16. Mckeon, G.M.; Hall, W.B.; Yee Yet, J.; Stone, G.S.; Crimp, S.J.; Peacock, A.; Richards, R.; Tynan, R.W.; Watson, I.W.; Power, S.B. Learning from history: Land and pasture degradation episodes in Australia's rangelands. In Proceedings of the Conference on Managing Australian Climate Variability (CVAP), Albury, Australia, 23–25 October 2000; Power, S.B., Della Marta, P., Eds.; Bureau of Meteorology: Melbourne, Australia, 2000; pp. 64–67.
17. Callaghan, J.; Helman, P. *Severe Storms on the East Coast of Australia 1770–2008, Griffith Centre for Coastal Management*; Griffith University: Gold Coast, Australia, 2008.
18. Nicholls, N. Climate outlooks: From revolutionary science to orthodoxy. In *A Change in the Weather: Climate and Culture in Australia, 2005*; Sherratt, T., Griffiths, T., Robin, L., Eds.; National Museum of Australia Press: Canberra, Australia, 2005; pp. 18–29.
19. Smith, A.W.; Jackson, L.A. Assessment of the past extent of cyclone beach erosion. *J. Coast. Res.* **1990**, *6*, 73–86.
20. Chapman, D.M.; Geary, M.; Roy, P.S.; Thom, B.G. *Coastal Evolution and Coastal Erosion in New South Wales*; Coastal Council of New South Wales: Sydney, Australia, 1982.
21. Zoulas, J.G.; Orme, A.R. Multidecadal-scale beach changes in the Zuma littoral cell, California. *Phys. Oceanogr.* **2007**, *28*, 277–300. [CrossRef]
22. Brenstrum, E. The cyclone of 1936: The most destructive storm of the Twentieth Century? *Weather Clim.* **2000**, *20*, 23–27.
23. Horton, B.P.; Rahmstorf, S.; Engelhart, S.E.; Kemp, A.C. Expert assessment of sea-level rise by AD 2100 and AD 2300. *Quat. Sci. Rev.* **2014**, *84*, 1–6. [CrossRef]

Journal of
Marine Science and Engineering

MDPI

Article

Potential Impacts of Sea Level Rise and Coarse Scale Marsh Migration on Storm Surge Hydrodynamics andWaves on Coastal Protected Areas in the Chesapeake Bay

Alayna Bigalbal, Ali M. Rezaie *, Juan L. Garzon and Celso M. Ferreira

Civil, Environmental, and Infrastructure Engineering, George Mason University, 4400 University Drive, MS 6C1, Fairfax, VA 22030, USA; abigalba@masonlive.gmu.edu (A.B.); jgarzon3@masonlive.gmu.edu (J.L.G.); cferrei3@gmu.edu (C.M.F.)
* Correspondence: arezaie@gmu.edu

Received: 16 May 2018; Accepted: 3 July 2018; Published: 16 July 2018

Abstract: The increasing rate of sea level rise (SLR) poses a major threat to coastal lands and natural resources, especially affecting natural preserves and protected areas along the coast. These impacts are likely to exacerbate when combined with storm surges. It is also expected that SLR will cause spatial reduction and migration of coastal wetland and marsh ecosystems, which are common in the natural preserves. This study evaluates the potential impacts of SLR and marsh migration on the hydrodynamics and waves conditions inside natural protected areas during storm surge. The study focused on four protected areas located in different areas of the Chesapeake Bay representing different hydrodynamic regimes. Historical and synthetic storms are simulated using a coupled storm surge (ADCIRC) and wave (SWAN) model for the Bay region for current condition and future scenarios. The future scenarios include different rates of local SLR projections (0.48 m, 0.97 m, 1.68 m, and 2.31 m) and potential land use changes due to SLR driven marsh migration, which is discretized in the selected preserve areas in a coarse scale. The results showed a linear increase of maximum water depth with respect to SLR inside the protected areas. However, the inundation extent, the maximum wave heights, and the current velocities inside the coastal protected areas showed a non-linear relationship with SLR, indicating that the combined impacts of storm surge, SLR, and marsh migration depend on multiple factors such as storm track, intensity, local topography, and locations of coastal protected areas. Furthermore, the impacts of SLR were significantly greater after a 1 m threshold of rise, suggesting the presence of a critical limit for conservation strategies.

Keywords: coastal protected areas; Chesapeake Bay; sea level rise; storm surge; marsh migration; ADCIRC+SWAN

---

1. Introduction

Both ocean water level records and satellite altimetry from the last century indicate a rise in global sea level [1–4]. In the next century sea level is expected to rise at a greater rate than during the past 50 years [5]. The Fifth Assessment Report (AR5) of the Intergovernmental Panel on Climate Change (IPCC) projected that from 1986–2005 to 2081–2100, the global mean sea level will rise by 0.26–0.55 m and by 0.45–0.82 m respectively, under the lowest (RCP2.6) and highest greenhouse-gas concentration scenarios (RCP8.5) [1]. The potential rise in sea level can largely affect coastal ecosystems through increased flooding, salinity, erosion, and loss of wetlands [6]. Although the loss of coastal wetlands can occur from various reasons [7], studies suggests that sea level rise (SLR) can reduce 22% of the world's coastal wetlands by the 2080s [8,9]. It is also estimated that 66% of coastal wetlands in 76 developing countries are at risk considering 1 m of SLR in the future [7]. Additionally, with projected hurricane intensification over the next century [10,11], the combined effects of storm surge and SLR are likely to increase flood impacts in coastal areas [12].

Furthermore, SLR is most likely to exacerbate the impacts of storm surge by amplifying the total inundated area and maximum water levels [12–15]. Higher surge elevations, along with increasing wave heights driven by SLR [16], may result in increased tidal current and changes in shorelines [17]. The gradual shoreline recession will also reduce the wave energy dampening and increase the long-term erosion rates [18]. This will largely affect coastal wetlands and salt marshes containing plants and vegetation that can only withstand within a limited tidal range, salinity [19,20], and elevation range to mean sea level [21]. The sustainability of these natural habitats depends on the accretion rate at which they are vertically rising with respect to the rate of SLR [22]. If the water elevation is rising at a faster rate than the marsh is able to build in order to sustain vegetation, the marsh will begin to migrate inland [23]. Although some tidal wetlands are capable of vertical movement with small changes in sea level [20,24], higher increase in SLR will cause submergence and landward movement of marshes across the coastal landscape [24,25]. The incapacity of a migration could lead to the loss of the entire marsh and cause major land cover changes. Nonetheless, the physical response of these coastal lands to SLR is complex [26], and thus sustainability of wetlands in these protected lands will depend on multiple factors such as local geomorphology, sediment supply, vertical accretion, subsidence, and interaction between biotic and hydrologic processes [22].

While several studies [13,27–31] evaluated the impacts of storm surge and SLR on coastal lands and communities, their specific effect on protected lands in coastal areas, such as the Federal and State preservation areas, are less discussed. A recent study [32] looked at the potential exposure of coastal protected areas to SLR at a macro level and estimated that about 95% of the protected areas in the eastern United States (US) will be affected by a 0.92 m (3 ft.) of SLR. The study also suggested that adaptation policies for the protected areas should focus on a local scale. However, the study only assessed the impacts of inundation due to SLR. Another study [33] investigated the combined inundation risks from both storm surge and SLR on two northeastern coastal National Parks and found that their vulnerability to inundation varied according to the site location. The results of the study showed that the natural habitat with high-elevation settings is less vulnerable to inundation than the low-lying National Park site. Although the spatial scale of the studies is different, both studies have indicated a higher flood exposure and inundation risk to the coastal protected areas. However, rather than applying coastal numerical models to address the coastal hydrodynamic processes in the protected areas, both studies used a "bathtub" approach to estimate the inundation within the protected areas.

The protected areas along the coasts of the US are natural, undisturbed lands, which contain wetlands, forests, marshes, etc. and provide a range of ecosystem services. This includes water purification, storm surge attenuation, fish nurseries, carbon sequestration, and protection of wildlife habitat [34,35]. With changes in sea level, these ecosystems can become more vulnerable and lead to changes in the hydrodynamic and hydrological regimes [36,37] in the coastal wetlands. In this study, a coupled hydrodynamic and waves model, ADCIRC+SWAN [38–40], is applied for the Chesapeake Bay region to evaluate potential impacts of storm surge, SLR, and marsh migration to the hydrodynamic and wave responses in four coastal protected areas. One historical hurricane and two synthetic storms are simulated for a baseline condition and future SLR scenarios (0.48 m, 0.97 m, 1.68 m and 2.31 m of SLR) to compare and contrast the inundation extent, maximum water elevation, current velocities, and wave heights in the preserve areas. Based on the National Climate Assessment [41] and regional land subsidence rate [42], four local SLR projections are included in the modeling approach. Additionally, projected land use changes due to potential reduction and migration of coastal wetlands and marshes are included in the analysis. The objective is to use regional SLR and publicly available large geographical scale marsh migration projection data in regional coastal numerical models to explore how these land cover changes will impact the hydrodynamics and waves regimes during storm surge events in coastal protected areas in the future.

2. Methodology

2.1. Study Area

The Chesapeake Bay is located within the Mid-Atlantic regions of the east coast of the US (Figure 1). The Bay is surrounded by the coastal counties of Maryland (MD) and Virginia (VA), which has been identified as one of the "hot spot" coasts for SLR [43]. In addition, the bay areas are experiencing a higher rate of land subsidence than the accretion rate (Boon, 2010). The lower accretion rate in the Bay areas can exacerbate the impacts for wetlands and marshes that are located in the low-lying coastal landscape. A study by Beckett et al. [44] used surface elevation and accretion measurements in freshwater and brackish marshes in the Nanticoke estuary of the Chesapeake Bay and demonstrated that, on average, the wetland elevation has decreased by 1.8 ± 2.7 mmyr^{-1}, which is at least 5 mmyr^{-1} below the rate at which global sea level is rising. Recent studies also suggest that due to SLR, Virginia can lose about 50% to 80% of its wetlands [45]. Additionally, a 0.92 m (3 ft.) rise in sea level can affect about 25.2% of the protected areas in Virginia and 24.3% in Maryland [32].

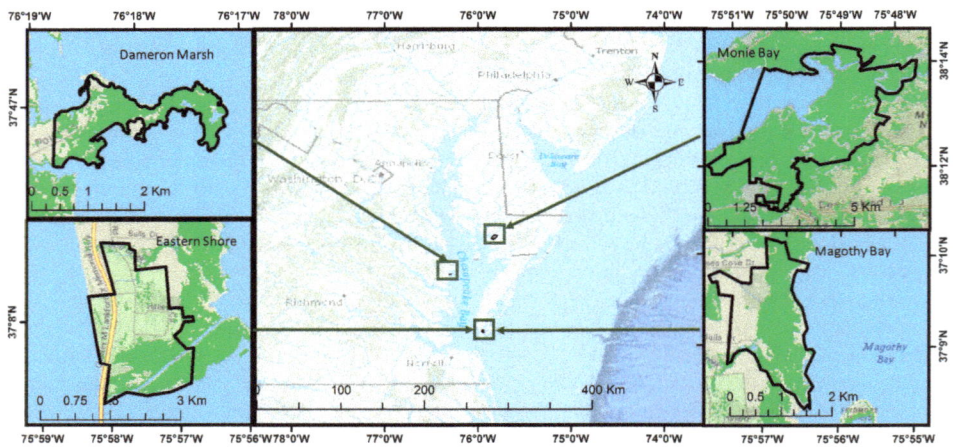

Figure 1. Location of the four study areas within the Chesapeake Bay (tidal wetlands and marshes areas are highlighted in green).

For this study, four protected areas in the Chesapeake Bay were selected due to their different exposure to the tides and surge along the Bay. Figure 1 shows the locations of the sites in the Bay and representative photographs collected during the study. The preserve areas are the Dameron Marsh Natural Area Preserve (DM), the Eastern Shore National Wildlife Refuge (ES), the Magothy Bay Natural Area Preserve (MGB), and the Monie Bay National Estuarine Research reserve (MB). Each site contains different types of wetlands and marshes presenting specific characteristics and increasing the research interest on these sites. For instance, Dameron Marsh is affected by an unbalanced sediment transport problem, highly eroding the north part and building up the southern portions. Eastern Shore and Magothy Bay are located at the southern portion of the Delmarva Peninsula and are highly exposed to storm surge. Monie Bay is the only site located at the mid-eastern side of the Bay, and additionally, it is used for numerous research projects on marsh ecology. It should be noted that terms such as "protected areas," "preserve areas," or "reserve areas" are interchangeably used in the literature referring to natural preserve areas. In this study, the term "preserve area" refers to each selected site, whereas "protected areas" is used to denote protected areas in general.

The selected preserve areas present the typical characteristics of tidal marshes in temperate regions, mainly composed of *Spartina alterniflora* in the lower marsh and *Spartina patens* in the upper marsh [46]. This provides the opportunity to examine the impacts on storm-induced waves, currents, and water levels by a marsh with vegetation typical for the Mid-Atlantic region. The mean elevation of these lands varied with their locations along the Chesapeake Bay. Preserve areas located at the mouth of the Bay (Eastern Shore and Magothy Bay) have relatively lower elevation than the other two sites, which are located near the middle of the Bay. The total area, location, mean elevation, and types of vegetation in each preserve area are summarized in Table 1.

Table 1. Summary chart of the four study sites.

Study Area				
Reserve	Location	Land Area	Mean Elevation *	Vegetation [5]
Dameron [1] Marsh	Northumberland County, VA	1.3 km² (316 Acre)	5.7 m	*Spartina alterniflora, Juncus roemerianus, Distichlis spicata, Scirpus robustus, Phragmites, Spartina patens*
Eastern [3] Shore	Delmarva Peninsula, Northampton County, VA	4.55 km² (1123 Acre)	1.7 m	*Spartina alterniflora, Juncus roemerianus, Distichlis spicata, Spartina cynosuroides, Spartina patens*
Magothy [4] Bay	Northampton, VA	1.16 km² (286 Acre)	1.7 m	*Spartina alterniflora, Juncus roemerianus, Distichlis spicata, Spartina cynosuroides, Spartina patens*
Monie [2] Bay	Somerset County, MD	13.87 km² (3426 Acre)	2 m	*Spartina alterniflora, Ruppia maritima, Spartina patens, Distichlis spicata, Juncus roemerianus*

* Mean Elevations are referred to NADV88. [1] (DCR, 2005); [2] (CBNERR-MD, et al.); [3] ("Easternshore", 2005); [4] ("Magothy Bay Natural Area Preserve", n.d.); [5] (Plant ES Natives Campaign).

2.2. Sea Level Rise in the Study Areas

In order to incorporate SLR in future scenarios, local SLR projections [42] for the Chesapeake Bay are used in this study. The local projections are derived from the synthesis and recommendations from National Climate Assessment (NCA) [41]. Based on global and regional assessment of past SLR trend and future IPCC emission scenarios, NCA prepared four possible SLR projections for managing the coastal resources in the US. Depending on different rates of SLR and ice sheet loss, these projections are considered as the lowest (or historic), low, high, and highest. Considering a constant regional subsidence rate of 2.7 mm/year, Mitchell et al. [42] derived four local SLR projections consistent to the national assessment. Due to low regional subsidence rate and higher rate of local SLR, the study [42] anticipated that subsidence rate for the region will be relatively constant [42]. Although this implies uncertainty in potential land use changes due to SLR, in order to capture a range of SLR and marsh migration impacts, we applied all four local projections in our study. The projected end of the century SLR values used in the study are provided in Table 2.

Table 2. Local sea level rise (SLR) projections at the end of the century for the Chesapeake Bay Regions and the nearest marsh migration scenario.

SLR Projections	Local SLR [1]	Marsh Migration Scenario [2]
Historic	0.49 m (1.6 ft.)	0.61 m (2 ft.)
Low	0.98 m (3.2 ft.)	0.92 m (3 ft.)
High	1.68 m (5.5 ft.)	1.53 m (5 ft.)
Highest	2.32 m (7.6 ft.)	1.83 m (6 ft.)

[1] (Mitchell et al., [42]); [2] (NOAA [47]).

2.3. Marsh Migration—Potential Land Cover Changes

The potential land use changes due to SLR driven marsh migration are collected from the NOAA SLR Viewer tool [47]. The tool estimates potential spatial reduction and migration of coastal wetlands and marshes through a "modified bathtub" approach that includes local and regional tidal range, tidal level and salinity [47]. The basic assumption is, with an increase in sea level, some marshes will move into the adjacent low-lying areas. Meanwhile, marshes unable to maintain their elevation relative to sea level will slowly submerge into open water or convert to an intertidal mudflat [20]. It also considers that, based on the varying frequency, salinity, and time of inundation, certain types of vegetation can exist and particular types of wetland will sustain within an established tidal range [48]. The projections allow wetlands and marshes to migrate into other vegetated canopies such as forested or agricultural lands. It should be noted that coastal physical processes such as erosion, subsidence, or ecological and geomorphologic changes are not included in the NOAA marsh migration projections, which oversimplify the complex processes and can impose uncertainty in their prediction. Although previous studies [49,50] have projected local scale distribution of marshes due to SLR using eco-geomorphologic models, there are trade-offs between acquiring fine scale marsh migration projection and simulating marsh evolution [51,52]. Therefore, this study utilized the NOAA projected marsh migration data for including the best publicly available regional scale coastal land use projection due to SLR induced marsh migration and reduction in a large-scale geographic region.

Also, note that the tool provides marsh migration projections from 0.31 m (1 ft.) to 1.83 m (6 ft.) of SLR with an increment of 0.3 m (1 ft.) of SLR. Therefore, to prepare future scenarios that integrates SLR projections with potential marsh migration, each SLR projection is combined with the closest SLR driven marsh migration projection. For example, in the end century "low" scenario, the projected rise in sea level is 0.96 m (3.2 ft.), while in the NOAA tool the land use change projection due to closet SLR value is available for 0.92 m (3 ft.). Thus, for modeling the "low" scenario, a 0.96 m of SLR is added in the model while the respective land use scenario is incorporated from a 0.92 m (3 ft.) SLR induced marsh migration projection. In Table 2, the correlation between each local SLR projections and nearest marsh movement or reduction scenario is outlined for the readers.

Additionally, Figure 2 demonstrates the projected land cover changes in the selected preserve areas due to marsh migration. The third column in Figure 2 shows the current land cover in the protected areas while the fourth and fifth columns show the projected land cover changes due to the "low" and "highest" SLR scenarios. It can be seen from the last two columns that with SLR, the existing salt-water marshes in the protected areas will either submerge into open water or convert to unconsolidated shore. For example, the second row on Figure 2 displays that with a 1.83 m (6 ft.) of SLR, the existing saltwater marshes in Dameron Marsh will submerge in the open water, while the freshwater wetlands can convert to brackish or transitional marshes. The current land use and land cover information is collected from NOAA's Coastal Change Analysis Program (C-CAP) [53]. The details of different land cover types can also be found from the C-CAP database.

2.4. Modeling Storm Surge and Waves

The coupled version of the hydrodynamic model, ADCIRC [38,39] and wind wave model, SWAN [54], was applied to simulate the impacts of storm surge and SLR in the coastal lands. ADCIRC is a numerical model that computes depth-averaged water levels through the Generalized Wave Continuity Equation (GWCE) and currents through vertically integrated momentum equations [55]. SWAN [54] is a third-generation wave model for estimating wave parameters based on the wave action balance equation. The coupled version of ADCIRC+SWAN simulates the interaction of wind waves and circulation on the same numerical mesh and thus shares the same model boundary. ADCIRC computes the water levels, currents, and wind speeds at each time step and passes it to SWAN. The information is used in SWAN to calculate the wave parameters and wave radiation stress gradients which are further applied to force ADCIRC in the next time step [55,56]. Further details about the coupling processes can be found in [55,56]. In this study, the FEMA Region III Mesh (R3) [57] is used to simulate storm surge in the Chesapeake Bay regions. The R3 Mesh was developed and validated by the United States Army Corps of Engineers [58,59]. It is composed of 1.8 million nodes and has a minimum resolution of 14 m in the Bay regions. The mesh is designed to study the storm surge impacts on the FEMA Region III areas, such as Washington DC, Maryland, Virginia, and thus, it has finer resolution in the Chesapeake Bay areas. The model domain extends from 60° W in the Atlantic Ocean to the 15 m contour line in the Mid-Atlantic coastal regions of the US. Figure 3 shows the model domain with selected storm tracks. The open ocean boundary of the model is forced by harmonic tides extracted from the Le Provost tidal database [60].

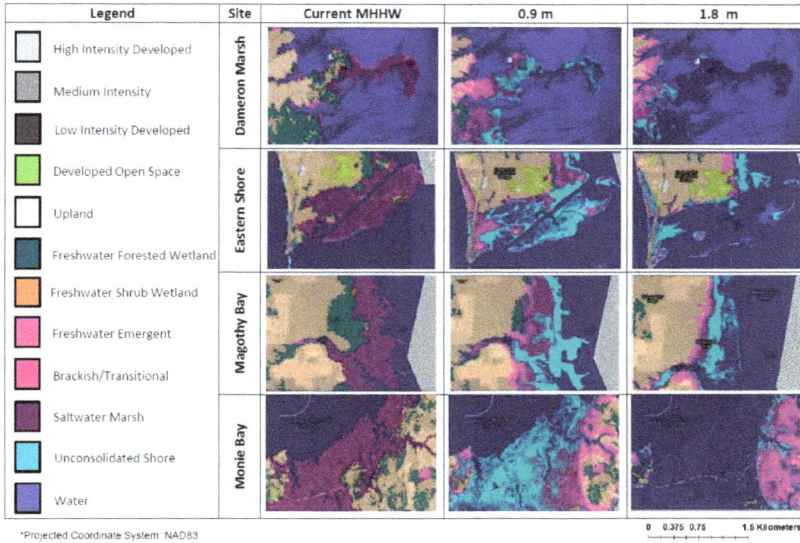

Figure 2. Land cover changes in each study area due to the different SLR–marsh migration scenarios.

Based on track locations and intensity three storms are selected to simulate the impacts of storm surge in coastal protected areas in the Chesapeake Bay. One recent historical storm, Irene (2011), and two synthetic storms (#35 and #145) developed under the North Atlantic Coast Comprehensive Study (NACCS) [61] are selected for this study. Based on statistical analysis of meteorological data and past historical storms tracks in the Mid-Atlantic region, NACCS generated symmetric synthetic storms for the Chesapeake Bay region [61]. The Synthetic145 storm travels parallel to the west side of the Bay, while the Synthetic35 cyclone travels through the Bay, crossing its main axis (Figure 3). Additionally, Hurricane Irene travels to the east of the Bay. In terms of storm intensity and forward speed, the selected storms represent a low to high strength hurricanes, including one of the major historical hurricanes, Irene that impacted the study area. Table 3 provides the minimum central pressure, maximum sustained wind speed and forward speed of the selected storms.

The storm parameters for Irene are collected from the National Hurricane Center (NHC) Hurricane Data 2nd generation (HURDAT2) database. To compute the meteorological forcing due to Hurricane Irene, ADCIRC uses the asymmetric vortex formulation [62,63] based on the Holland wind model [64], which generates the wind and pressure fields for each computational node in the model domain. Since the NACCS generated synthetic storms have symmetric wind field, the meteorological forcings for Synthetic35 and Synthetic145 storms in ADCIRC are calculated using the symmetric vortex formulation of the Holland Model [64]. Additionally, the wind stress over the free water surface is computed from the wind velocity using Garratt's drag formulation [24].

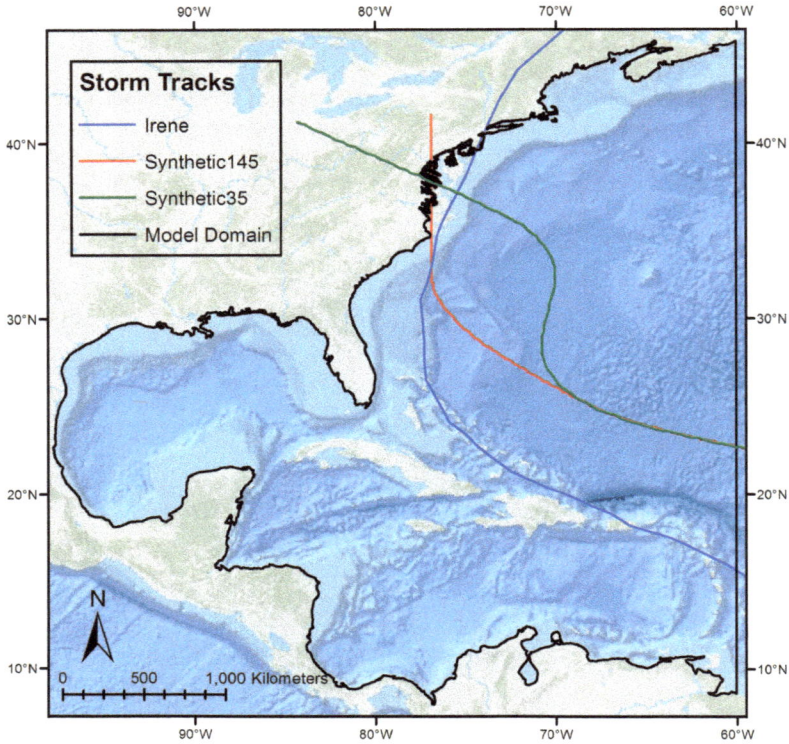

Figure 3. Numerical model domain and selected storm tracks.

Table 3. Wind and pressure intensity of the selected storms.

Storm	Min. Central Pressure (mb)	Max. Wind Speed (kt)	Forward Speed (m/s)
Hurricane Irene	942	105	8.76
Synthetic145	945	111	3.1
Synthetic35	985	64	9.18

The R3 mesh used in this study was calibrated and validated for multiple historical storms during the model development phase [58]. For this study, the model performance is estimated for Hurricane Irene by comparing the simulated peak water levels with observed maximum water levels at NOAA tidal stations located within the Chesapeake Bay areas. The model performance—difference between observed and modeled maximum water elevation—at most of the NOAA tide station locations across the Bay areas showed an overall variance or error (model minus observed) of 0.5 m. For example, the closest NOAA stations to Dameron Marsh and Monie Bay have approximate 0.1 m and −0.05 m errors, respectively (Figure A1: Appendix A). The averaged observed tidal range at the NOAA stations can vary from 0.9 m to 1.3 m [46,65–67]. For Eastern Shore and Magothy Bay, the error at the nearest observed locations are about 0.2 m. Thus, in the selected protected areas, the model performance can vary within a range of 0.2 m to 0.05 m. Although there are some differences in the modeled and observed peak water levels at some of the locations, the results are satisfactory for this study.

2.5. Preserve Areas and Sea Level Rise in the Model

The hydrodynamic and waves regimes in the preserve areas for the selected storms are calculated based on current conditions (i.e., without rise in sea level and using the existing land use information). The analysis is repeated considering the future scenarios incorporating the projected local SLR and potential land use change due to marsh migration. SLR is directly included in the models using eustatic method [29,31,68,69], in which mean sea water level is offset by the locally projected SLR values. Different land use and land cover is represented in the model through frictional drag coefficient (Manning's N) as bottom shear stress. Additionally, the dissipation of momentum transfer from wind to the water column by vegetation in the wetlands and marshes are delineated using surface canopy and land roughness length. Details about the land cover inclusion in ADCIRC is provided in Atkinson et al. [70] and Ferreira et al. [71]. The frictional parameters are assigned on each computational node of the mesh for different land cover types taken from C-CAP database. C-CAP divides wetlands and marshes as palustrine and estuarine categories where each category is subdivided into forested, shrub and emergent wetlands. Note that the frictional coefficients will change depending on the SLR and marsh migration scenarios according to the land cover type changes within the preserves. The frictional parameters in the storm surge model for each land cover types including wetlands and marshes are provided in Table A1. In addition, the average model mesh resolution within the selected preserve areas is approximately 200 m. Thus, the projected marsh migration in the preserve areas are represented in the model in a coarse scale.

Simulated maximum water level, maximum wave height, and maximum velocities for each storm and scenario were incorporated in ArcGIS using the Arc Storm Surge tool [72]. The hydrodynamic and waves regime is analyzed within each preserve area boundary. For calculating the hydrodynamic and wave responses in the study sites, each variable, such as maximum water level or wave heights, are averaged across all the numerical mesh nodes within the protected land boundaries.

3. Results

3.1. Impacts of Sea-Level Rise on Flooding Extent

In order to estimate the flooding extent in the protected areas, the percentage-flooded area due to the selected storms is plotted in Figure 4. The plot for each site shows the percent of inundated area due to storm surge for the current condition and different SLR scenarios. The zero SLR value in the *x*-axis of the plots represents the current condition.

As expected, the results show that SLR and land use change due to marsh migration will increase the inundation in all of the preserve areas. Figure 4 also shows that the percent flooded area for both current condition and the future SLR scenarios varies for each study sites. In addition, increase in flooded area substantially varied due to different rates of SLR. For example, depending on the storm intensity and location, in current condition, 7–10% of the land area of the Dameron Marsh is inundated due to storm surge, which rises up to inundating the entire reserve (100% inundation) with the "highest" SLR scenario regardless of the storm. On the other hand, in Monie Bay, regardless of SLR, the entire preserve area experienced storm surge inundation even under current conditions. The scenario is different for the Magothy Bay and Eastern Shore, which are located at the tip of the Chesapeake Bay. The results show a gradual increase in storm surge flooded areas due to SLR for both Eastern Shore and Magothy Bay preserve areas. None of the two preserve areas are expected to have a 100% flooded area for any storm, even with the highest rate of SLR. However, all preserve areas, except Eastern Shore, are likely to have more than 70% of the total area to be flooded with the highest SLR. Additionally, as expected for most of the cases, the storm intensity played an important role in storm surge inundation in the coastal protected areas. For instance, Hurricane Irene, which is the strongest of the selected storms, caused more coastal flooding than the Synthetic storms in the protected areas. The detailed percent of inundated area for the selected storms and protected areas are provided in Table A2.

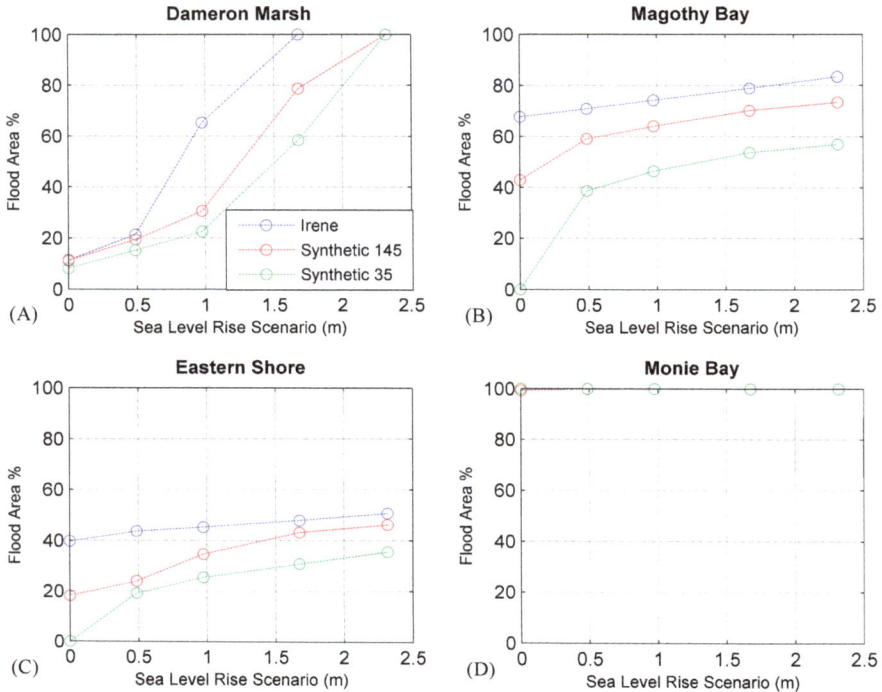

Figure 4. Percent of inundated area at each preserve areas for current condition and different SLR scenarios. The top panel shows the percent of inundated area for (**A**) Dameron Marsh and (**B**) Magothy Bay. In the bottom panel results are shown for (**C**) Eastern Shore and (**D**) Monie Bay Preserve Area.

3.2. Impact of Sea-Level Rise on Maximum Water Levels

In addition to increase in flooded extent, the results show an increase in surge induced maximum water levels in the preserve areas due to SLR. Similar to the Figure 5, the maximum water levels due to the selected storms are plotted in Figure 5, for each of the preserve areas for current conditions and future SLR scenarios.

The plots in Figure 5 clearly show that for both current and future conditions, Hurricane Irene has higher impact on maximum water levels than the other two storms. Although the maximum water levels in each of the preserve areas are distinct for different SLR scenarios, in general, the "highest" inundation height in the protected areas can rise up to 3.5 to 4.6 m, which is almost 1.5 to 2.5 m higher than the flood elevation in the current day storm surge flooding. The results also suggest that increase in water elevation tends to have a linear relationship with the increase in SLR. For example, observing the impact of Irene in Dameron Marsh, a SLR of 0.48 m raises the storm surge water level to 1.84 m. This is 0.42 m greater than the No SLR case, which is almost equal to the amount of SLR that was introduced into the system. Similar patterns are also observed in rest of the preserve areas. In terms of the maximum rise in water level, the Eastern shore and Magothy Bay preserve areas are likely to have higher surge induced flood elevation for both current condition and SLR scenarios than Dameron Marsh and Monie Bay. This could be due to the location of the preserve areas as both Eastern shore and Magothy Bay are situated in the mouth of the Bay and closest to the open ocean. Detailed maximum elevation values for the selected storms and scenarios for all protected lands are provided in Table A3.

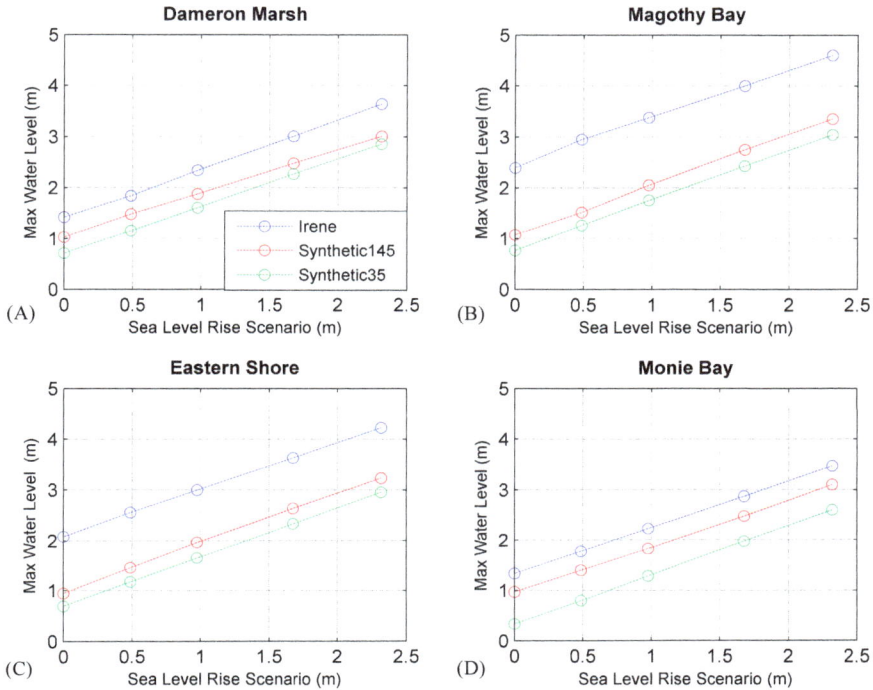

Figure 5. Maximum water levels at each preserve areas for current condition and different SLR scenarios. The top panel shows the simulated maximum water level in (**A**) Dameron Marsh and (**B**) Magothy Bay. In the bottom panel results are shown for (**C**) Eastern Shore and (**D**) Monie Bay Preserve Area.

3.3. Impact of Sea-Level Rise on Currents Velocities

In order to investigate flood propagation, erosion and potential vegetation damage in the protected areas, the study calculated the maximum current velocities in all preserve areas for the simulated storms. Figure 6 provides the maximum current velocities due to selected storms at each of the study sites for present-day condition and for the SLR projections.

The results suggest that the impacts of both storm surge and sea-level rise on currents in the protected areas are highly site specific, although a higher rate of SLR notably increases the current velocity at each preserve area. For example, depending on the storm intensity and track, the maximum current velocities in Monie Bay can reach up to a maximal of 0.35 m/s in the 'highest' SLR scenario. While for both Dameron Marsh and Eastern Shore, the "highest" maximum current velocities are higher than 1 m/s. Figure 6 also shows that current velocities are more sensitive to SLR in Dameron marsh and Eastern Shore than Monie Bay and Magothy Bay. Though both Magothy Bay and Eastern Shore are located close to the mouth of the Chesapeake Bay, Magothy Bay experiences significant lower current energy due to the protection from the surroundings Mockhorn Island and Smith Island. Similarly, Monie Bay is located in the mid Bay, confined by landmasses and exposed to relatively lower energy during hurricanes when compared to the other preserves. Maximum velocities substantially varied with different SLR scenarios for each preserve areas, presenting a non-linear response to SLR. For example, with a 0.49 m rise in sea-level, the maximum velocities in the Dameron Marsh and Magothy Bay decrease from the current condition, while Monie Bay shows increases from the baseline. However, with higher increase in SLR, regardless of the storms, all the preserve areas showed considerable increase in maximum currents velocities. Furthermore, the results show that hurricane Irene has higher impacts on currents velocities in the preserve areas than the Synthetic storms for both current condition and SLR scenarios. The only exception is at Dameron Marsh, where the impact of Synthetic Storm 145 is always higher than both Irene and Synthetic Storm 35. This indicates the significance of the location respect to the storm track on maximum current velocities in the protected areas. While Hurricane Irene travelled parallel to the east of the Bay, both Synthetic storms passed through the mid Bay region near the Dameron Marsh (Figure 3). Contrarily, the wind intensity of Synthetic35 is almost the half of the wind intensity of Synthetic145 and Irene. Therefore, in Dameron Marsh, Synthetic145 has the highest impacts on currents velocities than the other two storms. The detailed currents velocity values in all protected areas for the selected storms and scenarios are provided in Table A4.

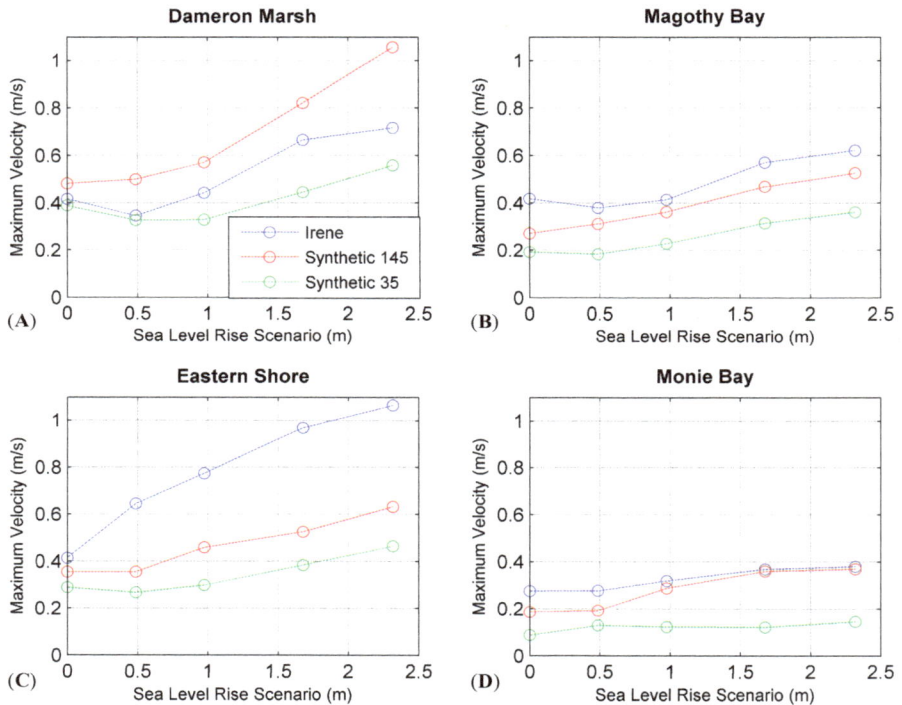

Figure 6. Maximum currents velocities at each preserve areas for current condition and different SLR scenarios. The top panel shows the simulated maximum currents velocity in (**A**) Dameron Marsh and (**B**) Magothy Bay. In the bottom panel results are shown for (**C**) Eastern Shore and (**D**) Monie Bay Preserve Area.

3.4. Impact of Sea-Level Rise on Wave Heights

In Figure 7, the maximum wave heights at each of the preserve areas are shown for current and future SLR scenarios. In terms of change in wave heights in the preserve areas, the results show an increase due to the rise in sea-level. Figure 7 shows that the impacts of wave heights in all preserve areas are higher for Hurricane Irene, which as expected, indicates that higher storm intensity will have higher impacts on wave heights in the protected areas.

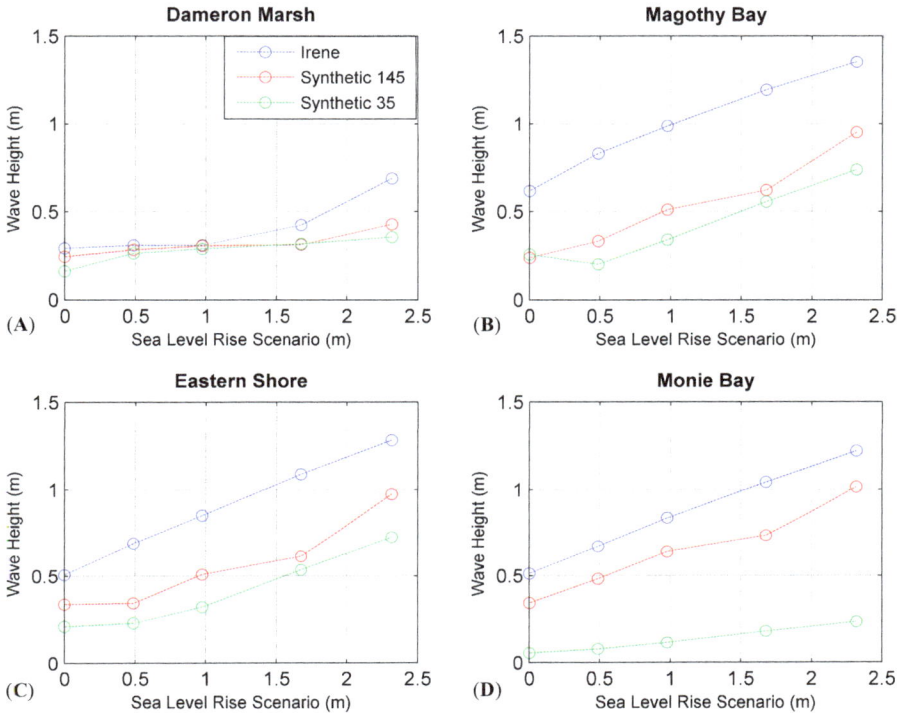

Figure 7. Maximum wave heights at each preserve areas for current condition and different SLR scenarios. The top panel shows the simulated maximum wave heights in (**A**) Dameron Marsh and (**B**) Magothy Bay. In the bottom panel results are shown for (**C**) Eastern Shore and (**D**) Monie Bay Preserve Area.

Furthermore, except for Dameron Marsh, SLR significantly increases the wave heights in the preserve areas. For instance, in Eastern Shore and Magothy Bay, a 0.97 m of SLR during hurricane Irene increases the maximum wave heights by 0.5 m, which is almost a 100% increase from the current conditions. For Dameron Marsh, until the "highest" scenario, no significant increase in wave heights is found with SLR. However, in the rest of the preserve areas, the current wave heights due to storm surge are considerably higher than Dameron Marsh. This indicates that preserve areas with higher wave heights without any SLR will have higher rate of increase due to SLR than the ones with lower wave impacts currently. Additionally, the "highest" SLR scenario leads to more than 100% increase in maximum wave heights from the current conditions during any storm events. For example, depending on the storms, a 2.3 m of SLR can increase the maximum wave heights in Monie Bay from a range of 0.05–0.5 m to 0.25–1.2 m. Detailed wave heights in the protected areas for the selected storms and scenarios are provided in Table A5.

4. Discussion and Implications

The results showed that the hydrodynamic responses inside the preserve area to storm surge, SLR, and marsh migration are site-specific. Therefore, in this section, the findings are summarized to contextualize the results for a regional scale and provide an overall understanding on the impacts of our results in coastal protected areas. Our results indicate that storm intensity plays a significant role in inundation, maximum water levels, and wave heights in the protected areas. The highest intensity storm, Irene, showed higher impacts on the study sites. Figure 8 shows the inundation area and the maximum water levels under different rates of SLR for hurricane Irene in the four preserve areas. The inundation maps in Figure 8 also indicates that the "highest" SLR can raise inundation height in the protected areas to an average of 3.5 to 5 m which is almost 1.5 to 2.5 m higher than the current day flood elevation. The study by Xia et al. (2008) [73] also found out the significance of storm track and intensity in storm surge inundation.

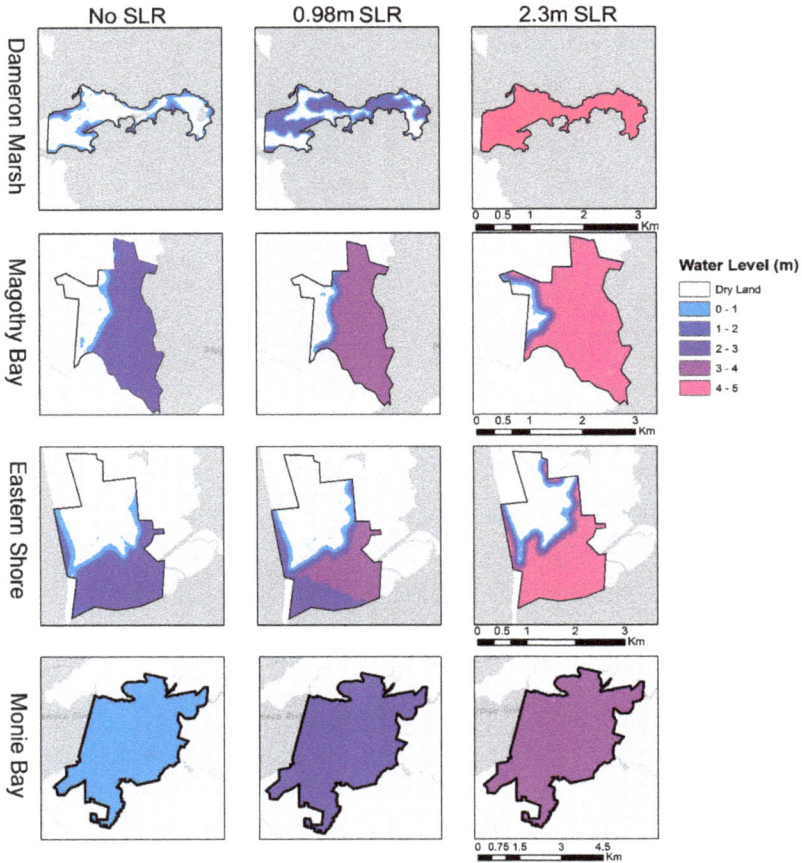

Figure 8. Current and future SLR scenario Inundation maps for the selected preserve areas in the Chesapeake Bay.

Figures 8 and 9 show that, regardless of the storm, almost 100% of Monie Bay land area is inundated by storm surge. In terms of increase in flood extent, results indicate that Dameron Marsh is the most sensitive preserve area to SLR. The average percent of storm surge flooded area in Dameron Marsh increases rapidly, as the rate of SLR increases while for Magothy Bay and Eastern Shore the flooded area gradually expands with the increases in SLR. This reflects the uniqueness of the coastal protected areas in terms of their location and topography. For example, the nearshore elevation at both Eastern Shore and Magothy Bay are around 0–2 m which rises up to an average 2–5 m in areas further inland. Therefore, even in the "highest" SLR scenario with an average maximum water elevation of 3–3.5 m, both preserve areas are not entirely inundated (Figure 8). In contrast, Dameron Marsh and Monie Bay have a relatively flat and constant slope, ranging between 0–2 m within the preserve areas. Thus, a "maximum" of 3 m height of flooding inundates the entire preserve areas. This higher exposure to coastal flooding in these preserve areas can consequently reduce the plant growth and organic matter input that decreases with excess inundation [20]. Our findings suggest that the increase in flooding extent inside the preserves is not linearly related to SLR. A study by Li et al. [74] (2012) in Norfolk, Virginia, also found a nonlinear relationship where a 50-year storm with 1 and 2 m SLR increased total inundated area 34% and 69%, respectively, that changed to 74% and 78% when analyzed for a 100-year storm for same SLR rates. Another study in the Galveston Bay and Jefferson County, Texas, found that the SLR and changing landscapes inundated three times more land when increasing from 0.402 m to 0.926 m [15]. However, the results demonstrated a linear relationship between surge induced water elevation in the protected areas with SLR. These results differ from Bilskie et al. [13], where for coastal areas in Alabama and Mississippi, they found that the increase in water levels was greater than the amount of water added due to SLR. It should be noted, that the focus of our study is on four coastal preserve areas that are spatially very small in compared to the coastal areas studied by Bilskie et al. [13]. These findings suggest that the preserves sizes are not large enough to allow for a fully developed interaction between storm surge hydrodynamics and friction; thus, the long wave associated to the storm surge is not significantly affected within these spatial scales, as observed by Bilskie et al. [13] in the much larger marshes of the Louisiana coast.

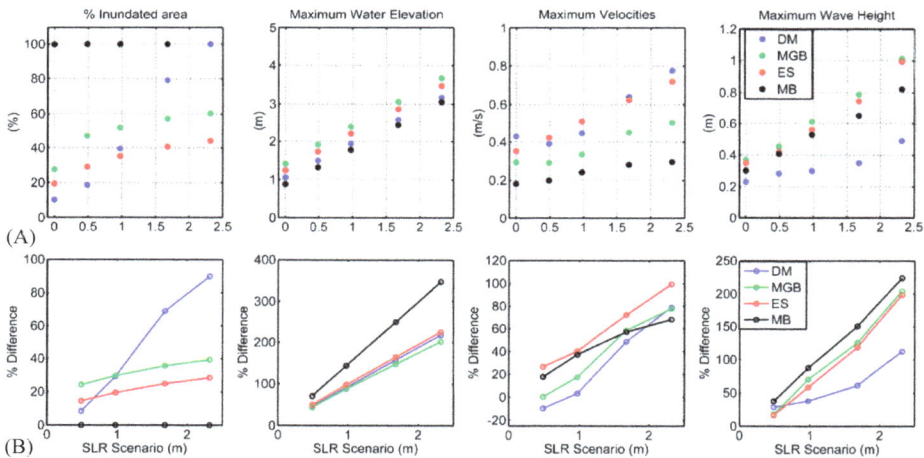

Figure 9. Overall impacts of SLR and marsh migration on Percentage of inundated area, maximum water levels, maximum currents and maximum wave heights at each preserve area. The top panel (**A**) shows the average of the simulated Percentage of inundated area, maximum water levels, maximum currents and maximum wave heights for the selected storms in both current and future scenarios while in the bottom panel (**B**) percentage increase of each of the parameters from the current condition are shown for the selected sites.

When compared with the current conditions, our results show that a maximum of 2.3 m rise in SLR can amplify the current maximum flood elevation in the protected areas by about 200%. This overall increase in maximum water elevation can largely affect the existing vegetation in coastal protected areas in Chesapeake Bay region. For current velocities, the increase is relatively small until 0.97 m of SLR, which significantly amplifies with a higher rate of SLR projections such as 1.68 m and 2.31 m. This implies that a higher rate of SLR can intensify the nearshore erosion in the coastal marshes in the preserve areas, though no linear relationship is found between SLR and current velocities in protected areas. Therefore, with increasing SLR, projected marsh migration and loss of wetlands, the potential shoreline erosion in the coastal protected areas are to likely intensify in future. Moreover, SLR has a significant effect on wave heights, where higher flood depths allow higher waves to propagate through the study area. Except at one site, Dameron Marsh, the increment in SLR showed an almost linear relationship with the increase in wave height in the preserve areas. A few recent studies [16,75] found a nonlinear trend between SLR and wave heights, although their focuses were not on the coastal protected areas. In terms of percent increase from current conditions, a 0.49 m of SLR increases maximum wave heights in the protected areas by less than 50% and a 2.3 m of SLR amplifies by more than 200%. The projected increase in wave heights and water levels in the protected areas can substantially affect the ecosystem service provided by the protected areas. For example, two recent studies [46,76] in the Chesapeake Bay regions applied observed field data in our study site to evaluate the surge and wave attenuation by coastal marshes. Both studies showed that marshes' capacity to attenuate surge and wave heights decreases with increasing inundation and water level. Thus, increase in inundation, water level, and wave heights in the preserve areas due to SLR and marsh migration will lower protected areas capacity to reduce surge level and wave heights, and provide flood protection service.

While the lower rates of sea level could result in adjustable hydrodynamic changes in the study sites, a higher increase of sea level has the potential to significantly alter the hydrodynamic responses to surge and waves and the hydrologic regime within the protected areas. Our results indicate that, for the coastal preserve areas in Chesapeake Bay, considerable increase in the hydrodynamics and waves is observed when SLR exceeds by 1 m. The increasing sea level will affect the distribution of salt marshes [77], and the losses of the saline wetlands are happening at a fast rate [8]. Our study indicates that, with storm surge, the inundation scenario will intensify in coastal wetlands and marshes. This implies that the sustainability of the marshes and wetlands in the protected areas are at a higher risk in the future. Larger currents and wave heights that are caused by the effects of SLR and storm surge might lead to increased coastal erosion. Higher current velocities and wave heights will transport more energy and momentum to the shore, which can cause a faster rate of erosion in the nearshore areas [78]. Therefore, results of the study can provide an improved understanding of the risks associated with SLR to support future management actions, policy, and practices to preserve the coastal protected areas. In addition, incorporating marsh migration in modeling future flooding can add further insights on how sea level can affect the coastal protected areas in the US. The methods applied in this study can also be implemented for other low-lying natural reserves in the coastal areas that are vulnerable to SLR and storm surge.

Though all sites showed higher flooding, each site revealed distinct responses to SLR in terms of faster intrusion of seawater and waves. This indicates two important implications in understanding the vulnerability of the coastal marshes in the protected areas. First, for improved interpretation of how the marshes and wetlands in the protected areas will respond to SLR, local scale analysis is required, since marsh dynamics are highly site specific. Second, the incorporation of marsh migration scenarios is essential when assessing the impacts of SLR on coastal protected areas.

5. Conclusions

Coastal protected areas serve as natural habitats to multiple ecosystems and offer a range of services from flood protection to recreation. Most of the coastal protected areas contain wetlands and marshes, which are unique in nature and are exposed to flooding due to storm surge and SLR. In this study, we combined the impacts of SLR, marsh migration and storm surge on four preserve areas located in different parts of the Chesapeake Bay to assess how coastal hydrodynamics and waves within the protected areas are likely to change in the future. Coupled surge and wave simulations are implemented to gain an improved understanding of the coastal inundation impacts to the Chesapeake Bay preserve areas. We included historical and synthetic storms in our simulations to capture a spectrum of storms and their impacts in the study areas. The simulations incorporated four different local SLR projections based on the National Climate Assessment and regional land subsidence rate. Potential land use changes due to SLR driven marsh migration are also included in our analysis to provide a more accurate representation of future land cover in the protected areas.

Comparing current and projected future inundation extent, maximum flood elevation, current velocity, and waves in four preserve areas in the Chesapeake Bay showed that SLR will increase the hydrodynamics and waves impacts in coastal protected areas. Our study indicates that protected areas responses to both storm surge and SLR are highly site-specific and depend on location, topography, and coastal features of the preserve areas. Therefore, adaptation strategies and restoration plans for the coastal protected areas should be site specific. In terms of the selected sites in the Chesapeake Bay, Monie Bay is found to be the most vulnerable preserve area to coastal flooding, while Dameron Marsh appears to be most sensitive site to SLR. Results also demonstrated that the hydrodynamic and wave responses of the protected areas significantly depend on storm intensity, track and proximity to the shore. Furthermore, findings on the preserve areas in the Chesapeake Bay suggest a linear relationship between SLR and surge induced water elevation in the protected areas. Moreover, the impacts of SLR were significantly greater after a 1 m threshold of SLR, suggesting the presence of a critical limit for conservation strategies.

The projected increase in the hydrodynamic and wave impacts on the preserves can affect the hydrologic regime, salinity, and local geomorphology in coastal protected areas. Higher increase in inundation and potential shoreline erosion can consequently change the ecology of the wetlands and marsh in these natural reserves. While the eco-morphological consequences in the protected areas due to storm surge and SLR are not investigated in this study, the results from the regional scale storm surge and wave models can be applied in local scale hydro-morphodyanmic model to quantify marshes vulnerability to SLR and to advance the understanding of the projected ecological changes in coastal protected areas. It is worth mentioning that fine scale site-specific morphodyanmic and eco-biological data are required to address the vulnerability of tidal wetlands and distribution of marshes due to SLR. Thus, our findings derived from the coarse scale representation of marsh migration should be qualitatively taken into account at best. Additionally, model parameterization needs improvement in representing the interaction between marsh vegetation and storm hydro- and wave dynamics. However, in this study, we provide a suitable method to estimate the potential changes in hydrodynamics and waves in coastal protected areas for storm surge, SLR, and marsh migration that may help to develop necessary adaptation plans for the long-term sustainability of coastal protected areas.

Author Contributions: The paper was conceived and designed by A.M.R., C.M.F., A.B., and J.L.G., A.B. has carried out the simulations, pre- and post-processed the results under the supervision of J.L.G., A.M.R., and C.M.F. Finally, results analysis, interpretation and manuscript preparation are carried out by all authors.

Funding: This research received no external funding.

Acknowledgments: This material is based upon work supported by the National Science Foundation under Grant No. SES-1331399. Any opinions, findings, and conclusions or recommendations expressed in this material are those of the authors and do not necessarily reflect the views of the National Science Foundation. This material is also based upon work supported by the National Fish and Wildlife Foundation and the U.S. Department of the Interior under Grant No. 43932. The views and conclusions contained in this document are those of the authors and should not be interpreted as representing the opinions or policies of the U.S. Government or the National Fish and Wildlife Foundation and its funding sources. Mention of trade names or commercial products does not constitute their endorsement by the U.S. Government, or the National Fish and Wildlife Foundation or its funding sources. Additionally, this research work was partially supported by Virginia Sea Grant Program Development funding (Project Number: R/71851B-PD) and the authors would like to thank Virginia Sea Grant for funding the open access publication of the manuscript. This work used the Extreme Science and Engineering Discovery Environment (XSEDE), which is supported by National Science Foundation grant number ACI-1053575. The authors would like to express sincere gratitude to NOAA Digital coast for providing the land cover projection data. The authors acknowledge the Texas Advanced Computing Center (TACC) at the University of Texas at Austin for providing HPC resources that have contributed to the research results reported within this paper (http://www.tacc.utexas.edu).

Conflicts of Interest: The authors declare no conflict of interest.

Appendix A

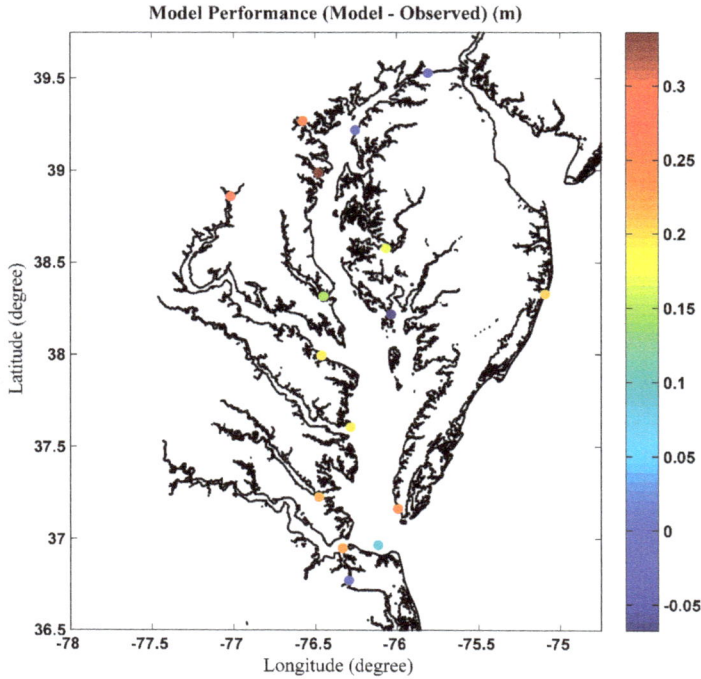

Figure A1. Model validation for Irene showing the error in maximum water level.

Table A1. Frictional parameter values applied in the storm surge and waves model.

Land Cover Type	Class Number	Manning's N	Surface Canopy Coefficient	Surface Directional Effective Roughness Length
Unclassified	1	0	1	0
Developed, High Intensity	2	0.12	1	0.3
Developed, Medium Intensity	3	0.12	1	0.3
Developed, Low Intensity	4	0.07	1	0.3
Developed, Open Space	5	0.035	1	0.3
Cultivated Crops	6	0.1	1	0.06
Pasture/Hay	7	0.055	1	0.06
Grassland/Herbaceous	8	0.035	1	0.04
Deciduous Forest	9	0.16	0	0.65
Evergreen Forest	10	0.18	0	0.72
Mixed Forest	11	0.17	0	0.71
Scrub/Shrub	12	0.08	1	0.12
Palustrine Forested Wetlands	13	0.2	0	0.6
Palustrine Scrub/Shrub Wetlands	14	0.075	1	0.11
Palustrine Emergent Wetlands	15	0.07	1	0.3
Estuarine Forested Wetlands	16	0.15	0	0.55
Estuarine Scrub/Shrub Wetlands	17	0.07	1	0.12
Estuarine Emergent Wetlands	18	0.05	1	0.3
Unconsolidated Shore	19	0.03	1	0.09
Barren Land	20	0.03	1	0.05
Open Water	21	0.025	1	0.001
Palustrine Aquatic Bed	22	0.035	1	0.04
Estuarine Aquatic Bed	23	0.03	1	0.04

Table A2. The percent of inundated land area for each marsh and the percent increase in flooded area shown in parenthesis.

			Irene		
Preserve Area	No SLR	0.48 m	0.97 m	1.68 m	2.31 m
DM	11.3	21.3 (10%)	65.3 (54%)	100 (88.7%)	100 (88.7%)
MGB	67.7	70.9 (3.2%)	74.3 (6.7%)	78.9 (11.3%)	83.4 (15.7%)
ES	39.7	42.6 (2.9%)	45.2 (5.5%)	47.9 (8.2%)	50.7 (11%)
MB	100	100 (0%)	100 (0%)	100 (0%)	100 (0%)
			Synthetic 145		
-	No SLR	0.48 m	0.97 m	1.68 m	2.31 m
DM	11.1	19.4 (8.3%)	30.5 (19.4%)	78.7 (67.7%)	100 (88.9%)
MGB	42.8	59.1 (16.3%)	64.1 (21.3%)	70.2 (27.4%)	73.4 (30.6%)
ES	18.2	24.0 (5.8%)	34.6 (16.4%)	43.1 (24.9%)	46.2 (28%)
MB	99.3	99.9 (0.6%)	100 (0.7%)	100 (0.7%)	100 (0.7%)
			Synthetic35		
-	No SLR	0.48 m	0.97 m	1.68 m	2.31 m
DM	8.08	15.2 (7.09%)	22.5 (14.4%)	58.6 (50.5%)	100 (91.9%)
MGB	14.6	53.2 (38.6%)	61.1 (46.5%)	68.4 (53.8%)	71.7 (57%)
ES	10.5	29.7 (19.1%)	36 (25.5%)	41.3 (30.8%)	46.1 (35.5%)
MB	100	100 (0%)	100 (0%)	100 (0%)	100 (0%)

Table A3. Modeled Data showing the maximum water levels for each Preserve Area. The percent increase in water level is shown in parenthesis.

			Irene		
Preserve Area	No SLR	0.48 m	0.97 m	1.68 m	2.31 m
DM	1.42	1.84 (29.5%)	2.34 (65%)	3.02 (112.7%)	3.64 (156.3%)
MGB	2.39	2.95 (23.2%)	3.38 (41.4%)	4.01 (67.5%)	4.60 (92.3%)
ES	2.08	2.56 (22.9%)	3.00 (44.1%)	3.63 (74.3%)	4.22 (102.9%)
MB	1.33	1.77 (33.2%)	2.23 (67.9%)	2.87 (116.5%)	3.47 (161.7%)
			Synthetic 145		
-	No SLR	0.48 m	0.97 m	1.68 m	2.31 m
DM	1.03	1.48 (43.7%)	1.87 (82.3%)	2.48 (141.3%)	3.01 (193.2%)
MGB	1.07	1.52 (41.4%)	2.05 (90.8%)	2.75 (156.1%)	3.36 (212.7%)
ES	0.94	1.45 (54.1%)	1.96 (108.8%)	2.64 (180.4%)	3.23 (243.9%)
MB	0.97	1.4 (43.2%)	1.8 (86.7%)	2.5 (154.8%)	3.1 (218.2%)
			Synthetic35		
-	No SLR	0.48 m	0.97 m	1.68 m	2.31 m
DM	0.71	1.16 (62%)	1.61 (125.5%)	2.27 (217.9%)	2.86 (301%)
MGB	0.76	1.26 (65%)	1.75 (129.4%)	2.43 (217.7%)	3.05 (298.3%)
ES	0.69	1.17 (69.5%)	1.65 (138.4%)	2.34 (237.5%)	2.95 (326.9%)
MB	0.34	0.80 (134.1%)	1.28 (275.7%)	1.97 (476.7%)	2.60 (662.2%)

Table A4. Modeled data showing the maximum currents velocities for each Preserve Area. The percent increase in maximum currents velocities is shown in parenthesis.

	Irene				
Preserve Area	No SLR	0.48 m	0.97 m	1.68 m	2.31 m
DM	0.42	0.34 (−17.1%)	0.44 (6.1%)	0.66 (59.9%)	0.71 (72.2%)
MGB	0.42	0.38 (−9.3%)	0.41 (−0.89%)	0.57 (36.5%)	0.62 (48.8%)
ES	0.41	0.64 (56.4%)	0.77 (87.4%)	0.97 (134.9%)	1.06 (158.2%)
MB	0.27	0.28 (0.71%)	0.32 (15.9%)	0.37 (34%)	0.38 (38%)
	Synthetic 145				
-	No SLR	0.48 m	0.97 m	1.68 m	2.31 m
DM	0.48	0.50 (3.8%)	0.57 (18.6%)	0.82 (70.6%)	1.06 (120%)
MGB	0.27	0.31 (15%)	0.36 (33.5%)	0.47 (73.8%)	0.52 (94.4%)
ES	0.35	0.46 (30.5%)	0.46 (29.1%)	0.52 (48.1%)	0.63 (78.3%)
MB	0.18	0.19 (3.53%)	0.29 (54.8%)	0.36 (93.4%)	0.37 (98.5%)
	Synthetic35				
-	No SLR	0.48 m	0.97 m	1.68 m	2.31 m
DM	0.39	0.33 (−15.8%)	0.33 (−15.4%)	0.44 (14.62%)	0.56 (44%)
MGB	0.19	0.18 (−4.86%)	0.23 (19.4%)	0.31 (65%)	0.36 (89.4%)
ES	0.29	0.27 (−7.52%)	0.30 (3.36%)	0.38 (33.1%)	0.46 (60.8%)
MB	0.09	0.13 (48.7%)	0.12 (40.8%)	0.12 (39.9%)	0.14 (67.4%)

Table A5. Modeled data showing the maximum wave heights for each Preserve Area. The percent increase in wave heights is shown in parenthesis.

	Irene				
Preserve Area	No SLR	0.48 m	0.97 m	1.68 m	2.31 m
DM	0.29	0.31 (5.79%)	0.31 (6.2%)	0.42 (45.6%)	0.69 (137.2%)
MGB	0.62	0.83 (34.6%)	0.99 (60.1%)	1.19 (93.6%)	1.35 (119.4%)
ES	0.50	0.69 (37%)	0.85 (68.7%)	1.09 (115.4%)	1.28 (153.7%)
MB	0.51	0.67 (30.9%)	0.84 (64%)	1.04 (104.1%)	1.22 (139.1%)
	Synthetic 145				
-	No SLR	0.48 m	0.97 m	1.68 m	2.31 m
DM	0.24	0.28 (16%)	0.30 (26%)	0.31 (29%)	0.43 (76.8%)
MGB	0.24	0.33 (39.4%)	0.51 (116.1%)	0.62 (163.4%)	0.95 (303.2%)
ES	0.33	0.34 (3.1%)	0.51 (51.8%)	0.61 (84%)	0.97 (192%)
MB	0.34	0.48 (39.5%)	0.64 (86%)	0.73 (113.3%)	1.01 (197.2%)
	Synthetic35				
-	No SLR	0.48 m	0.97 m	1.68 m	2.31 m
DM	0.16	0.26 (64%)	0.29 (80.6%)	0.32 (98%)	0.35 (122.2%)
MGB	0.25	0.20 (−21.6%)	0.34 (33%)	0.55 (118.4%)	0.74 (190.1%)
ES	0.21	0.23 (9.5%)	0.32 (53.5%)	0.53 (155.7%)	0.73 (248.1%)
MB	0.05	0.08 (42.6%)	0.11 (112%)	0.18 (233.2%)	0.24 (335.9%)

J. Mar. Sci. Eng. **2018**, *6*, 86

References

1. Church, J.A.; Clark, P.U.; Cazenave, A.; Gregory, J.M.; Jevrejeva, S.; Levermann, A.; Merrifield, M.A.; Milne, G.A.; Nerem, R.; Nunn, P.D.; et al. Sea level change. In *Climate Change 2013: The Physical Science Basis. Contribution of Working Group I to the Fifth Assessment Report of the Intergovernmental Panel on Climate Change*; Cambridge University Press: Cambridge, UK; New York, NY, USA, 2013; pp. 1137–1216. [CrossRef]
2. Church, J.A.; Gregory, J.M.; Huybrechts, P.; Kuhn, M.; Lambeck, C.; Nhuan, M.T.; Qin, D.; Woodworth, P.L. Changes in sea level. In *Climate Change 2001: The Scientific Basis: Contribution of Working Group I to the Third Assessment Report of the Intergovernmental Panel on Climate Change*; Houghton, J.T., Ding, Y., Griggs, D.J., Noguer, M., van der Linden, P.J., Dai, X., Maskell, K., Johnson, C.A., Eds.; Cambridge University Press: Cambridge, UK; New York, NY, USA, 2001; Chapter 11; pp. 639–694.
3. Jevrejeva, S.; Moore, J.C.; Grinsted, A.; Woodworth, P.L. Recent global sea level acceleration started over 200 years ago? *Geophys. Res. Lett.* **2008**, *35*, 8–11. [CrossRef]
4. Nerem, R.S.; Chambers, D.P.; Choe, C.; Mitchum, G.T. Estimating Mean Sea Level Change from the TOPEX and Jason Altimeter Missions. *Mar. Geod.* **2010**, *33*, 435–446. [CrossRef]
5. IPCC. *Climate Change 2013: The Physical Science Basis. Contribution of Working Group I to the Fifth Assessment Report of the Intergovernmental Panel on Climate Change*; Cambridge University Press: New York, NY, USA, 2013; p. 1535. [CrossRef]
6. Nicholls, R.J.; Cazenave, A. Sea Level Rise and Its Impact on Coastal Zones. *Science* **2010**, *328*, 1517–1520. [CrossRef] [PubMed]
7. Blankespoor, B.; Dasgupta, S.; Laplante, B. *Sea-Level Rise and Coastal Wetlands Impacts and Costs*; Policy Research Working Paper; No. 6277; World Bank: Washington, DC, USA, 2012.
8. Nicholls, R.J.; Hoozemans, F.M.J.; Marchand, M. Increasing flood risk and wetland losses due to global sea-level rise: Regional and global analyses. *Glob. Environ. Chang.* **1999**, *9*. [CrossRef]
9. Nicholls, R.J. Coastal flooding and wetland loss in the 21st century: Changes under the SRES climate and socio-economic scenarios. *Glob. Environ. Chang.* **2004**, *14*, 69–86. [CrossRef]
10. Knutson, T.R.; McBride, J.L.; Chan, J.; Emanuel, K.A.; Holland, G.; Landsea, C.; Held, I.; Kossin, J.P.; Srivastava, A.; Sugi, M. Tropical cyclones and climate change. *Nat. Geosci.* **2010**, *3*, 157–163. [CrossRef]
11. Elsner, J.B.; Kossin, J.P.; Jagger, T.H. The increasing intensity of the strongest tropical cyclones. *Nature* **2008**, *455*, 92–95. [CrossRef] [PubMed]
12. Lin, N.; Emanuel, K.; Oppenheimer, M.; Vanmarcke, E. Physically based assessment of hurricane surge threat under climate change. *Nat. Clim. Chang.* **2012**, *2*, 462–467. [CrossRef]
13. Bilskie, M.V.; Hagen, S.C.; Medeiros, S.C.; Passeri, D.L. Dynamics of sea level rise and coastal flooding on a changing landscape. *Geophys. Res. Lett.* **2014**, 927–934. [CrossRef]
14. Passeri, D.L.; Hagen, S.C.; Medeiros, S.C.; Bilskie, M.V. Impacts of historic morphology and sea level rise on tidal hydrodynamics in a microtidal estuary (Grand Bay, Mississippi). *Cont. Shelf Res.* **2015**, *111*, 150–158. [CrossRef]
15. ARCADIS. *ADCIRC Based Storm Surge Analysis of Sea Level Rise in the Galveston Bay and Jefferson County Area in Texas*; Prepared for The Nature Conservancy; Highlands Ranch, Colorado 80129. 2011. Available online: http://cbbep.org/publications/publication1306.pdf (accessed on 22 November 2017).
16. Cheon, S.H.; Suh, K.D. Effect of sea level rise on nearshore significant waves and coastal structures. *Ocean Eng.* **2016**, *114*, 280–289. [CrossRef]
17. Heather, S. Eshorance: An Online Methodology for Estimating the Response of Estuarine Shores From Sea Level Rise. In Proceedings of the NSW Coastal Conference, Batemans Bay, NSW, Australia, 10–12 November 2010; pp. 1–16.
18. Zhang, K.; Douglas, B.; Leatherman, S.; The, S.; July, N.; Zhang, K.; Douglas, B.; Leatherman, S. Do Storms Cause Long—Term Beach Erosion along the U.S. East Barrier Coast? *J. Geol.* **2002**, *110*, 493–502. [CrossRef]
19. The Nature Conservancy; NOAA National Ocean Service Center. *Marshes on the Move: A Manager's Guide to Understanding and Using Model Results Depicting Potential Impacts of Sea Level Rise on Coastal Wetlands*; Washington, DC, USA, 2011. Available online: https://coast.noaa.gov/data/digitalcoast/pdf/marshes-on-the-move.pdf (accessed on 22 November 2017).
20. Morris, J.T.; Sundareshwar, P.V.; Nietch, C.T.; Kjerfve, B.; Cahoon, D.R. Responses of coastal wetlands to rising sea level. *Ecology* **2002**, *83*, 2869–2877. [CrossRef]

21. Bertness, M.D.; Ellison, A.M. Determinants of Pattern in a New England Salt Marsh Plant Community. *Ecol. Monogr.* **1987**, *57*, 129–147. [CrossRef]

22. Cahoon, D.R.; Hensel, P.F.; Spencer, T.; Reed, D.J.; McKee, K.L.; Saintilan, N. Coastal Wetland Vulnerability to Relative Sea-Level Rise: Wetland Elevation Trends and Process Controls. *Wetl. Nat. Resour. Manag.* **2006**, *190*, 271–292. [CrossRef]

23. Kirwan, M.L.; Temmerman, S.; Skeehan, E.E.; Guntenspergen, G.R.; Fagherazzi, S. Overestimation of marsh vulnerability to sea level rise. *Nat. Clim. Chang.* **2016**, *6*, 253–260. [CrossRef]

24. Doyle, T.W.; Krauss, K.W.; Conner, W.H.; From, A.S. Predicting the retreat and migration of tidal forests along the northern Gulf of Mexico under sea-level rise. *For. Ecol. Manag.* **2010**, *259*, 770–777. [CrossRef]

25. Williams, K.; Ewel, K.C.; Stumpf, R.P.; Putz, F.E.; Thomas, W.; Workman, T.W. Sea-level rise and coastal forest retreat on the west coast of Florida, USA. *Ecology* **2007**, *80*, 2045–2063. [CrossRef]

26. Anderson, K.E.; Cahoon, D.R.; Gill, S.K.; Gutierrez, B.T.; Thieler, E.R.; Titus, J.G.; Williams, S.J. Executive summary. In *Coastal Sensitivity to Sea-Level Rise: A Focus on the Mid-Atlantic Region*; A Report by the U.S. Climate Change Science Program and the Subcommittee on Global Change Research; U.S. Climate Change Science Program and the Subcommittee on Global Change Research: Washington, DC, USA, 2009; pp. 1–8.

27. Frey, A.E.; Olivera, F.; Irish, J.L.; Dunkin, L.M.; Kaihatu, J.M.; Ferreira, C.M.; Edge, B.L. Potential impact of climate change on hurricane flooding inundation, population affected and property damages in Corpus Christi. *J. Am. Water Resour. Assoc.* **2010**, *46*, 1049–1059. [CrossRef]

28. Woodruff, J.D.; Irish, J.L.; Camargo, S.J. Coastal flooding by tropical cyclones and sea-level rise. *Nature* **2013**, *504*, 44–52. [CrossRef] [PubMed]

29. Irish, J.L.; Frey, A.E.; Rosati, J.D.; Olivera, F.; Dunkin, L.M.; Kaihatu, J.M.; Ferreira, C.M.; Edge, B.L. Potential implications of global warming and barrier island degradation on future hurricane inundation, property damages, and population impacted. *Ocean Coast. Manag.* **2010**, *53*, 645–657. [CrossRef]

30. Mousavi, M.E.; Irish, J.L.; Frey, A.E.; Olivera, F.; Edge, B.L. Global warming and hurricanes: The potential impact of hurricane intensification and sea level rise on coastal flooding. *Clim. Chang.* **2011**, *104*, 575–597. [CrossRef]

31. Bilskie, M.V.; Hagen, S.C.; Alizad, K.; Medeiros, S.C.; Passeri, D.L.; Needham, H.F.; Cox, A. Dynamic simulation and numerical analysis of hurricane storm surge under sea level rise with geomorphologic changes along the northern Gulf of Mexico. *Earth's Futur.* **2016**. [CrossRef]

32. Epanchin-Niell, R.; Kousky, C.; Thompson, A.; Walls, M. Threatened protection: Sea level rise and coastal protected lands of the eastern United States. *Ocean Coast. Manag.* **2017**, *137*, 118–130. [CrossRef]

33. Murdukhayeva, A.; August, P.; Bradley, M.; LaBash, C.; Shaw, N. Assessment of Inundation Risk from Sea Level Rise and Storm Surge in Northeastern Coastal National Parks. *J. Coast. Res.* **2013**, *291*, 1–16. [CrossRef]

34. Barbier, E.B.; Hacker, S.D.; Kennedy, C.; Koch, E.; Stier, A.C.; Silliman, B.R. The value of estuarine and coastal ecosystem services. *Ecol. Monogr.* **2011**, *81*, 169–193. [CrossRef]

35. Zedler, J.B.; Kercher, S. Wetland Resources: Status, Trends, Ecosystem Services, and Restorability. *Annu. Rev. Environ. Resour.* **2005**, *30*, 39–74. [CrossRef]

36. Burkett, V.; Kusler, J. Climate change: Potential impacts and interactions in wetlands of United States. *J. Am. Water Resour. Assoc.* **2000**, *36*, 313–320. [CrossRef]

37. Baldwin, A.H.; Egnotovich, M.S.; Clarke, E. Hydrologic change and vegetation of tidal freshwater marshes: Field, greenhouse, and seed-bank experiments. *Wetlands* **2001**, *21*, 519–531. [CrossRef]

38. Westerink, J.J.; Blain, C.A.; Luettich, R.A., Jr.; Scheffner, N.W. *ADCIRC: An Advanced Three-Dimensional Circulation Model for Shelves Coasts and Estuaries, Report 2: Users Manual for ADCIRC-2DDI, Dredging Research Program Technical Report DRP-92-6*; U.S. Army Engineers Waterways Experiment Station: Vicksburg, MS, USA, 1994.

39. Luettich, R.A., Jr.; Westerink, J.J. *Implementation and Testing of Elemental Flooding and Drying in the ADCIRC Hydrodynamic Model, Final Report, 8/95, Contract # DACW39-94-M-5869*; U.S. Army Crops of Engineers: Vicksburg, MS, USA, 1995.

40. Dietrich, J.C.; Bunya, S.; Westerink, J.J.; Ebersole, B.A.; Smith, J.M.; Atkinson, J.H.; Jensen, R.; Resio, D.T.; Luettich, R.A.; Dawson, C.; et al. A High-Resolution Coupled Riverine Flow, Tide, Wind, Wind Wave, and Storm Surge Model for Southern Louisiana and Mississippi. Part II: Synoptic Description and Analysis of Hurricanes Katrina and Rita. *Mon. Weather Rev.* **2010**, *138*, 378–404. [CrossRef]

41. Parris, A.; Bromirski, P.; Burkett, V.; Cayan, D.R.; Culver, M.; Hall, J.; Horton, R.; Knuuti, K.; Moss, R.; Obeysekera, J. *Global Sea Level Rise Scenarios for the United States National Climate Assessment*; Climate Program Office: Springfield, MD, USA, 2012.

42. Mitchell, M.; Hershner, C.; Julie, H.; Schatt, D.; Mason, P.; Eggington, E. *Recurrent Flooding Study for Tidewater Virginia*; Virginia Institute of Marine Science (VIMS) Center for Coastal Resources Management: Gloucester Point, VA, USA, 2013.

43. Sallenger, A.H.; Doran, K.S.; Howd, P.A. Hotspot of accelerated sea-level rise on the Atlantic coast of North America. *Nat. Clim. Chang.* **2012**, *2*, 884–888. [CrossRef]

44. Beckett, L.H.; Baldwin, A.H.; Kearney, M.S. Tidal marshes across a chesapeake bay subestuary are not keeping up with sea-level rise. *PLoS ONE* **2016**, *11*, e0159753. [CrossRef] [PubMed]

45. Watch, W. *Adapting to Climate Change in the Chesapeake Bay: Virginia's Experience*; STAC Workshop: Norfolk, VA, USA, 2011.

46. Paquier, A.E.; Haddad, J.; Lawler, S.; Ferreira, C.M. Quantification of the Attenuation of Storm Surge Components by a Coastal Wetland of the US Mid Atlantic. *Estuaries Coasts* **2017**, *40*, 930–946. [CrossRef]

47. NOAA. *Detailed Method for Mapping Sea Level Rise Marsh Migration*; NOAA: Silver Spring, MD, USA, 2017.

48. Marcy, D.; Brooks, W.; Draganov, K.; Hadley, B.; Herold, N.; Mccombs, J.; Pendleton, M.; Ryan, S.; Sutherland, M.; Waters, K. *New Mapping Tool and Techniques for Visualizing Sea Level Rise and Coastal Flooding Impacts. Copyright ASCE 2011 Solutions to Coastal Disasters*; American Society of Civil Engineers (ASCE): Charleston, SC, USA, 2011; pp. 474–490.

49. Kirwan, M.L.; Murray, A.B. A coupled geomorphic and ecological model of tidal marsh evolution. *Proc. Natl. Acad. Sci. USA* **2007**, *104*, 6118–6122. [CrossRef] [PubMed]

50. D'Alpaos, A.; Lanzoni, S.; Marani, M.; Rinaldo, A. Landscape evolution in tidal embayments: Modeling the interplay of erosion, sedimentation, and vegetation dynamics. *J. Geophys. Res. Earth Surf.* **2007**, *112*, 1–17. [CrossRef]

51. Schile, L.M.; Callaway, J.C.; Morris, J.T.; Stralberg, D.; Thomas Parker, V.; Kelly, M. Modeling tidal marsh distribution with sea-level rise: Evaluating the role of vegetation, sediment, and upland habitat in marsh resiliency. *PLoS ONE* **2014**, *9*, e0088760. [CrossRef] [PubMed]

52. Craft, C.; Clough, J.; Ehman, J.; Jove, S.; Park, R.; Pennings, S.; Guo, H.; Machmuller, M. Forecasting the effects of accelerated sea-level rise on tidal marsh ecosystem services. *Front. Ecol. Environ.* **2009**, *7*, 73–78. [CrossRef]

53. NOAA-CSC C-CAP Land Cover Atlas. Available online: https://coast.noaa.gov/digitalcoast/tools/lca.html (accessed on 24 May 2017).

54. Booij, N.; Ris, R.C.; Holthuijsen, L.H. A third-generation wave model for coastal regions: 1. Model description and validation. *J. Geophys. Res.* **1999**, *104*, 7649–7666. [CrossRef]

55. Dietrich, J.C.; Zijlema, M.; Westerink, J.J.; Holthuijsen, L.H.; Dawson, C.; Luettich, R.A.; Jensen, R.E.; Smith, J.M.; Stelling, G.S.; Stone, G.W. Modeling hurricane waves and storm surge using integrally-coupled, scalable computations. *Coast. Eng.* **2011**, *58*, 45–65. [CrossRef]

56. Dietrich, J.C.; Tanaka, S.; Westerink, J.J.; Dawson, C.N.; Luettich, R.A.; Zijlema, M.; Holthuijsen, L.H.; Smith, J.M.; Westerink, L.G.; Westerink, H.J. Performance of the unstructured-mesh, SWAN+ADCIRC model in computing hurricane waves and surge. *J. Sci. Comput.* **2012**, *52*, 468–497. [CrossRef]

57. Hanson, J.L.; Wadman, H.M.; Blanton, B.; Roberts, H. *Coastal Storm Surge Analysis: Modeling System Validation*; Report 4: Intermediate Submission No. 2.0. US Army Engineer Research and Development Center (ERDC). 2013. Available online: http://www.dtic.mil/dtic/tr/fulltext/u2/a583150.pdf (accessed on 22 November 2017).

58. Blanton, B.; Stillwell, L.; Roberts, H.; Atkinson, J.; Zou, S.; Forte, M.; Hanson, J.; Luettich, R. *Coastal Storm Surge Analysis: Computational System Coastal and Hydraulics Laboratory*; Report 2: Intermediate Submission No. 1.2. US Army Engineer Research and Development Center (ERDC). 2011. Available online: http://acwc.sdp.sirsi.net/client/search/asset/1000839;jsessionid=6B9DCD6BB30B93C5752FEB1B865E6A2E.enterprise-15000 (accessed on 22 November 2017).

59. Forte, M.; Hanson, J.; Stillwell, L.; Blanchard-Montgomery, M.; Blanton, B.; Luettich, R.; Roberts, H.; Atkinson, J.; Miller, J. *Coastal Storm Surge Analysis System: Digital Elevation Model*; Report 1: IDS No. 1.1; US Army Engineer Research and Development Center (ERDC). 2011. Available online: http://acwc.sdp.sirsi.net/client/search/asset/1000838;jsessionid=6B9DCD6BB30B93C5752FEB1B865E6A2E.enterprise-15000 (accessed on 22 November 2017).

60. Le Provost, C.; Genco, M.L.; Lyard, F.; Vincent, P.; Canceil, P. Spectroscopy of the world ocean tides from a finite element hydro dynamic model. *J. Geophys. Res.* **1994**, *99*, 24777–24797. [CrossRef]

61. Nadal-caraballo, N.C.; Melby, J.A.; Gonzalez, V.M.; Cox, A.T. *Coastal Storm Hazards from Virginia to Maine Coastal and Hydraulics Laboratory*; US Army Engineer Research and Development Center (ERDC). 2015. Available online: http://www.nad.usace.army.mil/Portals/40/docs/ComprehensiveStudy/Coastal_Storm_Hazards_from_Virginia_to_Maine.pdf (accessed on 22 November 2017).

62. Mattocks, C.; Forbes, C.; Ran, L. *Design and Implementation of a Real-Time Storm Surge and Flood Forecasting Capability for the State of North Carolina*; 2006. Available online: http://www.unc.edu/ims/adcirc/publications/2006/2006_Mattocks.pdf (accessed on 22 November 2017).

63. Mattocks, C.; Forbes, C. A real-time, event-triggered storm surge forecasting system for the state of North Carolina. *Ocean Model.* **2008**, *25*, 95–119. [CrossRef]

64. Holland, G.J. An analytical model of the wind and pressure profiles in hurricanes. *Mon. Weather Rev.* **1980**, *108*, 1212–1218. [CrossRef]

65. NOAA NOAA Tides & Currents. Available online: https://tidesandcurrents.noaa.gov/ (accessed on 26 April 2018).

66. Garzon, J.; Ferreira, C. Storm Surge Modeling in Large Estuaries: Sensitivity Analyses to Parameters and Physical Processes in the Chesapeake Bay. *J. Mar. Sci. Eng.* **2016**, *4*, 45. [CrossRef]

67. Xiong, Y.; Berger, C.R. Chesapeake Bay Tidal Characteristics. *J. Water Resour. Prot.* **2010**, *02*, 619–628. [CrossRef]

68. Atkinson, J.; McKee Smith, J.; Bender, C. Sea-Level Rise Effects on Storm Surge and Nearshore Waves on the Texas Coast: Influence of Landscape and Storm Characteristics. *J. Waterw. Port Coast. Ocean Eng.* **2013**, *139*, 98–117. [CrossRef]

69. Passeri, D.L.; Hagen, S.C.; Plant, N.G.; Bilskie, M.V.; Medeiros, S.C.; Alizad, K. Tidal hydrodynamics under future sea level rise and coastal morphology in the Northern Gulf of Mexico. *Earth's Futur.* **2016**, *4*, 159–176. [CrossRef]

70. Atkinson, J.; Ph, D.; Hagen, S.C.; Ce, D.; Wre, D.; Zou, S.; Bacopoulos, P.; Medeiros, S.; Weishampel, J. Deriving Frictional Parameters and Performing Historical Validation for an ADCIRC Storm Surge Model of the Florida Gulf Coast. *Florida Watershed* **2011**, *4*, 22–27.

71. Ferreira, C.M. Uncertainty in hurricane surge simulation due to land cover specification. *J. Geophys. Res. Ocean.* **2014**, *119*, 1812–1827. [CrossRef]

72. Ferreira, C.M.; Olivera, F.; Irish, J.L. Arc StormSurge: Integrating Hurricane Storm Surge Modeling and GIS. *J. Am. Water Resour. Assoc.* **2014**, *50*, 219–233. [CrossRef]

73. Xia, M.; Xie, L.; Pietrafesa, L.J.; Peng, M. A Numerical Study of Storm Surge in the Cape Fear River Estuary and Adjacent Coast. *J. Coast. Res.* **2008**, *4*, 159–167. [CrossRef]

74. Li, H.; Lin, L.; Burks-Copes, K. Numerical modeling of coastal inundation and sedimentation by storm surge, tides, and waves at Norfolk, Virginia, USA. In Proceedings of the 33rd International Conference on Coastal Engineering, Santander, Spain, 1–6 July 2012.

75. Zhang, K.; Li, Y.; Liu, H.; Xu, H.; Shen, J. Comparison of three methods for estimating the sea level rise effect on storm surge flooding. *Clim. Chang.* **2013**, *118*, 487–500. [CrossRef]

76. Glass, E.M.; Garzon, J.L.; Lawler, S.; Paquier, E.; Ferreira, C.M. Potential of marshes to attenuate storm surge water level in the Chesapeake Bay. *Limnol. Oceanogr.* **2017**. [CrossRef]

77. Kennedy, V.S.; Twilley, R.R.; Kleypas, J.A.; Cowan, J.H.; Hare, S.R. *Coastal and Marine Ecosystems and Global Climate Change: Potential Effects on U.S. Resources*; Prepared for the Pew Center on Global Climate Change; Center for Climate and Energy Solutions: Arlington, VA, USA, 2002; p. 64.

78. Glick, P.; Clough, J.; Polaczyk, A.; Couvillion, B.; Nunley, B. Potential Effects of Sea-Level Rise on Coastal Wetlands in Southeastern Louisiana. *J. Coast. Res. Spec. Issue 63 Underst. Predict. Charg. Coast. Ecosyst. North. Gulf Mex.* **2013**, *63*, 211–233. [CrossRef]

Journal of
Marine Science and Engineering

MDPI

Article

Significance of Fluvial Sediment Supply in Coastline Modelling at Tidal Inlets

Janaka Bamunawala [1,2,*], **Shreedhar Maskey** [1], **Trang Minh Duong** [1,3] **and Ad van der Spek** [3,4]

1   Department of Water Science and Engineering, IHE Delft Institute for Water Education, P.O. Box 3015, 2601 DA Delft, The Netherlands; s.maskey@un-ihe.org (S.M.); t.duong@un-ihe.org (T.M.D.)
2   Department of Water Engineering and Management, University of Twente, P.O. Box 217, 7500 AE Enschede, The Netherlands
3   Applied Morphodynamics, Deltares, P.O. Box 177, 2600 MH Delft, The Netherlands; Ad.vanderSpek@deltares.nl
4   Department of Physical Geography, Faculty of Geosciences, Utrecht University, P.O. Box 80115, 3508 TC Utrecht, The Netherlands
*   Correspondence: j.bamunawala@un-ihe.org; Tel.: +31-062-734-9307

Received: 31 May 2018; Accepted: 22 June 2018; Published: 3 July 2018

Abstract: The sediment budget associated with future coastline change in the vicinity of tidal inlets consists of four components; sea level rise-driven landward movement of the coastline (i.e., the Bruun effect), basin infilling effect due to sea level rise-induced increase in accommodation space, basin volume change due to variation in river discharge, and coastline change caused by change in fluvial sediment supply. These four components are affected by climate change and/or anthropogenic impacts. Despite this understanding, holistic modelling techniques that account for all the aforementioned processes under both climate change and anthropogenic influences are lacking. This manuscript presents the applications of a newly-developed reduced complexity modelling approach that accounts for both climate change and anthropogenically-driven impacts on future coastline changes. Modelled results corresponding to the year 2100 indicate considerable coastline recessions at Wilson Inlet (152 m) and the Swan River system (168 m) in Australia and Tu Hien Inlet (305 m) and Thuan An Inlet (148 m) in Vietnam. These results demonstrate that coastline models should incorporate both climate change and anthropogenic impacts to quantify future changes in fluvial sediment supply to coasts to achieve better estimates of total coastline changes at tidal inlets. Omission of these impacts is one of the major drawbacks in all the existing coastline models that simulate future coastline changes at tidal inlets. A comparison of these modelled future coastline changes with the predictions made by a relevant existing modelling technique (Scale Aggregated Model for Inlet-interrupted Coasts (SMIC)) indicates that the latter method overestimates total coastline recessions at the Swan River system, and the Tu Hien and Thuan An Inlets by 7%, 10%, and 30%, respectively, underlining the significance of integrating both climate change and anthropogenic impacts to assess future coastline changes at tidal inlets.

Keywords: coastline modelling; inlet-affected coastlines; fluvial sediment supply; climate change impacts; anthropogenic impacts

---

1. Introduction

Coastlines in the vicinity of tidal inlets are shaped and affected not only by oceanic processes like tides, waves and mean sea level changes, but also by terrestrial processes, such as river flow, fluvial sediment supply, land use pattern changes, and land management [1–3]. For a given combination of anthropogenic impacts, such as deforestation, damming of rivers, and changes in land use and management patterns and environmental forcing, like rainfall and temperature, a river catchment will produce a certain amount of sediment flux. Depending on river morphodynamics, all or part of this sediment will eventually enter the downstream estuary. Depending on whether the estuary is in a sediment importing or exporting phase (relative to the ocean side of the estuary), all or part of the fluvial sediment received by the estuary will either settle within the estuary or enter the nearshore zone. Subjected to contemporary wave and tide processes, sediment supplied into the nearshore zone may contribute to one or more morphodynamic processes (e.g., ebb-delta development, coastline progradation, deposition on the lower shoreface). Any substantial change(s) in anthropogenic impact(s), such as dam construction or demolition, changes in crop patterns, changes in land management or variation(s) in environmental forcing due to climate change, could result in considerable changes in these sediment transport pathways. This could, in turn, have significant implications on estuary morphology, tidal flats, wetlands, coastlines, ebb-deltas, shorefaces, and a myriad of other physical impacts.

Despite scientists, engineers and managers having been, for decades, cognisant of this fact, owing to the inherent divisions between traditional academic disciplines such as hydrology, geology, oceanography, and coastal engineering, have resulted in a fragmented approach towards assessing anthropogenic and climate change driven impacts on long-term evolution of catchment, estuarine, and coastal systems. Owing to this reason, most of the studies to date have concentrated only on one or two of the three main components of catchment-estuary-coastal systems (CEC systems), while ignoring the other(s) (e.g., [2–7]). Given that the foreshadowed population growth and predicted climate change by the 21st century and beyond, suitable modelling techniques that can effectively simulate the long-term evolution of catchment, estuarine and coastal system behaviour are (urgently) required by contemporary coastal zone planners and managers to investigate the probable impacts of the aforementioned system forcing on the holistic behaviour of CEC systems.

The Scale-aggregated Model for Inlet-interrupted Coasts (SMIC) presented by [8] is the first of its kind that treats CEC systems in a holistic manner, while giving consideration to the description of physics governing each of the three components of the integrated systems. Although SMIC provides a solid platform to holistically probe into CEC systems, the highly-simplified method adopted in quantifying the fluvial sediment supply can be identified as its major drawback in accurately estimating the long-term behaviour of CEC systems under climate change and anthropogenic forcing.

Given the vulnerabilities of estuary-coastal systems, their significance and laps in existing modelling techniques that can holistically simulate their future behaviour, a reduced complexity modelling (RC modelling) technique is currently being developed to assess probable local scale (~30 km alongshore) coastline changes in the vicinity of small tidal inlets, estuaries with low-lying margins and barrier island coasts at macro (50–100 year) time scales. This manuscript presents a part of this study, which scrutinizes the significance of fluvial sediment supply in assessing future coastlines changes at small tidal inlets. Case studies used in SMIC applications [8] were used in this work as well, making it convenient to compare this model performances with that of relevant existing method(s). Therefore, this study focuses on coastline variations at CEC systems in Western Australia (Wilson Inlet and Swan River) and Vietnam (Tu Hien and Thuan An Inlets). Inlet-estuary systems considered in this study are small barrier estuaries, which can be found along microtidal, wave-dominated sandy coasts (~50% of the world's coastlines [9]). These four sites represent seasonally (Wilson), intermittently (Tu Hien), and permanently open (Swan River and Thuan An) inlet systems. They are also attributed with different basin and tidal prisms sizes and annual river flow volumes [8]. Compared to the present conditions, the catchments of Tu Hien and Thuan An inlet systems would generate larger river discharge volumes by 2100, whereas these volumes from the Wilson and Swan river systems are expected to be smaller [10].

2. Methods

Total coastline change at a tidal inlet can be synthesized according to the following [8]:

$$\Delta C_T = \Delta C_{BE} + \Delta C_{BI} + \Delta C_{BV} + \Delta C_{FS} \tag{1}$$

where ΔC_T is the total coastline change (m), ΔC_{BE} is sea-level rise driven landward movement of the coastline, ΔC_{BI} is the basin infilling due to sea-level rise induced increase in accommodation space, ΔC_{BV} is the basin volume change due to variation in river flow and ΔC_{FS} is the coastline change due to change in the fluvial sediment supply.

For detailed insights regarding the derivations of above four components in equation [1], readers are referred to [8]. Given the focal point of this study hinges on the significance of fluvial sediment supply in assessing future coastline changes, brief descriptions of the relevant methods adopted are presented herewith.

2.1. Sea-Level Rise-Driven Landward Movement of the Coastline

Conceivably the most renowned climate-change driven impact on coastlines, described by [11]. The resulting coastline recession (ΔC_{BE}) is given by:

$$\Delta C_{BE} = \frac{\Delta S}{\tan \beta} \tag{2}$$

where ΔC_{BE} is the coastal recession (m), ΔS is sea level rise (m), and β is the slope of the active beach profile.

2.2. Basin Infilling due to Sea-Level Rise-Induced Increase in Accommodation Space

Accommodation space is the additional volume created within the basin due to a given increment in mean sea-level (ΔS). This volume ($\Delta S \times A_b$; where, A_b is the basin surface area (m^2)) results in a sediment demand that creates an additional coastline erosion (ΔC_{BI} (m)). Such a coastal recession results in a landward shift of the coastal profile from the coastline to the depth of closure (h_{DoC} (m)) along a certain alongshore length (affected coastline length (L_{AC} (m))).

There exists a timescale difference in sea-level rise (hydrodynamic forcing) and concomitant basin infilling (morphological response). This lag in morphological response is estimated to be about 50% of the volume required by the corresponding forcing conditions [8]. These relationships can be expressed as the following:

$$\Delta C_{BI} \times L_{AC} \times h_{DoC} = \text{fac} \times \Delta S \times A_b \tag{3}$$

2.3. Basin Volume Change due to Variation in River Flow

Basin volume would change due to the possible future changes in annual river discharge (ΔQ_R (m^3)). In order to maintain the basin cross-sectional velocities, a basin-inlet system would undergo changes in its bed level, so that the changes in basin volume is accommodated. This process will result in a certain change in the nearby coast (ΔC_{BV} (m)), which can be calculated as the following [8]:

$$\Delta C_{BV} \times L_{AC} \times h_{DoC} = \frac{\Delta Q_R \times V_B}{(P + Q_R)} \tag{4}$$

where Q_R is the present river flow into the basin during ebb (m^3), ΔQ_R is the climate-change driven variation in river flow during ebb (m^3), V_B is the present basin volume (m^3) and P is the mean ebb tidal prism (m^3).

2.4. Coastline Change due to Changes in Fluvial Sediment Supply

Anthropogenic and climate-change driven impacts that may occur in future would result in swift changes in fluvial sediment (ΔQ_S in m^3) supplied to coasts. Subsequently, these changes would affect the volume of sediment exchanged between the basin and its neighbouring coast. The relevant changes in coastline (ΔC_{FS} (m)) can be calculated as [8]:

$$\Delta C_{FS} \times L_{AC} \times h_{DoC} = \int_0^T \Delta Q_S(t) \cdot dt \tag{5}$$

where T is the time period considered (in years).

2.5. Factors Affecting Fluvial Sediment Supply to Coasts

Rivers contribute for about 95% of the sediment received by the oceans [12]. Despite the amount of soil eroded from catchments is increasing due to the combined effects of climate change and anthropogenic impacts [12–14], it is being reported that the amount of sediment received by the oceans is considerably reduced caused by anthropogenic sediment retention [3,7,12,15]. Therefore, it is generally accepted that anthropogenic factors overwhelm the impact of natural processes on fluvial sediment supply to the coasts [15].

Climatic factors such as temperature, mean and extreme rainfall, and river flow are the main factors that affect fluvial sedimentation [12,13,16]. Predicted climate changes in the future would most likely result in an increased temperature [17], thus influencing the rate of soil erosion (both chemical and mechanical) and storage and the release of water from the Earth's lithosphere [13]. The combined effect of reduced rainfall and increased temperature results in water stresses to plants, resulting in diminished growths and, hence, amplified soil erosion that results in a larger sediment yield from catchments. Contrariwise, high rainfall and low temperatures facilitate favourable conditions for plant growth, and hence reduce soil erosion, which in turn diminishes the sediment yield from catchments [16]. Global-scale studies indicate that the rivers contemporary discharge less water to the oceans due to both increased water usage and diminished precipitation [7,15,17–19], which, in turn, directly affects fluvial sediment supply to the coasts and oceans.

On the other hand, anthropogenic impacts on fluvial sediment supply vary over a wider spectrum. Land-use management practices (e.g., urbanization, changes in land use, land management and agricultural practices), changes in water management practices (e.g., dam constructions or demolition, streamflow regulation, introduction of flood control mechanisms) and river sand mining are the most prominent human-induced drivers that affect fluvial sediment supply [7,12,13,15,18]. Owing to the aforementioned factors, human activities have simultaneously increased catchment sediment yield via accelerated soil erosion, yet, have also significantly reduced the amount of sediment received by the coasts due to retention within reservoirs [3,7,13,15,16,20–22].

2.6. Assessment of Fluvial Sediment Supply to Coasts

The proposed new reduced complexity model by [23] utilizes the empirical BQART model presented by [13] to assess the annual fluvial sediment supply to the coasts. This empirical model is based on 488 globally-distributed datasets. For catchments with a mean annual temperature greater than or equal to 2 °C, the aforementioned model estimates annual sediment volume supplied to the coast by the following equation:

$$Q_s = \omega \times B \times Q^{0.31} \times A^{0.5} \times R \times T \tag{6}$$

where ω is 0.02 or 0.0006 for the sediment volume (Q_s), expressed in kg/s or MT/year, respectively, Q is the annual river discharge from the catchment considered $\left(km^3/year\right)$, A is the catchment area $\left(km^2\right)$, R is the relief of the catchment (km) and T is the catchment-wide mean annual temperature (°C).

Term 'B' in the above equation [6] represents the catchment sediment production and comprises glacial erosion (I), catchment lithology (L) that accounts for its soil type and erodibility, a reservoir trapping efficiency factor (T_E), and the human-induced erosion factor (E_h), which is expressed as follows:

$$B = IL(1 - T_E)E_h \tag{7}$$

Glacial erosion (I) in above equation [7] is expressed as:

$$I = 1 + (0.09A_g) \tag{8}$$

where A_g is the percentage of ice cover of the catchment area.

The human-induced erosion factor (E_h; anthropogenic factor) of the above equation [7] depends on land-use practices, socio-economic conditions and population density [13]. The authors of [13] have estimated this human disturbance potential based on Gross National Product (per capita) and population density and have also suggested its optimum rage to between 0.3 and 2.0.

Instead of using countrywide estimates of GNP/capita and population density to estimate the human-induced soil erosion factor (E_h), high-resolution spatial information published by [24] in the form of a human footprint index (HFPI) is used in this study, so that the anthropogenic influences on sedimentation is better represented [21,23]. The human footprint index is developed by using several global datasets such as population distribution, urban areas, roads, navigable rivers, electrical infrastructures and agricultural land use [25]. This dataset is available at a spatial resolution of 0.25×0.25 degrees and is regionally normalized to account for the interaction between the natural environment and human influences [25]. Figure 1 illustrates the human footprint indices across the catchments of the selected four case study sites. These HFPI values were used to determine the human-induced erosion factors for the respective catchment areas.

In the model presented by [8], future changes in fluvial sediment supply are calculated based on the Universal Soil Loss Equation (USLE):

$$A_{sl} = R_{sl} \times K_{sl} \times L_{sl} \times S_{sl} \times C_{sl} \times P_{sl} \tag{9}$$

where; A_{sl} is the annual soil loss (m^3) per unit area in the catchment, R_{sl} is the rainfall erosivity factor, K_{sl} is the soil erodibility factor, L_{sl} is the slope length factor, S_{sl} is the slope gradient factor, C_{sl} is the crop/vegetation management factor and P_{sl} is the support practice factor (all dimensionless).

Despite the fact that the USLE encapsulates both anthropogenic and climate-change driven impacts on fluvial sediment supply, in the Scaled Aggregated Model for Inlet-interrupted Coasts (SMIC), change in fluvial sediment supply is calculated only by considering the variations in annual rainfall from a present to future time horizon. This highly-simplified method adopted by [8] omits factors like changes in temperature, river discharge, land use, and land management practises, which can significantly influence the changes in future sediment supply to the coasts.

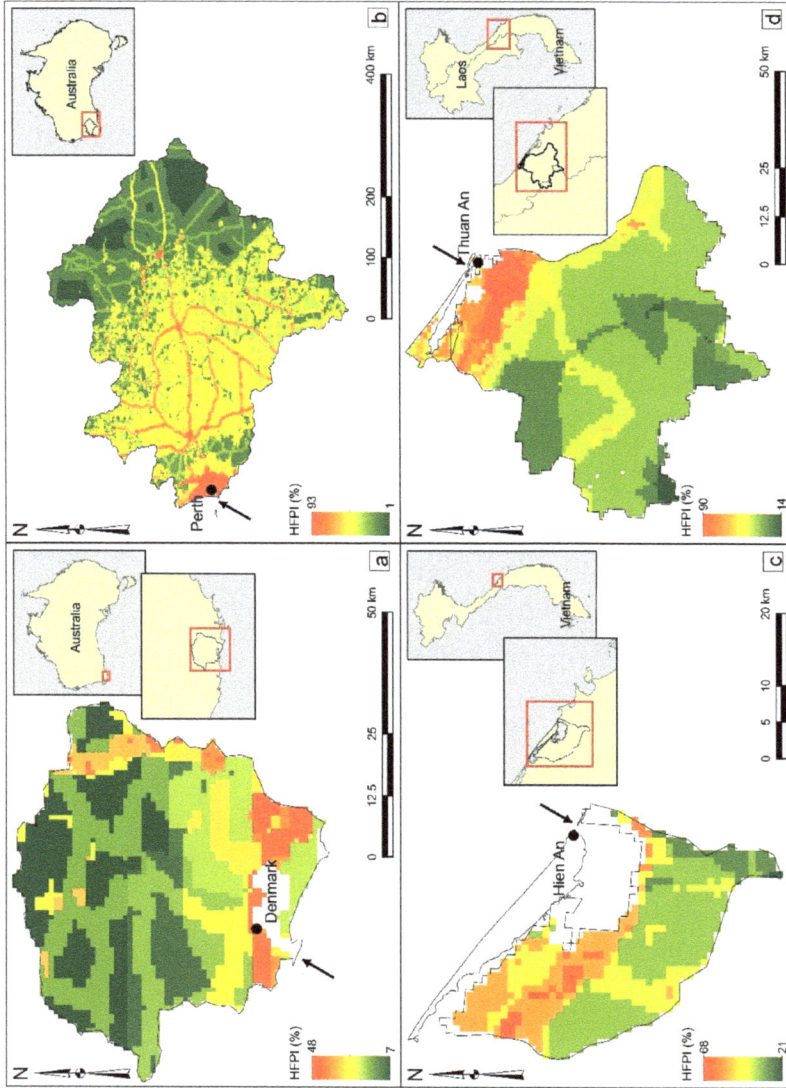

Figure 1. Human footprint index (HFPI) of the four systems considered; (**a**) Wilson Inlet system, Australia; (**b**) Swan River system, Australia; (**c**) Tu Hien Inlet system, Vietnam and (**d**) Thuan An Inlet system, Vietnam. Arrows indicate the locations of the inlets.

3. Model Applications and Results

The newly-developed reduced complexity model for small tidal inlets is applied at the selected four study sites to assess the potential coastline changes due to anthropogenic and climate change impacts by 2100. Model input parameters in Table 1 are obtained from published literature or by using standard calculations. River discharge (Q) values are obtained from the Supplementary Table 1 of [8]. Annual mean temperature values (T) are calculated by utilizing relevant daily temperature values, obtained from the ERA 40 dataset that was published by The European Centre for Medium-Range Weather Forecasts (ECMWF). Anthropogenic factors on soil erosion at catchments (E_h) are represented via HFPI data, obtained from [24]. Those HFPI values were linearly rescaled to be complied with the optimum range (0.3, 2.0) suggested by [13] for human induced erosion factor (E_h).

Table 1. Model input parameters (present conditions).

Parameter	Wilson Inlet	Swan River	Tu Hien Inlet	Thuan An Inlet
River discharge $\left(Q \text{ in km}^3/\text{year}\right)$	0.2	0.5	0.8	5.5
Temperature (T in °C)	23	23	25	25
Mean ebb tidal prism (P in 10^6 m^3) *	2.4	2.6	15	47
Basin surface area $\left(A_b \text{ in km}^2\right)$ *	48	52	100	110
Basin volume (V_b in 10^6 m^3) *	85	312	122	178
Catchment area $\left(A \text{ in km}^2\right)$ *	2263	121,000	600	3800
Catchment relief (R in km)	0.16	0.53	1.2	1.2
Lithology factor (L) **	0.5	0.5	1.0	1.0
Anthropogenic factor (E_h)	0.55	0.58	0.99	0.82
Beach profile slope (tan β) *	0.01	0.01	0.01	0.01
Depth of closure (h_{DoC} in m) *	20	20	15	15
Inlet − affected coastline length (L_{AC} in m) *	15	20	10	20

* according to the values adopted in [8], ** obtained from [13].

Temperature projections of Representative Concentration Pathway 8.5 (RCP 8.5) were obtained from the 5th Assessment Report of the Intergovernmental Panel on Climate Change [17]. The authors of [8] have adopted 0.8 m of sea-level rise as one of the climate-change driven model input parameters. Therefore, in this study, the future increment of temperature (ΔT) was selected to be the likely maximum increment predicted in [17]. Changes in future runoff (ΔQ) are the percentage values adopted in [8], which are obtained from Figure 3.5 of [10]. Selecting the corresponding increment for the anthropogenic erosion factor (E_h) was mainly governed by the rate of population growth, urbanization, and proposed development plans within the study areas. Due to the large uncertainties associated with those aforementioned factors, one value was set for all the four sites considered in this study. Thus, the baseline case simulation input parameters presented in Table 2 consist of the worst-case climate change and anthropogenic forcing combinations.

Table 2. Model input parameters for baseline case simulation (year 2100).

Parameter	Wilson Inlet	Swan River	Tu Hien Inlet	Thuan An Inlet
Sea level rise (ΔS in m)	0.8	0.8	0.8	0.8
Temperature increment (ΔT in °C)	4.8	4.8	4.8	4.8
Change in run off (ΔQ in %)	−30	−30	+15	+15
Change in anthropogenic factor (ΔE$_h$ in %)	+20	+20	+20	+20

Predicted coastline changes at the four study sites by 2100 are presented in Table 3 (herein, the baseline case simulation results). Positive and negative values in all the coastline changes are corresponding to coastline recession and progradation, respectively.

Table 3. Modelled coastline changes of the baseline case simulation.

Component	Potential Coastline Change (m) by 2100			
	Wilson Inlet	Swan River	Tu Hien Inlet	Thuan An Inlet
Bruun effect (ΔC_{BE})	80	80	80	80
Basin Infilling effect (ΔC_{BI})	64	52	266	146
Basin volume change effect (ΔC_{BV})	9	48	−4	−7
Fluvial sediment supply effect (ΔC_{FS})	−1	−12	−37	−71
Total (ΔC_T)	152	168	305	148

Above baseline case simulation results (Table 3) are compared with the modelled outcomes published by [8]. Since three of the four components (Bruun effect (ΔC_{BE}), Basin Infilling effect (ΔC_{BI}) and Basin volume change effect (ΔC_{BV}), highlighted in Table 3) of the total coastline change at tidal inlets are similar for both the models considered, only the relevant coastline changes due fluvial sediment supply variation and resulting total coastline changes by 2100 are presented in Table 4.

Table 4. Comparison of predicted coastline changes with SMIC predictions.

Component		Potential Coastline Change (m) by 2100			
		Wilson Inlet	Swan River	Tu Hien Inlet	Thuan An Inlet
Fluvial sediment supply effect (ΔC_{FS})	New RCM	−1	−12	−37	−71
	SMIC	0	0	−2	−7
Total (ΔC_T)	New RCM	152	168	305	148
	SMIC	153	180	340	212
Difference in Total [New RCM—SMIC]		−1 (~0%)	−12 (−7%)	−35 (−10%)	−64 (−30%)

RCM—Reduced Complexity Model.

In order to determine the sensitivity of climate change and anthropogenic forcing on fluvial sediment supply and subsequently their impacts on total coastline change at tidal inlets, different sets of model input parameters were also considered. For those model inputs, climate change projections relevant to the lower limits of the Representative Concentration Pathway 2.6 (RCP 2.6), presented in [17] were used. The corresponding value for anthropogenic impact was selected by considering the relentless population growth and myriad environmental pressures accompanied with it. These model input parameters, together with the relevant coastline changes are presented in Table 5.

Table 5. Modelled coastline changes by 2100 under different climate-change and anthropogenic forcing scenarios considered.

Parameter	Component	Potential Coastline Change (m) by 2100			
		Wilson Inlet	Swan River	Tu Hien Inlet	Thuan An Inlet
Baseline Simulation	Total (ΔC_T)	152	168	305	148
SMIC Results	Total (ΔC_T)	153	180	340	212
$\Delta T = 0.3\,°C$	FSS effect (ΔC_{FS})	~0	−4	−21	−39
	Total (ΔC_T)	152	176	321	180
	% Difference from BLS	-	5%	5%	22%
* $\Delta Q = \pm 5\%$	FSS effect (ΔC_{FS})	−1	−17	−34	−65
	BVC effect (ΔC_{BV})	1	8	−1	−2
	Total (ΔC_T)	144	123	311	159
	% Difference from BLS	−5%	−27%	2%	7%
$\Delta E_h = 10\%$	FSS effect (ΔC_{FS})	~0	−8	−28	−53
	Total (ΔC_T)	152	172	314	166
	% Difference from BLS	-	2%	3%	12%

* Coastline changes at Wilson Inlet and Swan River (Australia) are driven by $\Delta Q = -5\%$, whereas the same at Tu Hien and Thuan An Inlets (Vietnam) are driven by $\Delta Q = +5\%$. FSS—Fluvial sediment supply, BVC—basin volume change, BLS—baseline simulation.

Model results under baseline case simulation indicate that the four tidal-affected coastlines considered in this study are significantly eroding due to the combined effects of climate change and anthropogenic impacts (Table 4). Two arid/semi-arid systems in the Australian continent, Wilson Inlet and Swan River systems, and the larger system in Vietnam, Thuan An Inlet, would experience coastal recessions of similar magnitudes (~150 m), whereas the smaller system in Vietnam, Tu Hien Inlet, shows a coastline retreat as double as that of the other three systems (~300 m). At all these systems, the Bruun effect (ΔC_{BE}) and basin infilling effect (ΔC_{BI}) (i.e., oceanic processes) are the main contributors for the resulting total coastline recessions (Table 3).

As a consequence of their arid climate, predicted coastline changes at the two systems in Australia are less affected by climate change-driven terrestrial processes. Furthermore, estimated anthropogenic impact factors for these two systems are also small. Therefore, the combined effects of climatic and anthropogenic forcing yield very little changes in fluvial sediment supply to these coasts, and, thus, result in minor changes at the two CEC systems in Australia. Tu Hien and Thuan An Inlets in Vietnam will pass more fluvial sediment to their nearshore zones due to the combined effects of climate change and anthropogenic forcing, resulting in considerable coastline progradation which, in turn, reduces the total coastal recessions caused by the two oceanic processes.

The lower extreme temperature increment and river discharge changes considered in Table 5 illustrate the relevance of different climate parameters on fluvial sediment supply and, subsequently, on total coastline changes at tidal inlets. Under the least temperature increment considered (0.3 °C), arid CEC systems in Australia show trivial coastline changes due to variations in fluvial sediment supply (~0 m and −4 m, respectively) by 2100. This is quite different for the two systems in Vietnam, where the coastline changes for the same climatic condition were recorded as −21 m and −39 m at Tu Hien and Thuan An Inlets, respectively. These effects on the coastline are considerably less than the baseline case simulation results of −37 m and −71 m calculated for the respective locations. Therefore, when compared with the baseline case simulation, the reduced temperature increment has resulted in about a 5% and 22% increase in total coastline recessions at Tu Hien and Thuan An Inlets, respectively.

Climate change-driven changes in river discharges have trivial impacts on total coastline changes in all of the systems. Reduced future river discharges in the Australian CEC systems (−5%) would increase the estuarine sediment demands, thus resulting in coastal recessions at Wilson Inlet (1 m) and Swan River systems (8 m). Under the baseline case simulation, coastal recessions due to basin volume changes at the same locations were recorded as 9 m and 48 m, respectively. Slightly increased river discharge predictions (+5%) in Vietnam by 2100 would result in trivial coastline progradations at Tu Hien Inlet (1 m) and Thuan An Inlet (2 m).

Possible future changes in waves and currents due to climate change may change sediment transport patterns at tidal inlets. However, investigating future changes in sediment transport patterns due to climate change driven variations in waves and currents is beyond the scope of this study. Hence, it is assumed that the existing sediment transport conditions at the inlets remain invariant over the time period considered in this study. Climate change is also likely to alter both intensity and frequency of storms. Such changes in storm conditions would cause irrevocable changes to the inlet-estuary systems and its adjacent coasts. The reduced complexity modelling techniques adopted in this study cannot represent such extreme events. System response to such events would be better represented via insights obtained from detailed process-based modelling techniques.

When the anthropogenic impact factor is set to be 10% greater than the present conditions, all the systems indicate coastline recessions by 2100. Tu Hien Inlet (9 m) and Thuan An Inlet (18 m) systems in Vietnam exhibit considerably increased total coastline recessions, while the same increment at Swan River system in Australia is predicted to be 4 m. Therefore, when compared with the baseline case simulation results, the reduced anthropogenic influence factor has accounted for about 2% increment in total coastline change in the Swan River system, whereas the same for Tu Hien and Thuan An Inlets are found to be about 3% and 12%, respectively.

4. Discussion

A comparison of the model results of the new reduced complexity model for small tidal inlets with SMIC outcomes exemplifies that the latter method tends to overestimate total coastal recession. This can be attributed to the differences in the methods adopted to estimate future fluvial sediment supply changes. In SMIC, the variation of future fluvial sediment supply is considered to be depend only on the percentage change in precipitation. Contrarily, in the new reduced complexity model, the same future variations are attributed to temperature, river discharge, and anthropogenic influences. Wilson Inlet in Australia is the only instance where the two methods have yielded similar predictions for total coastal recessions by 2100. Arid conditions and less anthropogenic impact within its catchment can be attributed as the main reasons for such an outcome. SMIC has overestimated the total coastal recessions in the Swan River system, and the Tu Hien and Thuan An Inlet systems by 12 m, 35 m, and 64 m, respectively. These overestimations are about 7%, 10%, and 30% of the total coastal recessions estimated by SMIC at the respective CEC systems (Table 4). Except for the arid Wilson Inlet system in Australia, SMIC results overestimate the predicted total coastline changes corresponding to the least increments of temperature, river discharge, and anthropogenic factors at other CEC systems as well (Table 5).

5. Conclusions

This manuscript presents a part of a long-term study that is being undertaken to investigate the holistic behaviour of catchment-estuary-coastal systems under the influences of both climate change and anthropogenic impacts at macro time scales. The corresponding new modelling technique for small tidal inlets was applied to four case study sites considered in a previous study to compare the performance of the new model with the existing method.

Model simulations indicate that the four systems considered in this study would experience significant coastal recessions by 2100. In all the four case study sites, climate-change induced sea-level rise-driven oceanic processes (i.e., the Bruun effect and the basin infilling effect) overwhelm the climate-change and anthropogenic impacts driven terrestrial processes (i.e., the basin volume effect and fluvial sediment supply effect). Except for the arid catchment conditions in the Wilson Inlet system, future changes in fluvial sedimentation would result in extra sediment supply to the coastal zones, reducing the total coastal recessions, especially at the Tu Hien and Thuan An Inlets in Vietnam by the next century. Therefore, accurate estimation of future changes in fluvial sediment supply is essential to determine the total coastline changes at tidal inlets.

A comparison of the baseline case simulation results with the modelled results for the lowest predictions for climate change impacts and anthropogenic factors by 2100 indicate that for non-arid conditions, temperature, river discharge, and human influences make significant impacts on fluvial sediment supply to coasts. Neglecting these terms would significantly undermine the assessment of future changes in the fluvial sediment supply to the coasts and, subsequently, the total coastline changes.

Despite being a comprehensive technique of assessing holistic behaviour of the catchment-estuary-coastal system at macro time scales, the Scale Aggregated Model for Inlet-interrupted Coasts (SMIC) does not account for anthropogenic impacts and also omits some key climatic parameters, such as temperature and river discharge, in assessing future changes in fluvial sediment supply to the coasts. Therefore, SMIC applications at small tidal inlet-estuary systems with sandy coasts and non-arid catchments are most likely to underpredict future changes in fluvial sediment supply to the coasts and, subsequently, overpredict coastline recession.

Author Contributions: J.B. and S.M. conceived the conceptual design of the manuscript. J.B. carried out the model applications, analysis of the results, and drafted the original article. S.M. helped by revising the contents of the manuscript. T.M.D. helped with the conceptual design of the article and helped with revising the content of the manuscript. A.v.d.S. has contributed to the development of this study and provided the final and critical revision of the content of this manuscript.

Funding: This research received no external funding.

Acknowledgments: J.B. is supported by the Deltares research program 'Understanding System Dynamics; from River Basin to Coastal Zone'. Last of the Wild Project, Global Human Footprint, Version 2 data were developed by the Wildlife Conservation Society - WCS and the Center for International Earth Science Information Network (CIESIN), Columbia University and were obtained from the NASA Socioeconomic Data and Applications Center (SEDAC) at http://dx.doi.org/10.7927/H4M61H5F. Accessed 1 October 2015.

Conflicts of Interest: The authors declare no conflict of interest. The funding sponsors had no role in the design of the study; in the collection, analyses, or interpretation of data; in the writing of the manuscript; or in the decision to publish the results.

References

1. Cowell, P.J.; Stive, M.J.F.; Niedoroda, A.W.; De-Vriend, H.J.; Swift, D.J.P.; Kaminsky, G.M.; Capobianco, M. The Coastal-Tract (Part 1): A Conceptual Approach to Aggregated Modeling of Low-Order Coastal Change. *J. Coast. Res.* **2003**, *19*, 812–827.

2. Green, M.O. Catchment sediment load limits to achieve estuary sedimentation targets. *N. Z. J. Mar. Freshw. Res.* **2013**, *47*, 153–180. [CrossRef]

3. Syvitski, J.P.M.; Kettner, A.J.; Overeem, I.; Hutton, E.W.H.; Hannon, M.T.; Brakenridge, G.R.; Day, J.; Vörösmarty, C.; Saito, Y.; Giosan, L.; et al. Sinking deltas due to human activities. *Nat. Geosci.* **2009**, *2*, 681–686. [CrossRef]

4. Kettner, A.J.; Gomez, B.; Hutton, E.W.H.; Syvitski, J.P.M. Late Holocene dispersal and accumulation of terrigenous sediment on Poverty Shelf, New Zealand. *Basin Res.* **2009**, *21*, 253–267. [CrossRef]

5. Shennan, I.; Coulthard, T.; Flather, R.; Horton, B.; Macklin, M.; Rees, J.; Wright, M. Integration of shelf evolution and river basin models to simulate Holocene sediment dynamics of the Humber Estuary during periods of sea-level change and variations in catchment sediment supply. *Sci. Total Environ.* **2003**, *314–316*, 737–754. [CrossRef]

6. Smith, R.K. Poverty Bay, New Zealand: A case of coastal accretion 1886–1975. *N. Z. J. Mar. Freshw. Res.* **1988**, *22*, 135–142. [CrossRef]

7. Vörösmarty, C.J.; Meybeck, M.; Fekete, B.; Sharma, K.; Green, P.; Syvitski, J.P.M. Anthropogenic sediment retention: Major global impact from registered river impoundments. *Glob. Planet. Chang.* **2003**, *39*, 169–190. [CrossRef]

8. Ranasinghe, R.; Duong, T.M.; Uhlenbrook, S.; Roelvink, D.; Stive, M. Climate-change impact assessment for inlet-interrupted coastlines. *Nat. Clim. Chang.* **2013**, *3*, 83–87. [CrossRef]

9. Davis, R.A.; Hayes, M.O. What is a wave-dominated coast? *Mar. Geol.* **1984**, *60*, 313–329. [CrossRef]

10. IPCC. *Climate Change 2007: Synthesis Report. Contribution of Working Groups I, II and III to the Fourth Assessment Report of the Intergovernmental Panel on Climate Change*; IPCC: Geneva, Switzerland, 2007.

11. Bruun, P. Stability of Tidal Inlets—Theory and Engineering. *Dev. Geotech. Eng.* **1978**, *23*, 1–506.

12. Syvitski, J.P.M.; Peckham, S.D.; Hilberman, R.; Mulder, T. Predicting the terrestrial flux of sediment to the global ocean: A planetary perspective. *Sediment. Geol.* **2003**, *162*, 5–24. [CrossRef]

13. Syvitski, J.P.M.; Milliman, J.D. Geology, Geography, and Humans Battle for Dominance over the Delivery of Fluvial Sediment to the Coastal Ocean. *J. Geol.* **2007**, *115*, 1–19. [CrossRef]

14. Syvitski, J.P.M.; Saito, Y. Morphodynamics of deltas under the influence of humans. *Glob. Planet. Chang.* **2007**, *57*, 261–282. [CrossRef]

15. Overeem, I.; Kettner, A.J.; Syvitski, J.P.M. Impacts of Humans on River Fluxes and Morphology. *Treatise Geomorphol.* **2013**, *9*, 828–842. [CrossRef]

16. Shrestha, B.; Babel, M.S.; Maskey, S.; van Griensven, A.; Uhlenbrook, S.; Green, A.; Akkharath, I. Impact of climate change on sediment yield in the Mekong River basin: A case study of the Nam Ou basin, Lao PDR. *Hydrol. Earth Syst. Sci.* **2013**, *17*, 1–20. [CrossRef]

17. Stocker, T.F.; Qin, D.; Plattner, G.-K.; Tignor, M.; Allen, S.K.; Boschung, J.; Nauels, A.; Xia, Y.; Bex, V.; Midgley, P.M. *IPCC 2013: Climate Change 2013: The Physical Science Basis*; Contribution of Working Group I to the Fifth Assessment Report of the Intergovernmental Panel on Climate Change; Cambridge University Press: Cambridge, UK; New York, NY, USA, 2013.

18. Syvitski, J.P.M. Impact of Humans on the Flux of Terrestrial Sediment to the Global Coastal Ocean. *Science* **2005**, *308*, 376–380. [CrossRef] [PubMed]

19. Leonardi, N.; Kolker, A.S.; Fagherazzi, S. Interplay between river discharge and tides in a delta distributary. *Adv. Water Resour.* **2015**, *80*, 69–78. [CrossRef]

20. Syvitski, J.P.M.; Kettner, A.J.; Correggiari, A.; Nelson, B.W. Distributary channels and their impact on sediment dispersal. *Mar. Geol.* **2005**, *222–223*, 75–94. [CrossRef]

21. Balthazar, V.; Vanacker, V.; Girma, A.; Poesen, J.; Golla, S. Human impact on sediment fluxes within the Blue Nile and Atbara River basins. *Geomorphology* **2013**, *180–181*, 231–241. [CrossRef]

22. Walling, D.E. Linking land use, erosion and sediment yields in river basins. *Hydrobiologia* **1999**, *410*, 223–240. [CrossRef]

23. Bamunawala, J.; Ranasinghe, R.; van der Spek, A.; Maskey, S.; Udo, K. Assessing Future Coastline Change in the Vicinity of Tidal Inlets via Reduced Complexity Modelling. *J. Coast. Res.* **2018**, 636–640. [CrossRef]

24. Wildlife Conservation Society—WCS, and Center for International Earth Science Information Network—CIESIN—Columbia University. *Last of the Wild Project, Version 2, 2005 (LWP-2): Global Human Footprint Dataset (Geographic)*; NASA Socioeconomic Data and Applications Center (SEDAC): Palisades, NY, USA, 2005.

25. Sanderson, E.W.; Jaiteh, M.; Levy, M.; Redford, K.H.; Wannebo, A.V.; Woolmer, G. The Human Footprint and the Last of the Wild. *Bioscience* **2002**, *52*, 891–904. [CrossRef]

Journal of
Marine Science and Engineering

MDPI

Article

Modelling Hydrodynamic Impacts of Sea-Level Rise on Wave-Dominated Australian Estuaries with Differing Geomorphology

Kristian Kumbier *, Rafael C. Carvalho and Colin D. Woodroffe

School of Earth and Environmental Sciences, University of Wollongong, Wollongong 2522, Australia;
rafaelc@uow.edu.au (R.C.C.); colin@uow.edu.au (C.D.W.)
* Correspondence: kak609@uowmail.edu.au; Tel.: +61-432-595-813

Received: 20 April 2018; Accepted: 31 May 2018; Published: 5 June 2018

Abstract: Sea-level rise (SLR) will affect the hydrodynamics and flooding characteristics of estuaries which are a function of the geomorphology of particular estuarine systems. This study presents a numerical modelling of coastal flooding due to drivers such as spring-tides, storm surges and river inflows and examines how these will change under sea-level increases of 0.4 m and 0.9 m for two estuaries that are at different geomorphological evolutionary stages of infill. Our results demonstrate that estuarine response to SLR varies between different types of estuaries, and detailed modelling is necessary to understand the nature and extent of inundation in response to SLR. Comparison of modelling results indicates that floodplain elevation is fundamental in order to identify the most vulnerable systems and estimate how inundation extents and depths may change in the future. Floodplains in mature estuarine systems may drown and experience a considerable increase in inundation depths once a certain threshold in elevation has been exceeded. By contrast, immature estuarine systems may be subject to increases in relative inundation extent and substantial changes in hydrodynamics such as tidal range and current velocity. The unique nature of estuaries does not allow for generalisations; however, classifications of estuarine geomorphology could indicate how certain types of estuary may respond to SLR.

Keywords: inundation; coastal flooding; estuarine geomorphology; hydrodynamic modelling; barrier estuary; Australia

1. Introduction

The rise in global mean sea level (GMSL) and its expected consequences are undoubtedly one of the greatest challenges coastal communities are facing in the 21st century. The response of different coast types such as beaches, rocky shorelines, deltas or estuaries will be dynamic and include changes in hydrodynamics, coastal geomorphology and coastal ecology [1]. The fifth assessment report (AR5) of the Intergovernmental Panel on Climate Change (IPCC) indicates that average rises in GMSL (up to 0.82 m for RCP8.5) may differ from regionalised sea-level rise (SLR) projections [2]. More recent studies on GMSL reconstructions and satellite altimetry measurements confirm anticipated increases in sea level [3,4], whereas studies with an emphasis on icesheet contribution to SLR estimate that the rise in GMSL could exceed 2 m by 2100 [5,6]. Projections of SLR for the Australian coastline are similar to GMSL projections [7], even though regionalised studies for Australia indicate that SLR exhibits significant regional differences [8].

The consequences of SLR for coastal areas include immediate effects such as submergence, increased flooding and associated salt water intrusion of surface waters, as well as long-term effects such as increased erosion, changes in coastal ecosystems such as saltmarsh and mangrove habitats, and effects on tidal dynamics [9–13]. Estuarine environments may change considerably as the morphology of low-lying floodplains makes them particularly vulnerable to increased inundation by tides and storm surges. Assessing these changes is of great importance as many people live along estuarine shorelines and likely experience changes such as more frequent inundation, so-called nuisance flooding [14], or water backwashing up storm water drains. In addition, some estuaries are likely to experience a coincidence of coastal flooding with riverine discharge, so called compound flooding [15–17].

While bathtub models may be used to map flooding in certain estuarine environments [18], their application to assess estuarine response to SLR appears to be limited, because dynamics such as changes in tidal range or current velocity cannot be analysed. Bathtub models assume areas lower than a certain elevation to be inundated utilizing elevation data and a geographic information system (GIS) [19,20], but physical processes such as bottom friction or transfer of momentum are not considered. As the response of estuaries to SLR includes also changes in tidal range and current velocity, and responses appear to be more dynamic than bathtub models of SLR predict [21]; analysis of estuarine response to SLR requires detailed investigation using hydrodynamic modelling in order to assess changes in tidal range, current velocity, inundation extent or inundation depth. Estuarine response to SLR has been the focus of many studies [12,13,22–24]. Prandle and Lane [23] assessed how tide-dominated estuaries in the UK adapt to SLR and river flow using vulnerability indices. Yang et al. [24] investigated estuarine response to river flow and SLR under future climate change and land use changes in the Snohomish River, USA, using a coupled hydrologic–hydrodynamic model. Their simulations suggest that average water depth in inundated areas increases linearly with SLR, but at a slower rate than on the open coast, while average salinity in inundated areas increases linearly with SLR. Lee et al. [12] investigated tidal response to SLR in two coastal-plain estuaries in the USA, Chesapeake Bay and Delaware Bay, using the unstructured-grid finite volume Coastal ocean model (FVCOM) and hypothetical adaptation options (sea walls). Their simulations indicate a non-linear tidal response to SLR and a reduction in tidal range in up-stream locations due to tidal energy dissipation through inundation of low-lying areas. However, if inundation was restricted by sea walls, tidal range increased. Du et al. [13] assessed the tidal response to SLR for idealised and realistic estuaries (Chesapeake Bay and subestuaries) of different geomorphology using the semi-implicit cross-scale hydroscience integrated system model (SCHISM). Their simulations indicate that tidal response to SLR is spatially uneven and varies among different estuaries. While estuaries with a narrow channel and large floodplain are likely to experience decreased tidal ranges under high SLR, those with comparatively steep banks may experience an increase in tidal range.

However, none of the studies above compared estuarine response to SLR in wave-dominated barrier estuaries at different evolutionary stages. There are numerous estuaries along the wave-dominated coast of southeast Australia, which have been classified into several types and differ in terms of their geomorphological evolution [25]. Immature estuarine systems are incompletely infilled with sediments, whereas estuaries of mature evolutionary stage have been almost completely infilled with sediments. Whereas immature estuaries may experience considerable changes due to SLR, more mature estuaries that have developed broad alluvial floodplains may be particularly vulnerable to more frequent inundation when riverbank levees are overtopped or breached. A modelling study by Watterson et al. [22] in Lake Macquarie, an immature wave-dominated barrier estuary in central New South Wales (NSW), indicated a doubling of tidal range in response to SLR of 0.91 m. Hydrodynamic modelling can indicate the vulnerability of settlements and infrastructure to SLR in and around estuaries and contribute to studies of biophysical vulnerability of estuaries as presented by Rogers and Woodroffe [26].

This study investigates estuarine response to SLR in two wave-dominated barrier estuaries at different infilling stage (youthful and mature) by determining changes in inundation extent, inundation depth, and changes in the underlying hydrodynamics such as tidal ranges and current velocities, using hydrodynamic modelling. Measurements of tidal gauges and river discharge for a specific storm event were analysed to create model forcing data sets representing spring-tide, storm-tide and extreme river discharge conditions. These inputs were combined to model inundation extents, inundations depths, tidal ranges and current velocities for present day conditions as well as for SLR increases of +0.4 m and +0.9 m.

2. Study Sites

Lake Illawarra and the Shoalhaven Estuary are located approximately 80 km and 120 km south of Sydney on the wave-dominated microtidal southeast coast of Australia. Tides along the coastline are semi-diurnal with a significant diurnal inequality. Both estuaries have been categorized as wave-dominated barrier estuaries [25], even though they are at different evolutionary stage and display contrasting hydrodynamics. Barrier estuaries occur on wave-dominated coastlines and are separated from the open ocean by a sandy barrier. The opening of these estuaries comprises a mouth that can be temporarily sealed by a wave-generated berm on smaller systems, referred to as an intermittent entrance [27,28]. Mature wave-dominated barrier estuaries are defined by a channelised river that is surrounded by broad alluvial floodplains, whereas immature or youthful systems consist of a lake-like water body [25].

Whereas Lake Illawarra is at an immature infilling stage and characterised by a large water body that is bordered by comparatively narrow and small floodplains located around the main tributaries Mullet Creek and Macquarie Rivulet, the Shoalhaven Estuary has infilled its proto-lake with sediments over the past 6000 years [29] and is characterised nowadays by broad low-lying floodplains located mainly around Broughton Creek and south of the Shoalhaven River. The tidal regime of the Shoalhaven Estuary is charactersied by a tidal range of approximately 1.5 m at Crookhaven Heads during spring tides with only limited tidal attenuation (0.2 m) upstream towards Nowra. The highest tides of the year (HHWSS = highest high water summer solstice) reach approximately 0.95 m above MSL at the entrance in Crookhaven Heads [30]. In Lake Illawarra the tidal regime is defined by a tidal range of approximately 1.0 m at the entrance gauge with severe attenuation (0.8 m) once the tidal wave passes the entrance channel. The highest tides of the year (HHWSS) reach approximately 0.66 m above MSL at the entrance [30] and cause a pumping of sea water into the lake, so called spring tidal pumping [31]. Riverine flooding highly influences the lower Shoalhaven Estuary and causes inundation of the floodplain due to the comparatively large catchment area (7000 km^2) of the estuary. This contrasts with Lake Illawarra, where few peripheral areas are flooded, even though the two tributaries (of 75 and 100 km^2 catchment area) have been shown to respond to high discharge as well [18]. Both estuarine systems have been modified in the past. The Shoalhaven River has been redirected towards Crookhaven Heads with the construction of Berry's Canal in 1822 and in consequence the former opening at Shoalhaven Heads has transformed into an intermittent entrance that only breaches during the largest storm events [32]. At Lake Illawarra the formerly intermittent entrance was stabilized and permanently opened through construction of training walls in 2001 and 2007. Figure 1 illustrates the floodplain topography of both study sites given in metres relative to Australian height datum (AHD), which approximates mean sea level (MSL). Blue areas represent water at MSL whereas green areas (<1 m) represent mostly intertidal areas which are subject to regular inundation by the tide, even though several floodgates in the Shoalhaven Estuary restrict regular inundation.

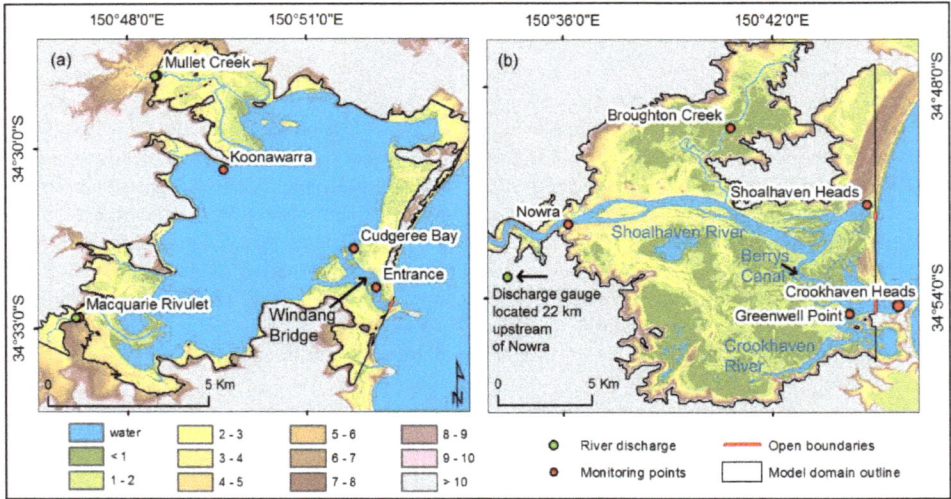

Figure 1. Topography and model setup at Lake Illawarra (**a**) and the Shoalhaven Estuary (**b**). LiDAR-derived topographic data of the floodplain is presented in m Australian height datum (AHD). Hydrodynamic model domains (black outline), open boundaries (red lines), river discharge locations (green dots) and monitoring points (red dots).

3. Materials and Methods

3.1. Input Data

Estuarine plain topography was determined from airborne LiDAR (Light Detection and Ranging) digital elevation data relative to AHD with a spatial resolution of 5 m (http://www.ga.gov.au/elvis/). Bathymetric data consisting of point measurements vertically referenced to AHD were downloaded from NSW Office of Environment and Heritage (OEH; http://www.environment.nsw.gov.au/estuaries/list.htm). Water level measurements at 15 min intervals for tidal gauges at the entrance to Lake Illawarra and along the Shoalhaven Estuary were provided for 1 year (June 2015 to July 2016) by OEH (distributed through Manly Hydraulics Laboratory). The entrance gauge, in Lake Illawarra, and Crookhaven Heads gauge, in the Shoalhaven Estuary, recorded a storm event in June 2016, which was used for the simulation of storm surge and compound flooding conditions. Discharge measurements for compound flooding simulations were provided at 15 min intervals for the Shoalhaven River at Tallowa Dam, which is located approximately 68 km upstream of the coast, and Macquarie Rivulet by NSW Water. Wind data of the same storm event measured at Port Kembla (5 km north of Lake Illawarra) were obtained from the server of the Bureau of Meteorology (http://www.bom.gov.au/oceanography/projects/abslmp/data/). Land use data were downloaded from OEH (http://data.environment.nsw.gov.au/dataset/nsw-landuseac11c).

3.2. Flood Drivers and SLR Scenarios

Modelling of present day hydrodynamic conditions used spring-tide forcing only (approximately HHWSS), storm surge only (HHWSS plus a positive surge resulting from the inverse barometric effect and wind set-up), and combined storm surge and riverine flooding, so called compound or coincident flooding. These three simulations for each estuary were then repeated for SLR increases of +0.4 m and +0.9 m to run 9 simulations in total per study site. SLR scenarios were selected in relation to the lower and upper boundary of the IPCC's AR5 SLR scenarios [2] and estimates of McInnes et al. [7],

Webb and Hennessy [33] and Zheng et al. [8] for the Australian coastline. Furthermore, values of +0.4 and +0.9 m in SLR have been advocated by the NSW state government as SLR planning benchmarks.

Time series of water-level changes for the storm surge simulation were taken from gauges at the entrance of Lake Illawarra and Crookhaven Heads in the Shoalhaven Estuary (red dots in Figure 1) for the previously mentioned storm event. According to Burston et al. [34], the coastal water level of this storm event, comprising storm surge and wave run-up, had an annual reoccurrence interval (ARI) of approximately 16 years for Lake Illawarra and 50 years for the Shoalhaven Estuary. The same time series of water-level changes (1 year of record) were analysed for predicted astronomical tides using UTide package for Matlab [35] in order to create time-series data sets for spring-tide-only forcing of model boundaries. River discharge measurements of the Shoalhaven River and Macquarie Rivulet were used to create time-series of discharge for normal (prior the storm) and extreme conditions (during peak of the storm). Spring-tide and storm surge simulations used discharge datasets for normal conditions while the compound flooding simulation used the original measurements of the storm. The same approach was used to create wind datasets for calm (for tidal simulations) and storm conditions (for storm surge and compound simulations). Scenarios for SLR of +0.4 m and +0.9 m were constructed through linear addition of the respective value to the spring-tide and storm surge water level time-series datasets described above.

3.3. Hydrodynamic Modelling

All hydrodynamic simulations were carried out using the Delft3D-Flow module of the open source numerical modelling software Delft3D [36]. The model was run in depth-averaged mode (2D) to solve the unsteady shallow water equations on a rectangular grid. For more information on the Delft3D code and the underlying unsteady shallow (2D) water equations the reader is referred to the software manual of Delft3D [36]. The general model setup comprising open boundaries, discharge locations and monitoring points is illustrated for both study sites in Figure 1.

Modelling at Lake Illawarra used a spatial resolution of 15 m, an open boundary at the entrance gauge and two river discharge locations at the tributaries Macquarie Rivulet and Mullet Creek. The modelling setup for tidal simulations at the Shoalhaven Estuary used one open boundary at Crookhaven Heads that was forced with data from the respective tidal gauge. Simulations of storm surge and compound flooding used an additional open boundary at Shoalhaven Heads, because the intermittent entrance opened in response to the simulated storm event in June 2016. All simulations at the Shoalhaven Estuary used a spatial resolution of 25 m. Spatially varying bottom friction with respect to different land use types was defined using Manning's friction coefficients taken from literature [37–39]. Adaptation measures such as tidal gates were only considered by their representation within the digital elevation data. Bathymetry and topography were assumed to be constant in time.

Model setups for both estuaries were validated for the previously mentioned storm event in June 2016. Therefore, observational data such as satellite imagery, aerial photographs, tidal gauges and water level logger measurements were used to compare against modelling results [17,18]. Table 1 summarises statistics derived from comparing modelled and observed water levels for several gauges shown in Figure 1.

Table 1. Statistical measures derived from comparison of modelled and observed water levels for the present-day storm surge simulation at Lake Illawarra and the Shoalhaven Estuary.

Lake Illawarra			Shoalhaven Estuary		
Tide Gauge	r^2	rmse (m)	Tide Gauge	r^2	rmse (m)
Entrance	0.97	0.12	Greenwell Point	0.98	0.09
Cudgegree Bay	0.98	0.12	Shoalhaven Heads	0.98	0.15
Koonawarra	0.97	0.14	Nowra	0.99	0.18
Macquarie R.	0.94	0.21			

At both sites, modelled and observed water levels correlated very well, whereas root mean square errors (rmse) were reasonably low, suggesting that the presented model setup is able to replicate inundation processes within the estuaries. This is further demonstrated by minor model underestimation of maximum current velocities (1 cm) at the Shoalhaven Estuary [17]. The reasonably good match between observed (satellite imagery and aerial photographs) and modelled inundation extents reinforces the accuracy of the modelling setup [17,18]. However, the absence of suitable data to validate current velocities at Lake Illawarra limits the credibility of velocity modelling and at this site and in consequence those results should be interpreted with caution.

Maps of maximum inundation extents were calculated in a GIS from the number of inundated pixels and known pixel dimensions. Outputs of maximum inundation depths per grid cell were used to calculate the average inundation depth (mean) for each simulation. Outputs of maximum inundation depth per computational grid cell were reclassified in a GIS into intervals of 0.25 m to enhance the comparison of modelling scenarios. Changes in tidal range were determined by comparison of maximum difference between consecutive high and low water at monitoring points indicated in Figure 1.

4. Results

4.1. Inundation Extents and Depths

Modelled maximum inundation extents using spring-tide, storm surge and compound flood drivers under different SLR scenarios for Lake Illawarra and the Shoalhaven Estuary are presented in Table 2. At Lake Illawarra, modelling for present-day conditions predicted an inundation extent of 1.5 km^2 for spring-tide forcing, 4.3 km^2 for storm surge (+186%) and 5.1 km^2 for compound flooding (+240%). Adding SLR of +0.4 m increased the inundation extent for all flood drivers by 60–90%, while differences in inundation extent between the drivers remained distinct. Simulations using a SLR of +0.9 m demonstrated another increase in inundation extent for all drivers by 50–100%. Differences in inundation extent between flood drivers remained considerable (100% between spring-tide and storm surge simulations), even though differences between storm surge and compound flooding decreased with an increase in sea level. Figure 2 presents spatial differences in inundation extents for spring-tidal simulations of present conditions and SLR scenarios at Lake Illawarra and the Shoalhaven Estuary. Spatial differences in inundation extents for storm surge and compound flooding simulations of present conditions and SLR scenarios at both estuaries are presented in Appendix A (Figure A1). At Lake Illawarra, the greatest increases in inundation extent were located around Mullet Creek, Macquarie Rivulet and the entrance of the lake.

Table 2. Modelled inundation extents and rounded average inundation depths at Lake Illawarra and Shoalhaven Estuary for flood driver and scenario combinations. SLR: Sea-level rise.

Study Site and Flood Driver	Present		+0.4 m SLR		+0.9 m SLR	
	Inundation Extent (km^2)	Average Inundation Depth (m)	Inundation Extent (km^2)	Average Inundation Depth (m)	Inundation Extent (km^2)	Average Inundation Depth (m)
L. Illawarra						
tide	1.5	0.4	2.9	0.7	5.8	0.9
storm-tide	4.3	0.8	7.3	0.9	11.6	1.1
compound	5.1	0.8	8.1	0.9	12.1	1.1
Shoalhaven						
tide	32	0.5	62	0.6	89	0.9
storm-tide	54	0.7	78	1.0	100	1.4
compound	75	0.8	91	1.1	104	1.6

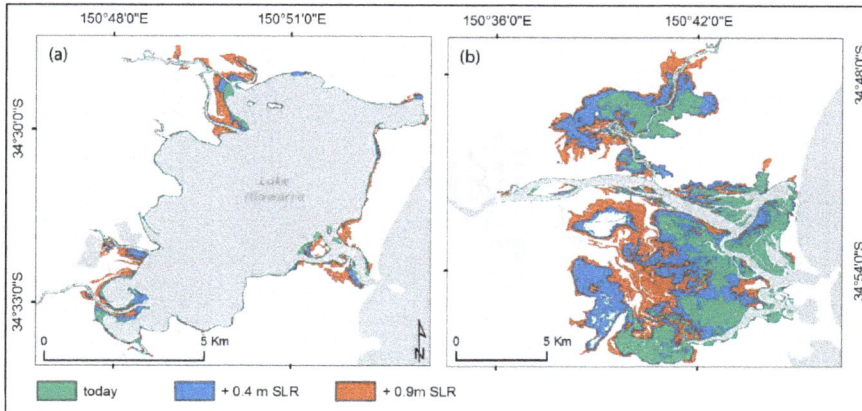

Figure 2. Maximum inundation extents of spring-tidal simulations for present conditions and SLR scenarios at Lake Illawarra (**a**) and the Shoalhaven Estuary (**b**).

At the Shoalhaven Estuary, spring-tidal forcing with today's sea level inundated an area of 32 km^2, whereas the storm surge simulation inundated an area of 54 km^2 (+70%) and compound flooding an area of 75 km^2 (+135%). These differences in inundation extents between flood drivers decreased under +0.4 m SLR, as the spring-tide inundated an area of 62 km^2, the storm surge an area of 78 km^2 (+25%) and the compound flooding an area of 91 km^2 (+45%). An increase in sea level by +0.9 m decreased differences in inundation extents between drivers even more. Spring-tidal forcing inundated an area of 89 km^2, whereas the storm surge and compound simulations inundated areas of 100 km^2 (+12%) and 104 km^2 (+17%) respectively. Inundation extents of spring-tidal simulations expanded mainly in the western part of the Crookhaven and Broughton Creek floodplain (Figure 2).

Average inundation depths at Lake Illawarra increased by approximately 0.1 m for the +0.4 m SLR simulations of storm surge and compound flooding and by 0.3 m for the spring-tidal simulation (Table 2). Simulations of +0.9 m SLR for storm surge and compound flooding increased average inundation depths by 0.3 m, whereas those from the spring-tidal simulation increased by 0.5 m.

Average inundation depths of storm surge and compound flooding under +0.4 m SLR at the Shoalhaven Estuary increased at a faster rate (0.3 m) than at Lake Illawarra, but were slower for spring-tidal simulations (+0.1 m for +0.4 m SLR). A similar pattern was observed for simulations of +0.9 m SLR, as inundation depths for storm surge and compound flooding increased by 0.7–0.8 m, whereas those for spring-tidal simulation were comparable to results at Lake Illawarra (+0.4 m).

Figure 3 illustrates the distribution of maximum inundation depths across the floodplain of Lake Illawarra and the Shoalhaven Estuary in classes of 0.25 m intervals for spring-tidal (red lines) and storm surge simulations (blue lines), as well as SLR scenarios (dashed and dotted) for both drivers. At Lake Illawarra, the areas per inundation depth class increased with SLR, but the distribution between classes remained similar. The inundation depth classes smaller than 0.5 m remained the dominating ones for all displayed simulations except for the storm surge simulation with +0.9 m SLR. In contrast, at the Shoalhaven Estuary inundation depths across the floodplain appear to shift from approximately 0.5–0.75 m towards 1.25 m (tide +0.9 m SLR) and even 1.5–2 m (storm surge +0.9 m SLR) with SLR.

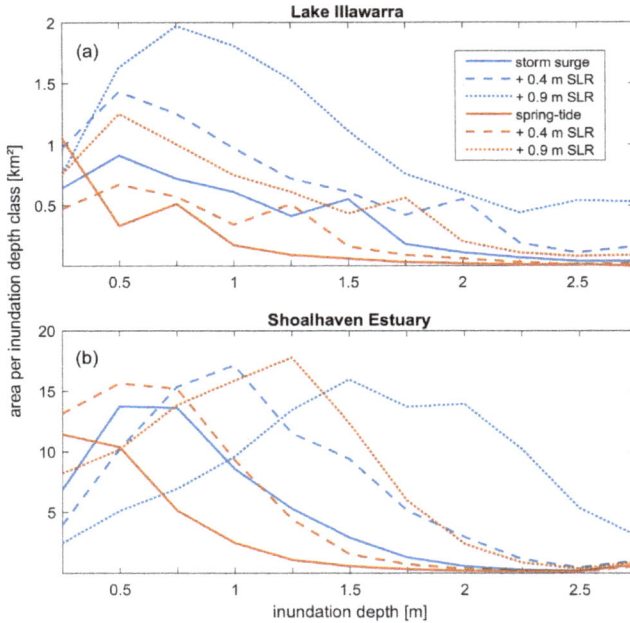

Figure 3. Distribution of maximum inundation depths in 0.25 m intervals across the floodplain of Lake Illawarra (**a**) and the Shoalhaven Estuary (**b**) for simulations of spring-tide (red) and storm surge (blue) in combination with present sea level as well as SLR of +0.4 m and +0.9 m. Note the difference in vertical scale due to the different size of the floodplains.

4.2. Hydrodynamics

Simulated water levels for spring-tidal conditions and SLR scenarios at Lake Illawarra are presented in Figure 4. Note that mean tidal level at both estuaries is elevated by approximately 0.25 m during simulation time due to the highest spring tides (summer solstice) of the year. Results indicate at all four monitoring locations inside Lake Illawarra an increase of approximately 0.4 m in MSL during simulation, which is independent of the applied SLR scenario and likely relates to spring-tidal pumping.

Changes in tidal range were very consistent between the monitoring points in Cudgeree, Koonawarra, Macquarie Rivulet and Mullet Creek. During present spring-tidal conditions the tidal range at the respective locations has a maximum of 0.25 m, except for the monitoring points in Mullet Creek, which is presently not subject to tides due to its upstream location behind a small weir. Adding SLR of +0.4 m to tidal simulations increases the tidal range at all places by 0.1 m (going to be 0.35 m). The simulation for SLR of +0.9 m demonstrated an increase of tidal range by 0.25 m (going to be 0.5 m). Similar increases in tidal range due to SLR were also observed for storm surge and compound flooding simulations. Comparison of tidal ranges at different locations in the entrance channel showed that the tide is mainly attenuated after is passes Windang Bridge, where the channel widens and the water depth decreases.

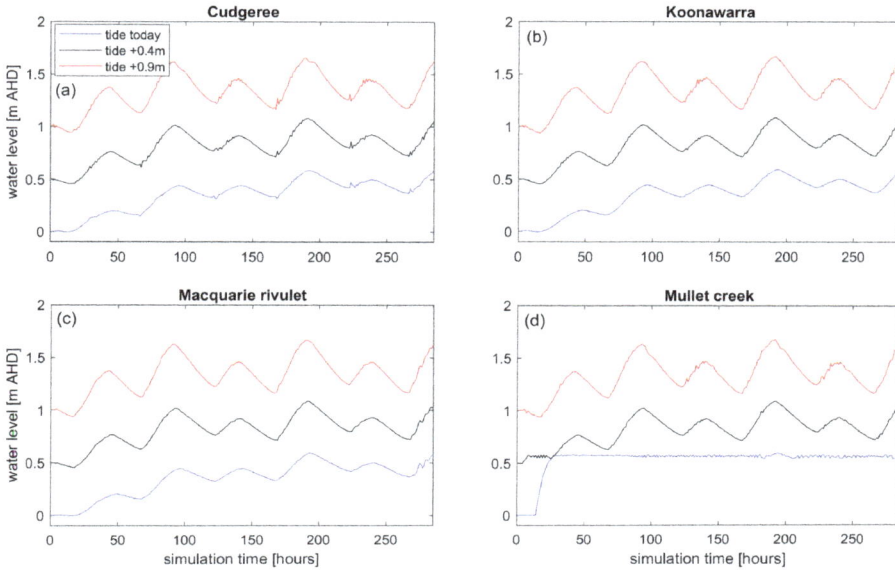

Figure 4. Simulated water levels for spring-tidal conditions and SLR scenarios over 6 tidal cycles at Cudgeree (**a**), Koonawarra (**b**), Macquarie Rivulet (**c**), and Mullet Creek (**d**) in Lake Illawarra. Mullet Creek monitoring point is not subject to tides at present sea level conditions due to its upstream location behind a tidal weir.

Figure 5 presents simulated water levels for spring-tidal conditions and SLR scenarios at the Shoalhaven Estuary. The maximum tidal range at Greenwell Point remained stable at 1.5 m independent of SLR scenario. Tidal range increased slightly further upstream. At Shoalhaven Heads, tidal range increased just slightly for SLR of +0.4 m and by 0.1 m for SLR of +0.9 m. At Nowra, tidal range increased by 0.1 m for SLR of +0.4 m and by 0.2 m for SLR of +0.9 m. Broughton Creek was the only monitoring location where tidal range decreased. While the maximum tidal range for present-day conditions was 0.95 m, it decreased to 0.8 m for SLR of +0.4m and to 0.5 m for SLR of +0.9 m. The steady increase in MSL during spring-tidal simulation with +0.9 m SLR further indicates tidal pumping during spring tides at Broughton Creek.

Modelling of depth-averaged current velocities indicated no considerable changes at locations inside the lake, but monitoring in the entrance channel at Windang Bridge (Figure 6a) indicated an increase by 0.3 m/s for SLR of +0.4 m (to 1.6 m/s) and an increase by 0.9 m/s for SLR of +0.9 m (to 2.1 m/s). Furthermore, Figure 6a illustrates the diurnal inequality in the tide and comparatively stronger flood current velocities, which seem to intensify with SLR. Peak current velocities in each scenario simulation correspond to rising tide towards the higher high water of the tidal cycle.

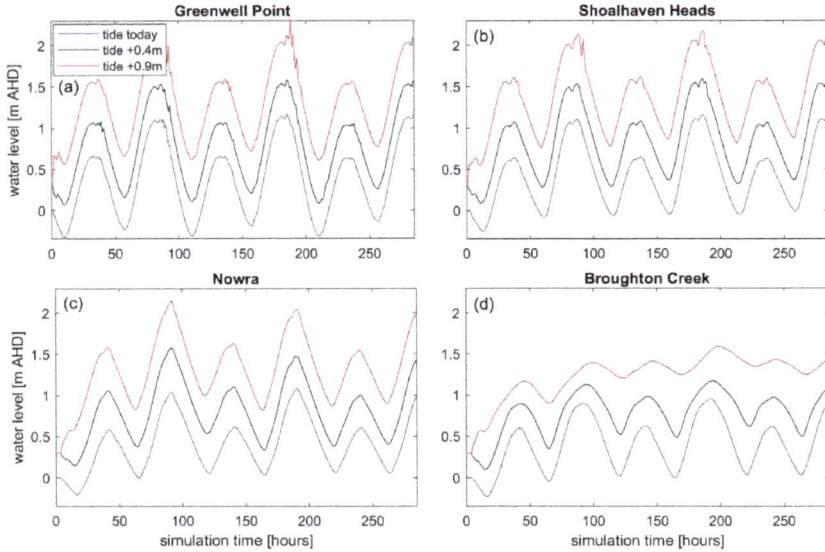

Figure 5. Simulated water levels for spring-tidal conditions and SLR scenarios over 6 tidal cycles at Greenwell Point (**a**), Shoalhaven Heads (**b**), Nowra (**c**), and Broughton Creek (**d**) in the Shoalhaven Estuary.

At the Shoalhaven Estuary depth-averaged current velocities only increased slightly in the main channel at Berrys Canal (0.1 m/s for +0.4 m and +0.9 m SLR). Changes at Nowra were just marginal. However, the lower Broughton Creek displayed a doubling of maximum current velocity (to then 0.3 m/s for SLR of +0.4 m and further strengthening during simulations of +0.9 m SLR (to then approximately 0.5 m/s) (Figure 6b). The monitoring of depth-averaged velocity in Figure 6b further shows the diurnal inequality in the tide and flood current dominance as observed already in the entrance channel of Lake Illawarra. The eastern Crookhaven River displayed also a doubling of maximum current velocities to 0.3 m/s for SLR of +0.9 m.

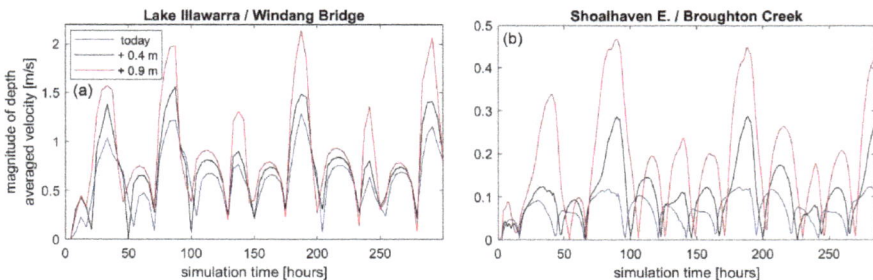

Figure 6. Simulated magnitude of depth-averaged current velocities for spring-tidal conditions and SLR scenarios over 6 tidal cycles at Windang Bridge in Lake Illawarra (**a**) and Broughton Creek in the Shoalhaven Estuary (**b**).

5. Discussion

5.1. Changes in Inundation Extents and Depths

The comparison of modelled inundation extents and depths due to different flood drivers and SLR conditions, whenever these will happen, revealed several differences between the two estuaries, which are discussed hereafter. At Lake Illawarra, modelled inundation extents due to spring-tide, storm surge and compound flooding were constantly increasing at high rates due to SLR (between 50–100%) and differences in extent between flood drivers remained distinct (e.g., 100% between simulated spring-tide and storm surge extent for 0.9 m SLR). This was different at the Shoalhaven Estuary, where inundation extents were increasing at lower rates (between 20% and 100%), especially for +0.9 m SLR scenarios (between 14% and 40%). Furthermore, differences in inundation extent between drivers became fairly small with increasing sea level (e.g., 12% between simulated spring-tide and storm surge extent for +0.9 m SLR). These differences in response to SLR are likely related to the morphological differences between the two sites. Most of the extensive floodplains of the Shoalhaven Estuary appear to overtop during flooding and in consequence, the inundation extent between drivers does not differ greatly. In contrast, at Lake Illawarra, SLR appears to elevate the lake water level and inundates low-lying areas surrounding the lake shoreline, even though tidal range is reduced in the entrance channel. This tidal attenuation is likely influencing the differences in inundation extent observed between flood drivers as peak water levels between storm surge and spring-tide differ.

At the Shoalhaven Estuary, the extent of inundation due to spring-tides, storm surge and compound flooding may become very similar in the future, because MSL appears to reach a tipping point that enables inundation of large parts of the floodplain independently of the flooding driver. In consequence, vast areas of the Shoalhaven floodplain are likely to be submerged and become tidal if no future adaptation measures are implemented. This is also confirmed by comparatively high increases in average inundation depths for the +0.9 m SLR scenarios. As inundation extents did not increase greatly between +0.4 m and +0.9 m SLR simulations, the additional floodwater added up nonlinearly in the vertical dimension. Figure 3 illustrates this shift from spring-tidal inundation depths being mainly between 0.5 and 0.75 m at present conditions, towards 1.5 m for future conditions. In consequence, floodplains in estuaries similar to the Shoalhaven Estuary may see considerable changes in the landscape through migration of tidal wetlands such as saltmarshes or mangroves. The broad floodplains may provide accommodation space for wetland migration [40], whereas estuaries with narrow floodplains or highly urbanised floodplains may prohibit the migration of tidal wetlands (coastal squeeze) and thus cause a loss of these valuable ecosystems [41,42]. Floodplain management in mature estuaries should consider the above and allow floodwater to enter low-lying areas in order to reduce the height of water levels in the channelised river and provide potential accommodation space for the migration and colonisation of tidal wetlands and their ecosystem services such as the mitigation of coastal flood risk [43,44].

In contrast, at Lake Illawarra, SLR caused considerable increases in inundation extent even for simulations of +0.9 m SLR. It appears that floodwater spreads horizontally, and in consequence average inundation depths increased at a smaller rate than observed in the Shoalhaven Estuary. Generally, changes in inundation depth in response to SLR in estuaries appear to be nonlinear. This was demonstrated by the presented differences in inundation depth between flood drivers and SLR scenarios at the two study sites. Our results corroborate findings by Yang et al. [24] who showed average water depth to increase linearly at roughly half of the SLR rate in the Snohomish River estuary in the United States. Estuarine geomorphology is too complex to come up with a universal formula as to how estuaries respond to SLR. Nevertheless, there appears to be a relationship between changes in inundation extent and depth due to SLR for estuaries: where inundation extents are increasing considerably, inundation depths increase at a comparatively slow rate as floodwater spreads horizontally. If inundated areas expand just insignificantly, inundation depths are going to increase at a similar rate to SLR. Similar findings were presented by Bilskie et al. [45].

The different responses of the estuaries to SLR can be further interpreted by investigating the floodplain elevation of the respective sites. Figure 7 shows the distribution of estuarine floodplain elevations (0 to 10 m AHD) in classes of 0.5 m at Lake Illawarra and the Shoalhaven Estuary. Areas given per elevation class were normalised to scale the data.

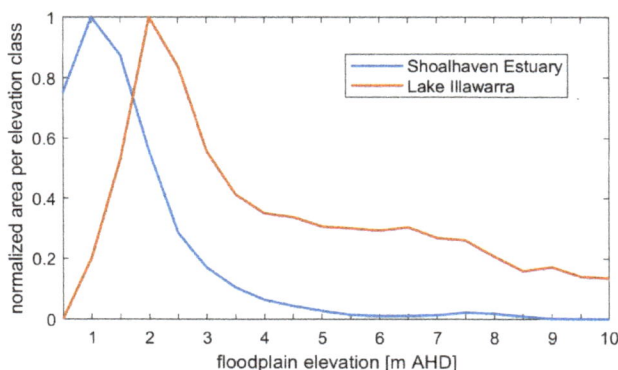

Figure 7. Distribution of estuarine floodplain elevation (0–10 m AHD) for Lake Illawarra and the Shoalhaven Estuary. Floodplain elevation was reclassified to intervals of 0.5 m and the corresponding areas per class normalised to scale the data.

Whereas the majority of low-lying areas in the Shoalhaven Estuary are below 2 m AHD and consequently subject to flooding during peak water levels higher than 2 m AHD (the peak high water during spring-tidal conditions with 0.9 m SLR was 2.1 m AHD at the entrance), the floodplain of Lake Illawarra is more heterogeneous with the majority of low lying areas being located between 1.5 and 2.5 m AHD. The distribution of floodplain elevation may contribute to understanding why inundation depths increase more rapidly in the Shoalhaven Estuary and why differences between inundation extents resulting from spring-tide, storm surge and compound flooding are more pronounced in Lake Illawarra. Data on floodplain elevation may be used to identify the most vulnerable estuarine systems and their tipping points as has been suggested by Rogers and Woodroffe [26] who recommend using geomorphology and elevation as indicators for vulnerability of estuaries to coastal and flood hazards. The Shoalhaven Estuary floodplains, for example, appear to be mostly inundated if water levels are higher than 2 m AHD. If the peak water level exceeds this tipping point (independently of the flood driver causing it), the extent of inundation increases only marginally, whereas inundation depths are likely to increase considerably.

5.2. Hydrodynamic Response to SLR

Changes in hydrodynamic parameters such as tidal range and current velocity differed between the two estuaries. Lake Illawarra displayed a doubling of tidal range for 0.9 m SLR at locations inside the estuary that most likely relates to decreased friction in the entrance channel and an increase in tidal prism due to the increase in MSL. As the entrance channel has comparatively steep banks, spreading of floodwater is restricted and thus leads to an increased tidal prism and tidal range in Lake Illawarra. Similar findings have been observed by Watterson et al. [22], who analysed the hydraulic response of Lake Macquarie to rising sea levels. An increase in tidal energy is also reflected in the modelled depth-averaged current velocities at Windang Bridge in the entrance channel of Lake Illawarra, which was shown to increase by 75%. The modelled velocities align well with observations by Wiecek et al. [46], but are limited in the sense that bathymetric conditions in the entrance channel are assumed to be static. As the entrance channel is subject to ongoing scouring [47], increases in tidal range are likely to be higher than those presented in this study. Significant increases in current velocity may even trigger additional erosion in the entrance channel of Lake Illawarra. These changes indicate how human modification has changed the estuarine system and also influences its future.

Tidal ranges in the Shoalhaven Estuary were demonstrated to increase only marginally within the Shoalhaven River, which most likely relates to the overtopping of floodwater onto the floodplain. The tidal pumping and increase in current velocity by more than 200% observed in Broughton Creek indicates that parts of the Shoalhaven Estuary floodplains may revert to becoming tidal in the future. Differences between the ebb and flow velocities displayed in Figure 6b resulting from variation in channel geometry over a tidal cycle may explain why tidal pumping is occurring in the constricted Broughton Creek in the future [31].

The modelling results of hydrodynamic response to SLR for Lake Illawarra and the Shoalhaven Estuary match well with recent studies by Lee et al. [12] and Du et al. [13]. Both studies suggested that tidal ranges may increase in those estuaries that are characterised by steep channel banks and that tidal ranges may remain unchanged or even decrease in estuaries that are characterised by broad floodplains, which may act as a sink for tidal energy. This kind of differentiation is very similar to estuarine characteristics given in Roy et al. [25], who showed that mature estuarine systems display a channelised river and broad floodplains as opposed to immature estuarine systems that are defined by a constricted entrance channel and narrow floodplains. Therefore, such classifications could indicate how hydrodynamics in certain estuaries may respond to SLR. This relation to estuarine geomorphology is of great importance as the southeast coast of Australia has at least 10 estuaries with similar geomorphological characteristics to Lake Illawarra and 25 estuaries similar to the Shoalhaven Estuary [25]. Variation in future exposure to tidal inundation for different types of estuaries in NSW was also shown by Hanslow et al. [48]. Wave-dominated estuaries in other parts of the world, such as New Zealand or South Africa, may respond in a similar way to SLR as the sites presented in this study.

Our results clearly demonstrate that estuarine response to SLR is variable and differs between estuarine environments. Nevertheless, the modelling is limited in the sense that adaptation measures such as tidal gates were only considered by their elevation (e.g., a tidal gate underneath a road is represented by the elevation of the road) and not by their functionality (one- or two-way gates). The Shoalhaven Estuary floodplain has been modified in the past 200 years of use (e.g., cleared for cattle farming, construction of Berrys Canal, construction of Tallowa Dam, etc.), and future modifications such as flow regulations or levee elevations may influence the way this estuarine system will respond to SLR. Furthermore, tidal ranges at the entrance of the estuaries were assumed to be constant in time, but tides have been shown to change in response to SLR [49–51]. For shallow areas such as the entrances of Lake Illawarra and the Shoalhaven Estuary, this could imply a SLR-induced increase or decrease in high water levels due to reduced bed friction on the shelf and changes in tidal characteristics [50] that are not considered in our estimates of SLR. Another simplification of our modelling is the assumption of a static morphology. Estuarine bathymetry and morphology are likely to adjust to SLR through processes such as erosion of river banks, sedimentary infilling or scouring of channels. Floodplain morphology may change in the long term through subsidence and changes in land use may alter the spatial variations in surface roughness used for modelling. Modelling with respect to morphological and land use changes would likely improve the quality of results, but come at high computational expenses.

6. Conclusions

Results of this study show how variables such as inundation extent, inundation depth, tidal range and current velocities relate to each other and how they can be used to assess the response and vulnerability of estuaries of different geomorphology to SLR.

Our comparison demonstrated that there seem to be certain thresholds in floodplain elevation that may be used to identify the most vulnerable systems and estimate how inundation extents and inundation depths, as well as estuarine hydrodynamics, may change in the future. Infilled mature estuaries with broad floodplains such as the Shoalhaven Estuary will be more vulnerable to drowning, regular saltwater intrusions, and considerable increases in inundation depth for vast areas of the floodplain. It appears that the majority of floodplains in these systems are drowned after a certain tipping point in elevation. Immature estuaries with comparatively narrow and steep floodplains may be vulnerable to substantial changes in estuarine hydrodynamics and extensive increases in relative inundation extent due to SLR. At the Shoalhaven Estuary, an increase in inundation extent mostly affects pasture land, but at Lake Illawarra, it may seriously threaten residential areas that are presently not protected against extreme water levels, because ocean driven floods were not a major concern before mankind permanently opened the entrance of the estuary. Therefore, settlements surrounding Lake Illawarra are likely to regularly experience an increase in flooding (nuisance flooding), if no adaptation measures are implemented.

Flood risk in estuaries will change considerably in the future. At both studies sites, inundation extents and depths of spring-tidal simulations with +0.9 m SLR were greater than those modelled for storm surge or compound flooding simulations at present sea level. Today's extreme water levels associated with storm events appear to be the usual conditions at spring tides, which occur naturally on several days in a month.

Our investigation demonstrates that estuarine response to SLR varies between different types of estuaries and detailed modelling is necessary to understand how systems may react to rise in MSL. The unique nature of each estuarine system does not allow for universal conclusions about how estuaries respond to SLR; however, classifications of estuarine geomorphology, as well as floodplain elevation, could indicate the nature of hydrodynamic adjustments in certain types of estuary.

Author Contributions: K.K., R.C.C. and C.D.W. conceived and designed the study; K.K. set-up the models and ran the simulations; K.K., R.C.C. and C.D.W. analysed and interpreted the results; K.K. wrote the paper with substantial input from R.C.C. and C.D.W.

Acknowledgments: This research was supported by Australian Research Council Linkage project LP130101025. K.K. expresses gratitude to Athanasios Vafeidis for supporting and supervising his MSc project. The authors thank four anonymous reviewers for their constructive reviews which helped to improve this manuscript. The provision of the following data sets is highly appreciated: bathymetric data provided by NSW Office of Environment and Heritage; water level data of tidal gauges provided by NSW Office of Environment and Heritage (distributed through Manly Hydraulics Laboratory); Land use data provided by NSW Office of Environment and Heritage; LiDAR elevation data provided by Commonwealth of Australia (Geoscience Australia); Discharge measurements provided by Water NSW.

Conflicts of Interest: The authors declare no conflict of interest.

Appendix A

Figure A1 presents spatial differences in inundation extents for storm surge and compound flooding simulations of present conditions and SLR scenarios at Lake Illawarra and the Shoalhaven Estuary.

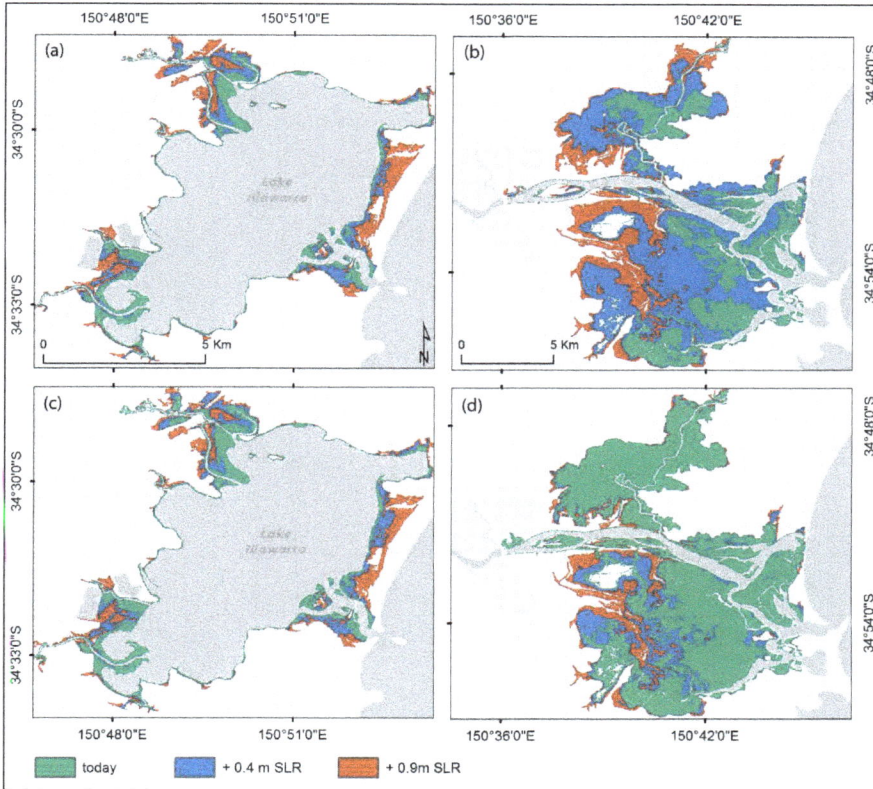

Figure A1. Maximum inundation extents of storm surge simulations for present conditions and SLR scenarios at Lake Illawarra (**a**) and the Shoalhaven Estuary (**b**), as well as the extent of compound flooding simulations (**c,d**) for the respective sites.

References

1. Nicholls, R.J.; Wong, P.P.; Burkett, V.R.; Codignotto, J.O.; Hay, J.E.; McLean, R.F.; Ragoonaden, S.; Woodroffe, C.D. Coastal systems and low-lying areas. In *Climate Change 2007: Impacts, Adaptation and Vulnerability. Contribution of Working Group II to the Fourth Assessment Report of the Intergovernmental Panel on Climate Change*; Parry, M.L., Canziani, O.F., Palutikof, J.P., Van der Linden, P.J., Hanson, C.E., Eds.; Cambridge University Press: Cambridge, UK, 2007.
2. Church, J.A.; Clark, P.U.; Cazenave, A.; Gregory, J.M.; Jevrejeva, S.; Levermann, A.; Merrifield, M.A.; Milne, G.A.; Nerem, R.S.; Nunn, P.D.; et al. Sea level change. In *Climate Change 2013: The Physical Science Basis. Contribution of Working Group I to the Fifth Assessment Report of the Intergovernmental Panel on Climate Change*; Stocker, T.F., Qin, D., Plattner, G.-K., Tignor, M., Allen, S.K., Boschung, J., Nauels, A., Xia, Y., Bex, V., Midgley, P.M., Eds.; Cambridge University Press: Cambridge, UK; New York, NY, USA, 2013.
3. Dangendorf, S.; Marcos, M.; Wöppelmann, G.; Conrad, C.P.; Frederikse, T.; Riva, R. Reassessment of 20th century global mean sea level rise. *Proc. Natl. Acad. Sci. USA* **2017**, *114*, 5946–5951. [CrossRef] [PubMed]
4. Nerem, R.S.; Beckley, B.D.; Fasullo, J.T.; Hamlington, B.D.; Masters, D.; Mitchum, G.T. Climate-change-driven accelerated sea-level rise detected in the altimeter era. *Proc. Natl. Acad. Sci. USA* **2018**, *115*, 2022–2025. [CrossRef] [PubMed]

5.   DeConto, R.M.; Pollard, D. Contribution of Antarctica to past and future sea-level rise. *Nature* **2016**, *531*, 591–597. [CrossRef] [PubMed]

6.   Oppenheimer, M.; Alley, R.B. How high will the seas rise? *Science* **2016**, *354*, 1375–1377. [CrossRef] [PubMed]

7.   McInnes, K.; Church, J.; Monselesan, D.; Hunter, J.; O'Grady, J.; Haigh, I.; Zhang, X. Information for Australian impact and adaptation planning in response to sea-level rise. *Aust. Meteorol. Oceanogr. J.* **2015**, *65*, 127–149. [CrossRef]

8.   Zhang, X.; Church, J.A.; Monselesan, D.; McInnes, K.L. Sea level projections for the Australian region in the 21st century. *Geophys. Res. Lett.* **2017**, *44*, 8481–8491. [CrossRef]

9.   Friedrichs, C.T.; Aubrey, D.G.; Speer, P.E. Impacts of Relative Sea-Level Rise on Evolution of Shallow Estuaries. *Coast. Estuar. Stud.* **1990**, *38*, 105–122.

10. Nicholls, R.J.; Cazenave, A. Sea-level rise and its impact on coastal zones. *Science* **2010**, *328*, 1517–1520. [CrossRef] [PubMed]

11. Geselbracht, L.L.; Freeman, K.; Birch, A.P.; Brenner, J.; Gordon, D.R. Modeled Sea Level Rise Impacts on Coastal Ecosystems at Six Major Estuaries on Florida's Gulf Coast: Implications for Adaptation Planning. *PLoS ONE* **2015**, *10*, e0132079. [CrossRef] [PubMed]

12. Lee, S.B.; Li, M.; Zhang, F. Impact of sea level rise on tidal range in Chesapeake and Delaware Bays. *J. Geophys. Res. Oceans* **2017**, *122*, 3917–3938. [CrossRef]

13. Du, J.; Shen, J.; Zhang, Y.J.; Ye, F.; Liu, Z.; Wang, Z.; Wang, Y.P.; Yu, X.; Sisson, M.; Wang, H.V. Tidal Response to Sea-Level Rise in Different Types of Estuaries: The Importance of Length, Bathymetry, and Geometry. *Geophys. Res. Lett.* **2018**, *45*, 227–235. [CrossRef]

14. Moftakhari, H.R.; AghaKouchak, A.; Sanders, B.F.; Matthew, R.A. Cumulative hazard: The case of nuisance flooding. *Earth's Future* **2017**, *5*, 214–223. [CrossRef]

15. Ikeuchi, H.; Hirabayashi, Y.; Yamazaki, D.; Muis, S.; Ward, P.J.; Winsemius, H.C.; Verlaan, M.; Kanae, S. Compound simulation of fluvial floods and storm surges in a global coupled river-coast flood model: Model development and its application to 2007 Cyclone Sidr in Bangladesh. *J. Adv. Model. Earth Syst.* **2017**, *9*, 1847–1862. [CrossRef]

16. Olbert, A.; Comer, J.; Nash, S.; Hartnett, M. High-resolution multi-scale modelling of coastal flooding due to tides, storm surges and rivers inflows. A Cork City example. *Coast. Eng.* **2017**, *121*, 278–296. [CrossRef]

17. Kumbier, K.; Carvalho, R.C.; Vafeidis, A.T.; Woodroffe, C.D. Investigating compound flooding in an estuary using hydrodynamic modelling: A case study from the Shoalhaven River, Australia. *Nat. Hazards Earth Syst. Sci.* **2018**, *18*, 463–477. [CrossRef]

18. Kumbier, K.; Carvalho, R.C.; Vafeidis, A.T.; Woodroffe, C.D. Comparing static and dynamic flood models in estuarine environments: A case study from southeast Australia. *Estuar. Coast. Shelf Sci.* **2018**. submitted.

19. Poulter, B.; Halpin, P.N. Raster modeling of coastal flooding from sea level rise. *Int. J. Geogr. Inf. Sci.* **2008**, *22*, 167–182. [CrossRef]

20. Van de Sande, B.; Lansen, J.; Hoyng, C. Sensitivity of coastal flood risk assessments to digital elevation models. *Water* **2012**, *4*, 568–579. [CrossRef]

21. Passeri, D.L.; Hagen, S.C.; Medeiros, S.C.; Bilskie, M.V.; Alizad, K.; Wang, D. The dynamic effects of sea level rise on low-gradient coastal landscapes: A review. *Earth's Future* **2015**, *3*, 159–181. [CrossRef]

22. Watterson, E.K.; Burston, J.M.; Stevens, H.; Messiter, D.J. The hydraulic and morphological response of a large coastal lake to rising sea levels. In Proceedings of the 19th NSW Coastal Conference, Batemans Bay, Australia, 10–12 November 2010.

23. Prandle, D.; Lane, A. Sensitivity of estuaries to sea level rise: Vulnerability indices. *Estuar. Coast. Shelf Sci.* **2015**, *160*, 60–68. [CrossRef]

24. Yang, Z.; Wang, T.; Voisin, N.; Copping, A. Estuarine response to river flow and sea-level rise under future climate change and human development. *Estuar. Coast. Shelf Sci.* **2015**, *156*, 19–30. [CrossRef]

25. Roy, P.S.; Williams, R.J.; Jones, A.R.; Yassini, I.; Gibbs, P.J.; Coates, B.; West, R.J.; Scanes, P.R.; Hudson, J.P.; Nichol, S. Structure and function of south-east Australian estuaries. *Estuar. Coast. Shelf Sci.* **2001**, *53*, 351–384. [CrossRef]

26. Rogers, K.; Woodroffe, C.D. Geomorphology as an indicator of the biophysical vulnerability of estuaries to coastal and flood hazards in a changing climate. *J. Coast. Conserv.* **2016**, *20*, 127–144. [CrossRef]

27. Thom, B. *Coastal Geomorphology in Australia*; Academic Press: Sydney, Australia, 1984; ISBN 0126878803.

28. Short, A.D.; Woodroffe, C.D. *The Coast of Australia*; Cambridge University Press: Melbourne, Australia, 2009; ISBN 0521873983.

29. Woodroffe, C.D.; Buman, M.; Kawase, K.; Umitsu, M. Estuarine infill and formation of deltaic plains, Shoalhaven River. *Wetlands* **2000**, *18*, 72–84.

30. MHL NSW Tidal Planes Analysis 1990–2010 Harmonic Analysis, Report MHL2053. Available online: http://new.mhl.nsw.gov.au/docs/oeh/tidalplanes/mhl2053%20OEH%20tidal%20planes%20analysis%20final%20report.pdf (accessed on 10 October 2017).

31. Hinwood, J.; McLean, E.; Trevethan, M. Spring Tidal Pumping. In Proceedings of the Coasts and Ports, Coastal Living-Living Coast: Australasian Conference, Adelaide, Australia, 20–23 September 2005.

32. Carvalho, R.C.; Woodroffe, C.D. Rainfall variability in the Shoalhaven River catchment and its relation to climatic indices. *Water Resour. Manag.* **2015**, *29*, 4963–4976. [CrossRef]

33. Webb, L.B.; Hennessy, K. Projections for Selected Australian Cities. Available online: https://www.climatechangeinaustralia.gov.au/media/ccia/2.1.6/cms_page_media/176/CCIA_Australian_cities_1.pdf (accessed on 14 September 2017).

34. Burston, J.; Taylor, D.; Garber, S. Contextualizing the return period of the June 2016 East Coast Low: Waves, water levels and erosion. In Proceedings of the 25th NSW Coastal Conference, Coffs Harbour, Australia, 9–11 November 2016.

35. Codiga, D.L. Unified Tidal Analysis and Prediction Using the UTide Matlab Functions, Technical Report 2011-01. Available online: http://www.po.gso.uri.edu/pub/downloads/codiga/pubs/2011Codiga-UTide-report.Pdf (accessed on 20 October 2016).

36. Deltares. Delft3D-Flow. Simulation of Multi-Dimensional Hydrodynamic Flows and Transport Phenomena, Including Sediments, User Manual. Available online: https://oss.deltares.nl/documents/183920/185723/Delft3D-FLOW_User_Manual.pdf (accessed on 4 October 2016).

37. Chow, V.T. *Open Channel Hydraulics*; Blackburn Press: New York, NY, USA, 1959; ISBN 1932846182.

38. Fisher, K.; Dawson, H. *Reducing Uncertainty in River Flood Conveyance, Roughness Review*; DE-FRA/Environmental Agency Flood and Coastal Defense R&D Program: Lincoln, UK, 2003.

39. Kaiser, G.; Scheele, L.; Kortenhaus, A.; Loevholt, F.; Roemer, H.; Leschka, S. The influence of land cover roughness on the results of high resolution tsunami inundation modelling. *Nat. Hazards Earth Syst. Sci.* **2011**, *11*, 2521–2540. [CrossRef]

40. Woodroffe, C.D. Mangrove response to sea level rise: Palaeoecological insights from macrotidal systems in northern Australia. *Mar. Freshwater Res.* **2018**, *69*, 917–932. [CrossRef]

41. Phan, L.K.; van Thiel de Vries, J.S.M.; Stive, M.J.F. Coastal mangrove squeeze in the Mekong Delta. *J. Coast. Res.* **2015**, *31*, 233–243. [CrossRef]

42. Woodroffe, C.D.; Rogers, K.; McKee, K.L.; Lovelock, C.E.; Mendelssohn, I.A.; Saintilan, N. Mangrove Sedimentation and Response to Relative Sea-Level Rise. *Annu. Rev. Mar. Sci.* **2016**, *8*, 243–266. [CrossRef] [PubMed]

43. Temmerman, S.; Meire, P.; Bouma, T.J.; Herman, P.M.J.; Ysebaert, T.; De Vriend, H.J. Ecosystem-based coastal defence in the face of global change. *Nature* **2013**, *504*, 79–83. [CrossRef] [PubMed]

44. Van Coppenolle, R.; Schwarz, C.; Temmerman, S. Contribution of Mangroves and Salt Marshes to Nature-Based Mitigation of Coastal Flood Risks in Major Deltas of the World. *Estuaries Coasts* **2018**, *13*. [CrossRef]

45. Bilskie, M.V.; Hagen, S.C.; Medeiros, S.C.; Passeri, D.L. Dynamics of sea level rise and coastal flooding on a changing landscape. *Geophys. Res. Lett.* **2014**, *41*, 927–934. [CrossRef]

46. Wiecek, D.; Regena, C.; Laine, R.; Williams, R.J. Quantifying change and impacts to Lake Illawarra from a permanent opening. In Proceedings of the 25th NSW Coastal Conference, Coffs Harbour, Australia, 9–11 November 2016.

47. Couriel, E.; Young, S.; Jayewardene, I.; McPherson, B.; Dooley, B. Case study: Assessment of the entrance stability of the Lake Illawarra Estuary. In Proceedings of the Coast and Ports 2013: Australasian Port and Harbour Conference Coasts and Ports 2013: 21st Australasian Coastal and Ocean Engineering Conference and the 14th Australasian Port and Harbour Conference, Sydney, Australia, 11–13 September 2013.

48. Hanslow, D.; Morris, B.D.; Foulsham, E.; Kinsela, M.D. A regional scale approach to assessing current and potential future exposure to tidal inundation in different types of estuaries. *Sci. Rep.* **2018**, *8*, 1–13. [CrossRef] [PubMed]

49. Mudersbach, C.; Wahl, T.; Haigh, I.D.; Jensen, J. Trends in high sea levels of German North Sea gauges compared to regional mean sea level changes. *Cont. Shelf Res.* **2013**, *65*, 111–120. [CrossRef]
50. Idier, D.; Paris, F.; Le Cozannet, G.; Boulahya, F.; Dumas, F. Sea-level rise impacts on the tides of the European Shelf. *Cont. Shelf Res.* **2017**, *137*, 56–71. [CrossRef]
51. Pickering, M.D.; Horsburgh, K.J.; Blundell, J.R.; Hirschi, J.J.M.; Nicholls, R.J.; Verlaan, M.; Wells, N.C. The impact of future sea-level rise on the global tides. *Cont. Shelf Res.* **2017**, *142*, 50–68. [CrossRef]

Journal of
Marine Science and Engineering

MDPI

Article

Projected 21st Century Coastal Flooding in the Southern California Bight. Part 1: Development of the Third Generation CoSMoS Model

Andrea C. O'Neill [1,*]**, Li H. Erikson** [1]**, Patrick L. Barnard** [1]**, Patrick W. Limber** [1]**, Sean Vitousek** [2]**,
Jonathan A. Warrick** [1]**, Amy C. Foxgrover** [1] **and Jessica Lovering** [1]

[1]   U.S. Geological Survey, Pacific Coastal and Marine Science Center, 2885 Mission Street,
     Santa Cruz, CA 95060, USA; lerikson@usgs.gov (L.H.E.); pbarnard@usgs.gov (P.L.B.);
     plimber@usgs.gov (P.W.L.); jwarrick@usgs.gov (J.A.W.); afoxgrover@usgs.gov (A.C.F.);
     jlovering@usgs.gov (J.L.)
[2]   Civil and Materials Engineering, University of Illinois at Chicago, 2095 Engineering Research Facility,
     842 W. Taylor Street (M/C 246), Chicago, IL 60607-7023, USA; vitousek@uic.edu
*   Correspondence: aoneill@usgs.gov; Tel.: +1-831-460-7586

Received: 24 April 2018; Accepted: 11 May 2018; Published: 24 May 2018

Abstract: Due to the effects of climate change over the course of the next century, the combination of rising sea levels, severe storms, and coastal change will threaten the sustainability of coastal communities, development, and ecosystems as we know them today. To clearly identify coastal vulnerabilities and develop appropriate adaptation strategies due to projected increased levels of coastal flooding and erosion, coastal managers need local-scale hazards projections using the best available climate and coastal science. In collaboration with leading scientists world-wide, the USGS designed the Coastal Storm Modeling System (CoSMoS) to assess the coastal impacts of climate change for the California coast, including the combination of sea-level rise, storms, and coastal change. In this project, we directly address the needs of coastal resource managers in Southern California by integrating a vast range of global climate change projections in a thorough and comprehensive numerical modeling framework. In Part 1 of a two-part submission on CoSMoS, methods and the latest improvements are discussed, and an example of hazard projections is presented.

Keywords: sea-level rise; coastal storm flooding; coastal hazards

---

1. Introduction

With over 600 million people living in the coastal zone worldwide [1], changes in sea-level and atmospheric conditions, including winds, sea-level pressures (SLPs), and precipitation [2], represent significant potential hazards. As such, changes will affect coastal erosion and flood patterns [3], increasing development and populations in these areas [1] will exasperate vulnerabilities and exposure unless methods to identify future hazards are developed and appropriate mitigation and adaptation strategies implemented.

Global Climate Models (GCMs) are often the best tools to evaluate potential changes in large-scale conditions and hazards. The coarse resolution and inability of GCMs to represent meso-scale conditions essential for local coastal impact studies [4], however, make downscaling of GCMs necessary [5] for community-scale coastal hazard identification. Several studies have conducted regional downscaling of GCMs for evaluation of changes in future storm surges and wave conditions of interest to coastal communities [6–10]. However, only a few have translated that work to the coastal zone and developed flood hazard maps from the combined impacts of projected sea-level rise (SLR), wave setup and runup, storm surge, and other coastal water level contributors.

One such study is the Coastal Storm Modeling System (CoSMoS) [11,12], a physics-based numerical modeling approach developed to comprehensively assess future coastal flooding risk by integrating SLR, dynamic water levels, and shoreline change. CoSMoS yields detailed hazards projections through a series of nested hydrodynamic models of increasing resolution that are applied to a 2 m resolution digital elevation model (DEM) to depict high-resolution storm-induced coastal flooding over large geographic regions. It has already been used by local emergency managers and planning councils to assess the exposure of 95% of the 26 million coastal residents in the state of California. The prototype system, developed for the California coast by USGS in collaboration with Deltares, uses swell computed with the global WaveWatchIII numerical wave model, the TOPEX/Poseidon satellite altimetry-based global tide model, and atmospheric forcing data from GCMs, to determine regional wave and water-level boundary conditions. These physical processes are dynamically downscaled using a series of nested SWAN (Simulating Waves Nearshore, Delft University of Technology) and Delft3D-FLOW numerical models that are linked at the coast to tightly spaced XBeach (eXtreme Beach) cross-shore profile models [11]. That first iteration of CoSMoS focused on evaluating flood hazards associated with historical storms and two SLR scenarios as well as a hypothetical extreme storm [13]. That initial work was expanded upon across the greater San Francisco Bay Area and up to Pt. Arena by including 40 SLR and storm scenarios, and incorporating downscaled atmospheric forcing and river discharge within San Francisco Bay [14–16]. Since then, further improvements and enhancements to the overall modeling system have been made to incorporate long-term shoreline change, better represent spatially-variable coastal storm responses, and integrate additional contributions to coastal flood levels.

Building upon earlier implementations, we present work showcasing the development of the third generation of CoSMoS in Southern California. This latest generation of explicit, numerically-derived flooding includes several new developments and modeling improvements: (1) high resolution grids for better representation of harbors, lagoons, bays, estuaries, and overland flow; (2) fluvial discharges that might locally impede and amplify flooding associated with coastal storms; (3) long-term morphodynamic change (i.e., beach change and cliff/bluff retreat) and its effect on coastal flooding projections; (4) uncertainty associated with terrain models, numerical model errors and vertical land motion; and (5) alterations to coastal storm intensity and frequency associated with a changing climate. Resulting model projections include flood extent, depth, duration, uncertainty, water elevation, wave run-up, maximum wave height, maximum current velocity, and long-term shoreline change and bluff retreat.

The objectives of this manuscript, Part 1 of 2, are to present the global-to-local scale downscaling methodology of the latest generation of CoSMoS, used to define flood hazards and assess the exposure of people, property, and infrastructure to future SLR and coastal storms in the Southern California Bight. This includes presentation and discussion of: (1) the modeling architecture; (2) the required model system inputs; (3) the inclusion of fluvial discharge; (4) the incorporation of long-term shoreline change and cliff retreat; and (5) the resultant generation of local-scale coastal hazards. Evaluation of the resulting hazards, exposure and vulnerability associated with various SLR scenarios combined with the full spectrum of plausible future coastal storms, variations in coastal response, and ecological and economic exposure that are available in web tools are detailed in Part 2 [17].

2. Materials and Methods

2.1. Study Area

Five counties comprise the coast of Southern California, extending from the U.S./Mexican Border to Point Conception: Santa Barbara, Ventura, Los Angeles, Orange, and San Diego Counties (Figure 1). The 500 km stretch of coast is an active, complex tectonic setting along the Pacific and North American plate boundary, resulting in much of the Southern California Bight (SCB) being fronted by a narrow continental shelf, a series of islands, beaches often backed by semi-resistant bedrock sea cliffs, and a highly irregular complex bathymetry that hosts a plethora of submerged seamounts, troughs, and canyons [18,19]. The deep-water wave climate is transformed to a more complicated nearshore wave field by the numerous seamounts, knolls, canyons, and the Channel Islands [20–23]. The islands block waves approaching from many directions, yielding a large shadow zone of wave energy. Additionally, complex shallow water bathymetry adjacent to the islands, seamounts and canyons, scatters, focuses, and dissipates wave energy, resulting in highly variable wave energy distribution patterns along the coast. Although swell dominates nearshore wave energy, locally-generated seas contribute as much as ~40% to the total wave energy spectrum [24,25].

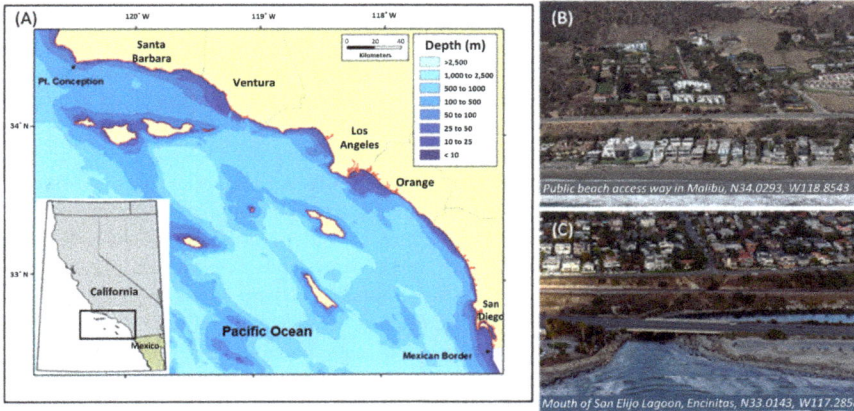

Figure 1. Overview of study area: (**A**) Southern California Bight and coastal counties; (**B**) aerial oblique photograph of Malibu; and (**C**) aerial oblique photograph of Encinitas. Both oblique photos highlight the urban infrastructure common throughout the study area. Image source: California Records Project, http://www.californiacoastline.org/.

Tides are mixed, semi-diurnal, with a mean diurnal range of 1.7 m [26]. Offshore significant wave height can reach ~8 m during the most extreme events [27], and therefore even with dissipation across the shelf, wave-driven water levels (i.e., set-up and runup) are still the dominant contributors to extreme coastal water levels across the region, contributing as much as 3 m to the total water level, while storm surge and El-Niño-driven water level anomalies rarely contribute more than ~20–30 cm each [28,29].

The heavily urbanized region hosts one of the largest economies in the United States, with a coastal county population of 17 million. Many vulnerable coastal areas are presently protected by sea walls or other flood and erosion defenses designed to withstand present-day storm impacts. Thus, the efficacy of these defenses against SLR or the combined effects of storms and SLR is questionable.

The coincident occurrence of storm-driven elevated water levels with high astronomic tides yield the greatest flooding [30,31]. Whereas astronomic tide ranges along the open coast are well predicted and not expected to significantly vary over the 21st century compared to historical levels [32], rates of SLR and the frequency and magnitude of storm-generated water levels are less well-constrained, and thus are the main foci of this study.

2.2. Modeling Framework

The third generation of CoSMoS, CoSMoS v. 3.0, is comprised of one global-scale wave model and a suite of regional and local scale models that simulate coastal hazards in response to projections of 21st century waves, storm surge, anomalous variations in water levels, river discharge, tides, and SLR (Figure 2). A total of 40 scenarios, resulting from the combination of ten sea levels, three storm conditions, and one background condition (i.e., daily average wave conditions), were simulated. Because scientific consensus on the magnitude and rate of SLR projections is constantly evolving [33–39], CoSMoS does not use a SLR rate projection or curve. Rather, CoSMoS uses discrete amounts of SLR, ranging from 0 m to 2 m, at 0.25 m increments, plus an additional 5 m extreme. Conceptually, SLR is not tied to a specific period in CoSMoS, but rather a range of discrete SLR values are chosen to illustrate conditions under various climate scenario assumptions and time periods. Future storm conditions represent the 1-year, 20-year, and 100-year return level coastal storm events, as derived and downscaled from winds, sea-level pressures (SLPs), and sea-surface temperatures (SSTs) of the RCP (Representative Concentration Pathway) 4.5, GFDL-ESM2M global climate model (GCM). Thus, the full suite of projections spans plausible SLR and 21st century storm conditions for an array of planning horizons and uses.

Summarizing the CoSMoS approach, at the global scale, GCM wind fields developed for the Coupled Model Intercomparison Project Phase 5 (CMIP5) [40] are fed into the WaveWatchIII (WWIII) [41] global wave model (see Section 2.2.1 for details). A higher-resolution, nested WWIII model produces a regional time-series of 21st century wave conditions across a range of models and climate scenarios at the edge of the continental shelf for the western U.S. coast [42].

Scaling down, select wave conditions from the regional scale are subsequently fed into a series of nested, higher resolution, coupled hydrodynamic and wave models (i.e., Delft3D-FLOW and SWAN) [43,44] that dynamically-downscale waves across the shelf to the nearshore in conjunction with astronomic tides and storm surge (see Section 2.2.3 for details). Nested model simulations include atmospheric pressures and winds [14,45], local river discharge, and seasonal water-level anomalies. Highest resolution grids (~10–20 m) are used to simulate overland flows in areas surrounding protected embayments. Along the open coast, cross-shore XBeach [46,47] models are used every 100–200 m in the along-shore direction to explicitly simulate wave set-up and swash (i.e., run-up) due to infragravity waves, a key driver of extreme water levels during storm events on dissipative beaches [48]. Modeled flood levels are ultimately interpolated onto regularly spaced grids and differenced from a 2 m resolution DEM [49] to generate local-scale flood projections and identify vulnerable areas not hydraulically connected to the open ocean but wetted by the numerical model.

The explicit CoSMoS downscaling approach summarized above, from a global to local scale, is computationally expensive and thus does not lend itself to simulating extensive, 100-year-long continuous time-series. Instead, the model system is run for pre-determined storm scenarios of interest, such as the 1-year or 100-year storm event, in combination with SLR amounts. These storms are first identified from time-series of total water level proxies (TWL_{px}) at the shore (see Sections 2.2.2 and 2.2.3). The 21st century TWL_{px} time series are computed using a linear super-position of the major processes contributing to the overall total water level, to efficiently estimate coastal water level impacts and identify extreme events. The identified events define the boundary conditions for subsequent detailed numerical modeling with CoSMoS.

TWL_{px} time series are also used to force a cliff recession and shoreline change model, both of which were developed for this study (Section 2.2.2). The data-driven sandy beach evolution [50,51] and cliff retreat [52,53] models are run at thousands of cross-shore transects spaced approximately 100 m apart along the coast. Both models use shoreline positions and oceanographic forcing to calibrate a suite of equations and develop robust relationships between forcing parameters and coastal response. Results from the two models provide time-varying beach shoreline and cliff positions, defined as the mean-high-water (MHW) line and top of the cliff, respectively, that are used to evolve cross-shore profiles [54] (Section 2.2.2.5) extracted from the underlying 2-m resolution DEM [49]. Evolved profiles illustrating projected cliff retreat and beach change are used to modify the three-dimensional DEM prior to running the thirty-six scenarios that incorporate future SLR (SLR 0.25 m and higher) using the approach of the CoSMoS system described above.

Each of the aforementioned components is detailed in the following subsections, with emphasis on the newest inclusions for Southern California.

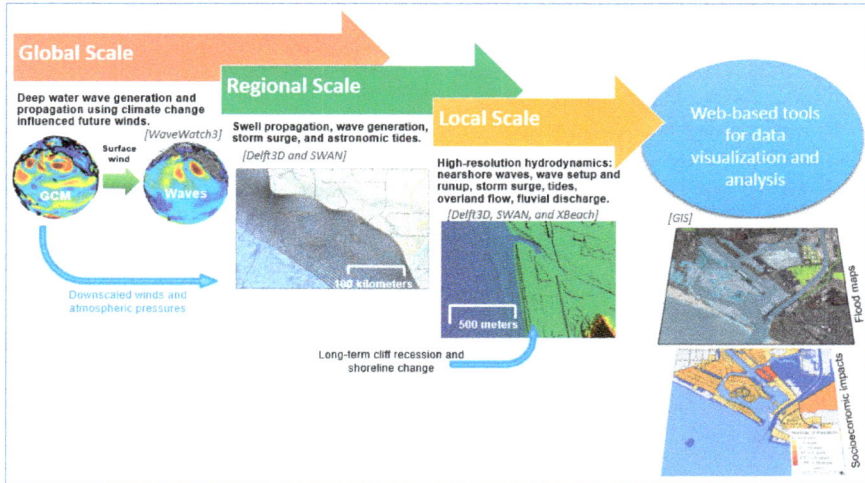

Figure 2. CoSMoS model system overview. Models and tools used in each step are shown with gray text. GIS-enabled web-tools for data visualization are detailed in Part 2 [17].

2.2.1. Inclusion of Swell Waves and GCM Selection

Previous versions of CoSMoS [11,12] included swell wave projections as boundary forcing, derived from four GCMs [42]. However, with the additional complexity and inclusion of local wave generation, storm surge, and efforts to capture the range of storm impact responses, it was necessary to identify one GCM for use in this version of CoSMoS. Thus, internal physical consistency (between derived swell projections, simulated wind waves, simulated storm surge, etc.) within the model system is maintained for all included processes. Therefore, an appropriate GCM was selected based on comparisons to wave buoy observations (using GCM winds from 1976 through 2010). The Earth-system model from National Ocean and Atmospheric Administration (NOAA) Geophysical Fluid Dynamics Laboratory (GFDL-ESM2M) was shown to best represent observed wave conditions in the extremes (winter H_s rmsd of 7–17 cm, extreme H_s bias < 5 cm, $D_p < 7° \pm 0.5°$) based on wave simulations along the California coast [42]. Lower percentile waves were somewhat underestimated with a bias of about 25 cm, commensurate with other global scale models. As GCM-driven wave projections associated with the RCP4.5 emissions scenario resulted in higher waves, and more extreme coastal wave climates, compared to RCP8.5 for the study region [8,42], GFDL-ESM2M RCP4.5 was selected to illustrate representative storm events. All further GCM-downscaling across the Southern California Bight used the GFDL-ESM2M RCP4.5 model scenario.

Downscaled wind and SLP fields used in this version of CoSMoS were produced and obtained from Scripps Institution of Oceanography [14,45,55]. For model validation and hindcast runs, data from California Reanalysis Downscaling at 10 km (CaRD10) was used; CaRD10 is a reconstruction of the high-spatial resolution/high-temporal scale analysis of atmosphere and land covering the state of California for global change studies [55]. For future periods, data fields from Localized Constructed Analogs (LOCA, 2011–2100 at 2.5° × 1.5°, 3-hourly resolution, derived from GFDL-ESM2M RCP 4.5) was used. Unique to LOCA's approach, a multiscale matching scheme [45] is used to select appropriate analogs from historical libraries, constructed from CaRD10 datasets.

To obtain global-scale wave projections, a WWIII model [41] was applied over a near-global grid (NWWIII, latitude 80° S–80° N) with 1° × 1.25° spatial resolution, and a one-way nested Eastern North Pacific (ENP) grid of 0.25° spatial resolution (~23.6 to 27.7 km at latitude 32° N). Details of model settings are found in Appendix A.

For further use in CoSMoS v 3.0, GFDL-ESM2M RCP4.5 wave projections [42] were extracted at the Scripps Institution of Oceanography California Data Information Program (CDIP) buoy 067 (33.221° N, 119.881° W) offshore the SCB (Figure 3). CDIP 067 is located offshore the shelf and has a robust historical record. Conditions at this location served as regional offshore conditions for deriving nearshore wave time series, nearshore TWL_{px}, and boundary conditions in higher-resolution nested models, in Section 2.2.2, Section 2.2.3, and Section 2.2.3.3, respectively. Return periods of offshore wave height are summarized in Table 1. Of note, these values are ~1 m lower than measured waves at the same location where a maximum significant wave height of 7.76 m (wave period, T_p = 14.3 s) was observed on 28 December 2006. While swell waves are projected to be lower in the 21st Century for this region compared to the recent past, the wave period is projected to increase and the incidence angle to be more southerly [42,56]. More southerly incidence angles and longer wave periods are related to the intensification of Southern Ocean wave generation, a consistent feature in global climate model predictions [57,58]. The projected decrease of extreme wave heights is thought to be related to a poleward shift in North Pacific extra-tropical storm tracks [8,31,59].

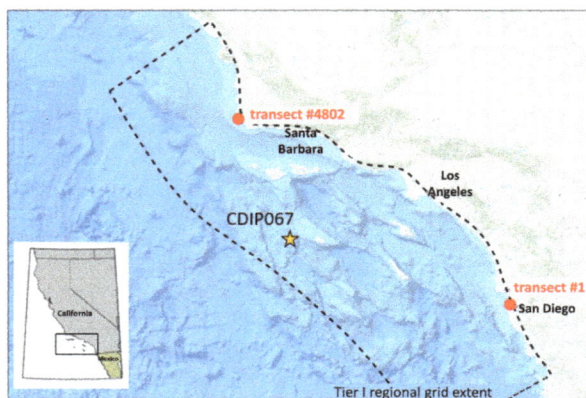

Figure 3. Deep-water conditions at CDIP067 in relation to nearshore transect locations. Dashed line shows the extent of the regional CoSMoS model grid (see Section 2.2.3.3).

Table 1. Projected and observed deep-water waves at buoy CDIP067. A Generalized Pareto Distribution (GPD) was fit through the wave data and in the case of the observation data, extrapolated for return periods greater than the length of the time-series. Listed T_p and D_p are the means of all instances for which the shown $H_s \pm 0.1$ m occurred.

Parameter	1-year	5-year	10-year	20-year	50-year	100-year
GFDL-ESM2M (full 100 years up to year 2100) (T_p and D_p are means of all $H_s \pm 0.1$ m)						
H_s (m)	4.93	5.93	6.25	6.5	6.76	6.91
T_p (s)	16	16	17	17	16	17
D_p (deg)	294	292	291	282	282	284
Observed (September 1996 through December 2017)						
H_s (m)	6.04	7.21	7.47	7.65	7.79	7.86
T_p (s)	15	16	14	17	No data	No data
D_p (deg)	299	296	305	306	No data	No data

2.2.2. Modeling Long-Term Shoreline Change

Two data-driven models to simulate cliff retreat and sandy beach evolution were developed for this study. The two models and supporting wave data are briefly described in the following subsections to illustrate function and position within the CoSMoS approach. Several different management scenarios involving beach nourishment and the existence and maintenance of hard structures to limit erosion were simulated with both the cliff recession and sandy shoreline change models. Further details can be found in respective references for individual models [50–53].

2.2.2.1. Initial Elevation

The 2-m CoNED DEM was constructed from recent bare-earth topographic and bathymetric lidar and multi- and single-beam sonar datasets [49]. The seamless topobathymetric DEM was constructed to define the shape of nearshore, beach, and cliff surfaces as accurately as possible, utilizing dozens of the most recent bathymetric and topographic datasets available at the time of generation. The vast majority of the data were derived from the Coastal California Data Merge Project which includes lidar data collected from 2009 to 2011 and multi-beam bathymetry collected between 1996 and 2011 extending out to the three nautical mile limit of California's state waters [60]. Harbors and some void areas in the nearshore were filled in with bathymetry from either more recent multi-beam surveys, 1/3 arc-second NOAA coastal relief model data, or single-beam bathymetry. The DEM is used as initial conditions and calibration data in two long-term coastal change models (Sections 2.2.2.3 and 2.2.2.4) that are run prior to the CoSMoS flood model, in addition to defining grid elevations for hydrodynamic and wave models (Section 2.2.3). Combined long-term shoreline changes are applied two-dimensionally to alter the DEM for SLR scenarios and are used in local-scale Tier III transect models (Sections 2.2.3.4 and 2.2.3.5).

2.2.2.2. Nearshore Wave Time Series and Water Level Proxies

Both cliff and shoreline change models were forced with hindcast (1980–2010) and projected (2010–2100) wave time series developed by Hegermiller et al. [61,62]. The hindcast was generated from high resolution SWAN model runs that capture changes in the wave field due to wave refraction across complex bathymetry and shadowing, focusing, diffraction, and dissipation of wave energy by islands. The model was forced at the open boundaries by intermediate-depth Wave Information Study (WIS) [63] wave time-series located landward of the Channel Islands, and by CaRD10 near-surface wind fields [55]. Three-hourly wave parameters (significant wave heights, mean wave period, peak wave period, mean wave direction, and peak wave direction) were output at 4802 points along the 10 m bathymetric contour every ~100 m in the alongshore direction, coincident with offshore ends of cross-shore transects used in the long-term cliff/sandy beach models and XBeach simulations (Figure 2). The hindcast was validated against 23 co-located CDIP buoys and found to behave reasonably well with an RMSE of 27 cm. See Hegermiller et al. [61] for further details and data.

To relate nearshore wave parameters to deep-water wave conditions, the 30-year hindcast time-series was correlated with deep-water waves at the CDIP067 to generate a look-up-table [61]. The look-up-table, in conjunction with the dynamically downscaled waves and winds (GFDL-ESM2M RCP 4.5) as the forcing fields (WWIII; Section 2.2.1), was used to derive nearshore wave parameters for the aforementioned nearshore points at three-hourly intervals out to the year 2100.

To identify significant coastal storm events, in terms of coastal flooding impacts at the shore, a proxy of total water levels (TWL_{px}) was developed from the nearshore wave time series. To efficiently estimate TWL_{px}, wave runup ($R_{2\%}$) was empirically computed from the hindcast and projected wave time-series [61,62], and then linearly super-imposed onto empirically derived time-series of storm surge (SS) [64] and other water level anomalies (SLA):

$$TWL_{px} = R_{2\%} + SS + SLA \tag{1}$$

Further details on generation and limitations of TWL_{px} can be found in Erikson et al. [64]; see Section 2.2.3.1 for use of TWL_{px} for storm selection. Conditional dependencies are accounted for using internally-consistent boundary conditions from a single GCM (GFDL-ESM2M RCP 4.5), the choice of which is justified above in Section 2.2.1.

While CoSMoS outputs use discrete amounts of SLR (see Section 2.2), specific SLR rates were used in long-term cliff and sandy beach models, with simulation end times (representing long-term morphodynamic change) matching overall CoSMoS SLR amounts. These rates are necessary as long-term cliff and sandy beach models were run for the entire 21st century period, unlike CoSMoS simulations. SLR scenarios for the coastal cliff/beach projections were represented with a second-order polynomial curve (see [51] for details) that reached 1 m or greater by the year 2100, relative to 2000 (Figure 4, modified from [51]). To align with CoSMoS SLR amounts of 0.25 m, 0.50 m and 0.75 m, long-term morphodynamic change simulations were simulated until 01 January 2044, 2069, and 2088, respectively, based on the National Research Council values developed for Southern California [34]. Cliff/beach SLR scenarios are in line with the range of long-term predictions from the 4th California Climate Assessment [37] (Figure 4) and other studies [33,65,66]. The 0.93 m scenario is a regional sea-level projection developed specifically for Southern California by the National Research Council [34].

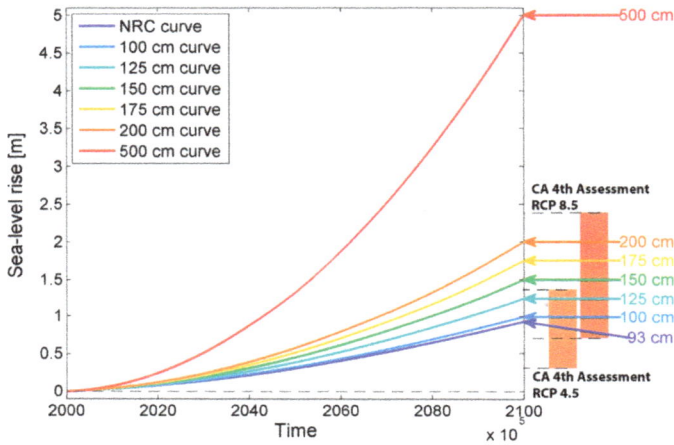

Figure 4. SLR curves used for the cliff retreat and shoreline change models (modified from [51]). Bands (right side) illustrate the range of SLR projections at 2100 from the 4th California Climate Assessment [37].

2.2.2.3. Coastal Cliff Retreat Model

Cliff retreat, defined as the landward movement of the cliff-top edge, is projected at each applicable transect in SCB using a multi-model ensemble of up to seven models that relate sea-cliff retreat to wave impacts using the hindcast and projected wave time series (Section 2.2.2.2), SLR, historical cliff behavior, and cross-shore geometry [67–71]. Mean long-term historic cliff retreat rates during ~1930–2000 calculated by the USGS National Shoreline Assessment were used to calibrate the models. The multi-model ensemble mitigates limitations of a single model and, therefore, can develop more robust predictions. Included models can be classified into two general groups: (1) simple 1-D models relating wave impacts to cliff retreat; and (2) 2-D transect models that include a discretized, evolving profile of nearshore and cliff morphology. The ensemble approach uses a weighted mean among all model outputs, giving preference to models with the least uncertainty [52,53].

The cliff-retreat model results in time-averaged cliff edge positions and rates for each SLR scenario, at all applicable SCB transects. Projections include two management scenarios considering the existence or non-existence of hard structures that limit cliff erosion. Considered structures and armoring locations were obtained from the California Coastal Commission [72] and aerial photographs from the California Coastal Records Project [73].

Included models do not explicitly distinguish between soft rock and hard rock coasts because they represent only basic physical interactions between waves and cliffs that are common to both morphologies. Smaller-scale details, such as vertical variations in rock strength on the cliff face [74] and seasonal variations in beach width and height, are not explicitly represented. Neither are dynamics related to seasonal beach erosion [75] and talus deposition and subsequent removal [76,77]. Finally, while rainfall can affect sea cliff evolution in parts of Southern California [78], CoSMoS cliff-retreat projections focus on wave-driven erosion, because relationships between rainfall, groundwater, and cliff failures are not well established. Rainfall-induced cliff erosion and other factors that might affect cliff retreat rates, such as jointing, fractures, and groundwater flow, are, however, implicitly included in the historical cliff retreat rates used to calibrate the models.

2.2.2.4. Sandy Beach Shoreline Change Model

The CoSMoS-COAST sandy shoreline change model [51] combines geographic information, management scenarios, and forcing conditions (due to waves and SLR) with three process-based models that compute: (1) wave-driven longshore transport [50]; (2) cross-shore transport due to waves [75]; and (3) cross-shore transport due to SLR [79]. CoSMoS-COAST integrates the process-based models with historical shoreline observation via an Extended Kalman Filter data assimilation method [80]. The model uses historical shoreline positions and oceanographic observations to calibrate a suite of equations and develop robust relationships between forcing parameters and shoreline response and project these relationships into the future [51].

The projected time-series of nearshore waves [61,62] and water levels (Section 2.2.2.2), combined with sea-level rates of change, were used to model one-dimensional shoreline change (represented by position of the mean high water (MHW) line) to the year 2100 at all applicable SCB transects. Four different management scenarios, representing the combinations of existing/no beach nourishment and the existence/non-existence of hard structures that limit erosion (i.e., "hold the line" on erosion) were considered. In the latter scenario, erosion was limited to a fixed polyline digitized from aerial photos accessible in Google Earth [81] representing the division of beach and urban infrastructure.

As the process-based models referenced above assume an equilibrium beach profile, changes in the profile shape are not computed in CoSMoS-COAST, which outputs shoreline positions and uncertainty (see [51] for details). Contributing to uncertainty, natural and anthropogenic sediment supply for management scenarios is also estimated from limited shoreline data throughout the region.

2.2.2.5. Combining Long-Term Morphodynamic Change

Projections of shoreline change and cliff retreat were combined to illustrate long-term morphodynamic change within CoSMoS. This was done by evolving the present-day (0 m SLR) cross-shore profiles to reflect the projected shoreline change (as illustrated by the MHW contour change) and cliff retreat for each SLR scenario. Details of the profile evolution scheme were presented by Erikson et al. [54]. Only projections from one management scenario were used for incorporating of long-term morphodynamic change: beach nourishment would cease but that existing cliff armoring and flood/beach protection infrastructure remains in place (i.e., the "hold-the-line" scenario). The resulting "evolved" profiles were used in simulations of runup and event-based morphodynamic change with the Tier III XBeach model (Section 2.2.3.5). Additionally, the 4000+ evolved profiles along the SCB were merged together to modify the 3-dimensional DEM for each CoSMoS SLR scenario. Projected elevation changes (ΔZ at each geographic coordinate) along each transect were blended at 10 m spacing between each profile, to interpolate the results of discrete beach/cliff responses along shore. These higher-density elevation changes were spatially interpolated to create a two-dimensional elevation adjustment, altering the present-day DEM to reflect long-term shoreline change for each SLR scenario. This DEM was used in final fine-scale flood projections and mapping potential hazards (Section 2.3).

2.2.3. Core Flood Model Architecture

In contrast to the TWL proxies (Section 2.2.2.2) that were computed to aid in identification of extreme storms (following subsection) and to provide temporally continuous boundary conditions for the cliff recession and shoreline change models (Sections 2.2.2.3 and 2.2.2.4), flood hazard modeling is done explicitly and deterministically with a suite of numerical models accounting for changes in water levels, waves, currents and resulting interactions between them (Figure 2). For all flood hazard simulations, projected deep water wave conditions, computed with the global scale wave model (Section 2.2.1), are propagated to shore with a suite of regional (Tier I) and local (Tiers II and III) models that additionally simulate regional and local wave growth (seas) in combination with event-driven morphodynamic change and water level changes due to astronomic tides, winds, sea-level pressure, steric effects, and sea-level rise. The nested framework (Tier I–Tier III) was run separately for multiple identified storm events (Section 2.2.3.1) and SLR.

The regional Tier I model consists of one Delft3D hydrodynamic FLOW grid for computation of currents and water level variations (astronomic tides, storm surge, and steric effects) and one SWAN grid for computation of wave generation and propagation across the continental shelf. The highest grid resolution is 1.2 km × 2.5 km nearshore. Wave conditions from the global wave model are applied at the open-boundaries of the SWAN model. The FLOW and SWAN models are two-way coupled, so that wave-current interaction is accounted for.

Employing high-resolution grids for simultaneous fine-scale modeling of the entire study area was not feasible given model and computational limitations; Tier II simulations were therefore segmented into 11 sections covering the SCB. Each sub-model consists of two SWAN grids and multiple FLOW grids of variable resolution, with the finest resolution of 5 m × 15 m in nearshore regions. Wave and water level time-series from the Tier I model are applied at the open boundaries of each Tier II sub-model (Section 2.2.3.4). Tier II simulations also included river discharges to capture impacts of potential riverine backflow at the coast.

Tier III consists of over 4000 cross-shore XBeach transect models (highest resolution of 5 m) simulating event-driven morphodynamic change, water level variations, and infragravity wave runup every ~100 m alongshore (Section 2.2.3.5). Wave runup is the maximum vertical extent of wave uprush on a beach or structure above the still water level, and in cases where infragravity waves exist, the reach of wave runup can be significantly further inland compared to wave runup driven by shorter incident waves [46,47]. The U.S. west coast is particularly susceptible to infragravity wave runup due to the prevalence of breaking long-period swell (low wave steepness) across wide, mildly sloping (dissipative) beaches that result in a shoreward decay of incident wave energy and accompanying growth of infragravity energy.

2.2.3.1. Storm Event Selection

The CoSMoS framework is computationally expensive, therefore a full 21st century simulation is infeasible. Additionally, as major contributors to coastal water levels such as astronomic tides and storms are independent processes, a century-long simulation may not fully characterize potential flood risk, as the timing between high tide occurrences and storm impacts would be necessary. Therefore, future storm events are identified a priori in CoSMoS; these events are then run with Tiers I–III, synched to representative spring tide cycles.

The storm selection process employs the same TWL proxies used as forcing for the cliff recession and shoreline change models (Section 2.2.2). Proxies do not account for variations in water levels due to astronomic tides and SLR, as they are independent of atmospheric conditions. It is recognized, however, that nearshore wave heights and $R_{2\%}$ are affected by tidal stage and currents, and non-linear interactions can be important [82]. As runup is the dominant contribution to TWL_{px} (over 85%), it is unlikely that selection of extreme storm events would be significantly impacted. Additionally, these components are accounted for in the more detailed CoSMoS model simulations.

In keeping with the approach of identifying coastal storms with specific recurrence intervals (Section 2.2.1), the ranked 1-year, 20-year, and 100-year future coastal storm events, derived from the 100-year TWL time series at the shore, were identified at each nearshore transect location (4802 sites). Event results were clustered with a k-means algorithm to delineate coherent coastal segments which had similar return period water levels for specific storms; see Erikson et al. [64] for more details on identification methodology. Clustering of extreme events showed that the more severe but rare coastal flood events (e.g., the 100-year event) occur for most of the region from the same storm event. In contrast, different storms from varying directions were responsible for the less severe, but more frequent, local coastal flood events (Figure 5). To this end, two 100-year storms were identified (February 2044 and March 2059), two 20-year storms (February 2025 and February 2095), and three 1-year storms (March 2020, December 2056, and January 2097). Upon completion of 1-year storm simulations using the entire train of models (Tier I–Tier III resolving detailed flow dynamics and wave-current interaction) for a range of SLRs, results showed a single 1-year storm (March 2020) consistently yielded the highest water levels throughout the SCB; thus, 1-year projections use contributions from only this event. Event conditions are summarized for each of the identified storms in Table 2. For each identified event, associated event conditions are applied as follows: deep water wave conditions (H_s, T_p, and D_p) applied at all open boundaries of the Tier I wave grid; SLA applied uniformly to all model domains; and downscaled time- and space-varying SLP and wind fields applied to all domains synched such that minimum SLP in the region coincided with high tide water levels.

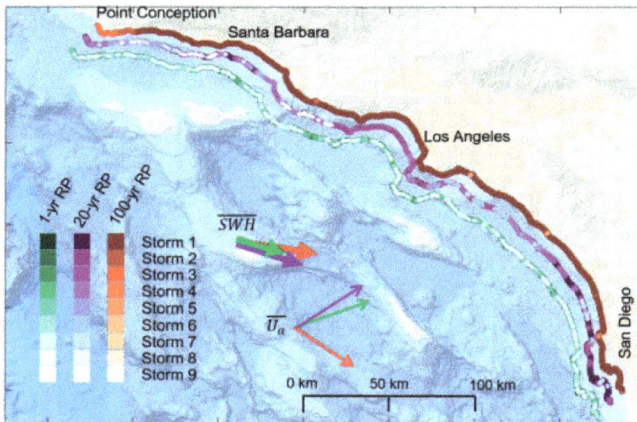

Figure 5. Map showing coastal regions that respond similarly to region-wide storms as determined through k-means clustering of nearshore total water level proxies [64]. Large colored arrows show the weighted mean offshore wave heights and winds for the 1-year, 20-year and 100-year return period coastal storms.

Table 2. Boundary conditions associated with each modeled scenario.

Scenario	H_s, Meters	T_p, Seconds	D_p, Degrees	SLA, Meters	Minimum SLP (Kilopascal)	Maximum Wind Speed, Meters/Second
Background	1.75	12	286	0	NA	NA
1-year storm 1	4.39	16	284	0.16	100.56	22.8
20-year storm 1	5.86	18	281	0.18	100.79	22.3
20-year storm 2	6.13	18	292	0.24	100.41	28.7
100-year storm 1	6.20	16	264	0.19	100.43	26.6
100-year storm 2	6.80	18	287	0.23	98.67	30.3

2.2.3.2. Fluvial Discharge Model

CoSMoS incorporated discharges from 37 coastal rivers in the SCB considered most relevant in influencing coastal flooding (Table 3; Figure 6); discharges are included in Tier II sub-models. At the time of this study, no available time series of 21st century discharge rates associated with this project's RCP scenario exist, and therefore discharge projections were developed. The approach does not erroneously assume that a 100-year fluvial discharge event coincides with a 100-year coastal storm event for the simulation period [83], but instead employs atmospheric patterns common to both events with the aim to obtain more realistic joint occurrences.

Rivers were initially separated into two groups: (1) primary gauged rivers for which relationships between peak flows and an independent atmospheric variable (from downscaled GCM) are identified; and (2) subordinate river for which relations with assigned primary rivers were used to estimate future flows.

Seven gauged rivers for which identifiable relationship between peak flows and SLPs were attainable (Table 3) were labeled as "primary/parent representations" within the SCB. As many as 15 sub-ordinate rivers were assigned to each of these primary rivers using previous studies that have evaluated similar relationships [84]. Exceptions are the Rio Hondo and Santa Ana Rivers, for which sub-ordinate rivers and tributaries were assigned as shown in Figure 6 due to the urbanized lowland nature of the former and contrasting mountainous and dammed Santa Ana watershed. See Appendix A for details on discharge data sources.

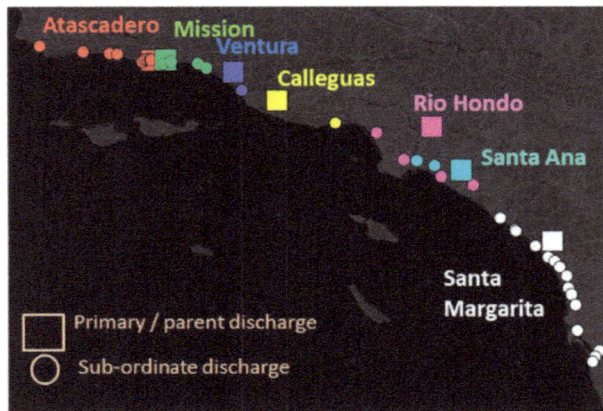

Figure 6. Locations and delineation of primary and sub-ordinate rivers and tributaries modeled in Southern California CoSMoS. Area-normalized runoff rates of primary (squares) discharge points were used to estimate discharge rates at sub-ordinate (circles of corresponding color) point locations.

Table 3. Primary and sub-ordinate rivers within the Southern California study area.

Primary Stream	Sub-Ordinate Stream
Atascedero	Jalama, Gaviota, Refugio, El Capitan, Devereux, Goleta
Mission Creek	Arroyo Burro, Mission, Carpinteria, Rincon
Ventura	Santa Clara
Calleguas	Malibu
Santa Margarita	San Juan, San Mateo, San Onofre, Los Flores, San Luis Rey, Buena Vista, Agua Hedionda, Batiquitos, San Elijo, Del Mar, Pensaquitos, San Diego, Sweetwater, Otay, Tijuana
Rio Hondo	Ballona, Dominguez, Bolsa Chica, Newport Bay
Santa Ana	Los Angeles, San Gabriel

To estimate future peak discharge rates, and relate them to coincident storm events, observation-based least-squares linear regression equations relating peak discharges to SLPs were developed. Using these relationships with future SLPs, from the GFDL-ESM2M RCP4.5 GCM as the predictor, future rates consistent with storm event physics are established. Several variants of SLPs were tested against observed peak discharge rates (99.95th percentile flow rate) measured at the seven primary USGS gauging sites. Reasonably strong linear relationships ($0.50 \le r \le 0.99$, $0.001 \le p\text{-value} \le 0.076$) were found between maximum SLP gradients (ΔSLP) and peak discharge. To find the most appropriate predictor, ΔSLP (from CaRD10 reanalysis) were compared over one, three and five days prior to peak discharge and within 0.667°, 1°, and 5° radii of the gauging station. Best fits were obtained with the three-day window and 0.67° search radius for all but two gauging sites (Santa Ana and Santa Margarita; best fit of 1.0° radius). The greater search radius of the Santa Ana and Santa Margarita Rivers is consistent with the larger watershed areas (~>3 times) associated with each of these rivers compared to the five other watersheds.

Discharge projections of smaller, subordinate rivers within the SCB watersheds were derived using area-normalized flow rate relationship [16]

$$Q_i^{21} = R_{Primary}^{21} \cdot A^i \tag{2}$$

where and Q_i^{21} is the projected discharge rate (m³/s) of subordinate river i, A^i is the watershed area (m²) associated with river i, and $R_{Primary}^{21}$ is the area-normalized projected flow of the corresponding primary river.

An idealized dimensionless hydrograph was developed from nine gauging station discharge records throughout the study area. These stations had data available at 15 min or smaller sampling resolution and at least three events exceeding the 99.95th percentile during the record period. Events exceeding the 99.95th percentile were selected, normalized by the peak flow, $Q^{Historical}$, and fit with a lognormal distribution. Lognormal distributions are often used to develop unit hydrographs as they have been shown to predict peak flows and time to peak well [85]. The mean of the mean and mean variance of all nine fitted distributions were used to define the idealized hydrograph in Figure 7. The hydrograph is skewed toward rapid initial increases in flow and subsequent slower rates of decreasing discharge rates. The total duration is on the order of 0.7 days (17 h) for flows that exceed 10% of the peak discharge.

Simulations were constructed such that peak discharge was synched with peak tide levels and surge (i.e., minimum SLP) conditions. Winds fields, resultant waves, and fluvial discharge parametrization from the same downscaled GCM were thus dictated by the timing of the SLP low. Most USGS gauging stations are upstream of tidal influence; model discharge points were placed at gauge locations within applicable CoSMoS Delft3D domains (Figure 7). Discharge points were placed in the upmost position in the domain for the few cases where gauge locations were outside model boundaries.

2.2.3.3. Regional-Scale Wave and Hydrodynamic Model (Tier I)

The WAVE and FLOW modules of the Delft3D version 4.01.00 were used to simulate waves and hydrodynamics, respectively. The WAVE module allows for two-way coupling (communication) between wave computations and FLOW hydrodynamics simulating waves with the numerical model SWAN. SWAN is a commonly used third-generation spectral wave model specifically developed for nearshore wave simulations that account for propagation, refraction, dissipation, and depth-induced breaking [44,86]. Delft3D-FLOW, developed by WL/Delft Hydraulics and Delft University of Technology, is a widely used numerical model that calculates non-steady flows and transport phenomena resulting from tidal and meteorological forcing [43].

Tier I SWAN and FLOW models consist of identical structured curvilinear grids that extend from shore to ~200 km offshore in water depths >1000 m and range in resolution from 1.2 km × 2.5 km in the nearshore to 3.5 km × 5 km in the offshore (dashed black line in Figure 3; black line in Figure 8). The two-way coupled model was run in a spherical coordinate system and with FLOW in a vertically-averaged mode (2DH). Spatially varying astronomic tidal amplitudes and phases derived from the Oregon State University TOPEX/Poseidon global tide database [87] were applied along all open boundaries.

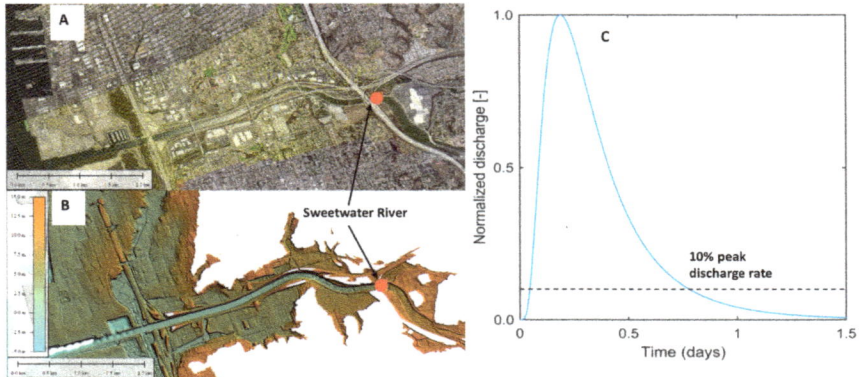

Figure 7. Illustration of discharge inclusion in Tier II hydrodynamic models: (**A**) high resolution model grid in San Diego, CA with discharge location shown as red dot; (**B**) elevation of same location with gauge location shown as red dot; and (**C**) normalized hydrograph used in discharge time-series included in each hydrodynamic simulation. Hydrograph was scaled by a river's estimated peak discharge, Q^{21}.

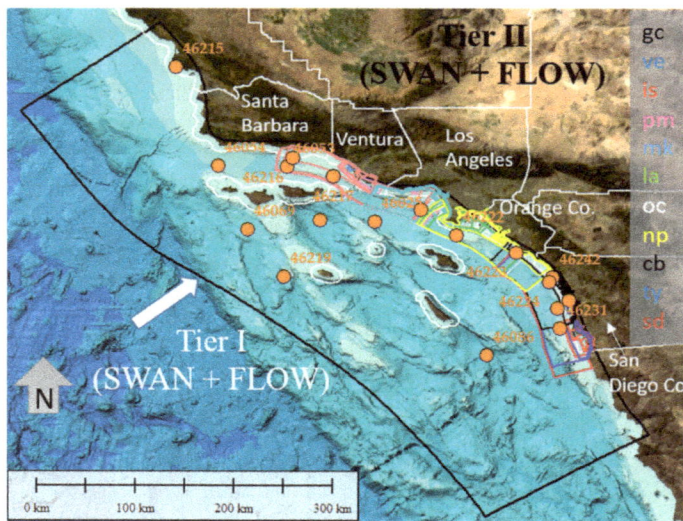

Figure 8. Map showing Tier I (black) and Tier II (colored) model grid extents. Wave observation buoys used in model validation (Section 3 and Appendix B) are shown with orange circles.

SLA due to large-scale meteorological and oceanographic processes unrelated to storms, were applied along all open boundaries of the Tier I FLOW grid. Elevated SLA are often observed in conjunction with El Niño events [28,29,88] and yield water levels of 10–20 cm above normal for several months [89]. In an effort to maintain simplicity, measured SLAs were correlated with sea surface temperature anomalies (SSTAs) for the same time periods. These relationships were used with projected GFDL-ESM2M SSTs to estimate future variations in SLAs, as described in [34].

Space- and time-varying wind (split into eastward and northward components) and SLP fields (see Section 2.2.1) were applied to all grid cells at each model time-step. The wind and SLP fields were input as equidistant points spaced 10 km apart and interpolated within the Delft3D model to the SWAN and FLOW grids. An average pressure of 101.3 kilopascals was applied to the open boundaries of the meteorological grid.

Deep water wave parameters (H_s, T_p, and D_p), obtained with the WWIII model for the CDIP067 buoy and each identified storm event (Section 2.2.3.1), were applied along all open boundaries of the Tier I SWAN grid. Wave direction is critical to accurate simulations of wave propagation from deep-water to the nearshore in the complex SCB region [22]. However, as alongshore variations in deep-water wave forcing available with the WWIII model outputs were small, particularly with respect to incident wave directions, non-varying wave boundary forcing is applied to the Tier I model.

2.2.3.4. Local-Scale 2D Wave and Hydrodynamic Model (Tier II)

Tier II consists of 11 local-scale sub-models to cover the SCB, each consisting of two SWAN grids and multiple FLOW grids (Figure 8). Physical overlap exists between sub-models along-shore extents to avoid erroneous boundary effects in regions of interest.

Each Tier II hydrodynamic FLOW sub-model consists of one "outer" grid and multiple two-way coupled "domain decomposition" (DD) structured grids. DD allows for local grid refinement where higher resolution (~10–50 m) is needed to adequately simulate the physical processes and resolve detailed flow dynamics and overland flood extents. Communication between the grids takes place along internal boundaries where higher-resolution grids are refined by three or five times that of the connected grid. This DD technique allows for two-way communication between the grids and for simultaneous simulation of multiple domains.

In the landward direction, Tier II DD FLOW grids extend to the 10 m topographic contour; exceptions exist where channels (e.g., the Los Angeles River) or other low-lying regions extend multiple kilometers inland. The number of DD FLOW grids ranges from 4 to 13, depending on local geography, bathymetry, and overall landscape. Grid resolution ranges from approximately 130 m × 145 m (across and along-shore, respectively) in the offshore region to as fine as 5 m × 15 m in the nearshore and overland regions.

Wave computations are done with the SWAN model using two grids for each Tier II sub-model: one larger grid covering the same area as the "outer" FLOW grid and a second finer-resolution two-way coupled nearshore nested grid. The nearshore SWAN grids extend from at least the 30 m isobath to well inland of the present day shoreline. The landward extension is included to allow for wave computations of the higher SLR scenarios. See Appendix A for details on settings.

Water level and Neumann time-series extracted from Tier I simulations were applied to the shore parallel and lateral open boundaries of each Tier II "sub-model outer" grid, respectively (one-way nesting). Several of the sub-models proved to be unstable with lateral Neumann boundaries; for those cases, one or both of the lateral boundaries were converted to water level time-series or left unassigned. The extracted time series from Tier I included variations due to tides, SLAs and spatially-varying storm surge. To account for further, local contributions of winds and SLPs to storm surge related wind-setup at the shore and local inverse barometer effects (IBE, rise or depression of water levels in response to atmospheric pressure gradients), the same 10 km hourly-resolution winds used in Tier I are also applied to each grid cell in the Tier II sub-models.

2.2.3.5. Local-Scale 1D Wave and Hydrodynamic Model (Tier III)

Nearshore hydrodynamics, wave setup, total wave runup and event-based erosion were simulated with the XBeach (eXtreme Beach) version 1.21.3667 (2014) model [46,47]. XBeach is a morphodynamic storm impact model specifically designed to simulate beach and dune erosion, overwash, and flooding of sandy coasts. XBeach was run in a profile mode, at 4466 cross-shore transects numbered consecutively from 1 at the U.S./Mexico border to 4802, north of Point Conception (Figure 3). Profiles across harbor mouths, inlets, etc. were excluded from simulations. Each of the profiles extend from the approximate −15 m isobath to at least 10 m above NAVD88 (truncated in cases where a lagoon or other waterway exists on the landward end of the profile). Cross-shore profiles obtained from the DEM (see Section 2.2.2.1) were re-sampled using an algorithm that evaluates long wave resolution at the offshore boundary, depth-to-grid-size ratio, and grid size smoothness constraints to obtain optimum grid resolution while reducing computation times. Final profile grid resolutions are between 25 m and 35 m in the offshore and 5 m in shallow nearshore and land regions. Further details on settings are provided in Erikson et al. [54] and Appendix A.

Time-series of water levels (hourly) and waves (20-min intervals) extracted from completed Tier II runs were applied at the seaward ends of each of the transect models. Water-level variations represented the cumulative effect of astronomic tides, storm surge (including inverse barometric effects and wind setup), SLAs, and SLR. Neumann boundaries set to zero were used along the lateral boundaries: a condition that has been shown to work well with quasi-stationary situations where the coast can be assumed to be uniform alongshore outside the model domain [46,47].

Within each simulation, morphodynamic change due to the storm event is computed along the profile transect. The event-based erosion extent simulated by XBeach is dependent on the hydrodynamics across the entire active and wetted profile, bordered on the landward side by the runup extent.

2.3. Determination of Flood Extents and Uncertainty

Flood extents were determined in two ways: (1) from the landward-most wet grid cell in the high-resolution Delft3D grids in embayments; and (2) from maximum sustained water levels calculated with XBeach cross-shore models along the open coast. The frequency-filtered sustained water levels (constant water levels of durations longer than 1 min) are intended to capture the wave setup at the shore, which is the increase in mean water level above the still water line due to the transfer of momentum by breaking waves. Maximum runup, computed with the Tier III XBeach model, are also output as part of the CoSMoS results, but are mapped as single points and are not included in the flood footprint (see Section 4). Mapped outputs are done in this way to distinguish between shorter duration wave runup which, depending on the beach slope, may only constitute a couple of centimeters of intermittent standing water. Except where overtopping occurs or at a narrow beach that fronts a near vertical cliff or wall, sustained water levels and flood extents are seaward of the maximum runup.

Melding of flood extents simulated with the XBeach and Delft3D high resolution models was done by interpolating (linear Delaunay Triangulation) resulting water level elevations onto a common 2 m resolution square mesh (within the Mathworks Matlab environment). In some areas, such as Mission Beach in San Diego County, where both XBeach and high resolution grids exist to capture flooding from either or both the landward or seaward side, XBeach results were given precedence. Because of Tier III transect model spacing, areas between transects contain interpolated flood depths and extents and are not exact representations of model outputs.

This post-processing step was done for all storms simulated as part of a given scenario. For the 20-year storm for example, two individual storm events were modeled to ensure that local effects, such as shoreline orientation with respect to incident storm direction, were taken into account. For those cases where more than one storm was modeled, all resulting 2 m gridded flood maps were overlain and maximum water levels saved at each grid cell to generate a single, composite flood map for a given scenario.

Resulting water-elevation surfaces were differenced from the high resolution DEM to isolate areas where the water level exceeds topographic elevations, indicating flooding. For scenarios that include SLR, 2 m DEMs that incorporate long-term morphodynamic changes were used (Section 2.2.2.5).

The resulting flood maps were then processed to exclude isolated wetted areas not hydraulically connected to the ocean; these disconnected areas were flagged as low-lying vulnerable areas below the flood elevation. To note, culverts or other manmade and natural underground pathways, not physically captured in the DEM, between coastal waters and land are not accounted for in the modeling and flood projection.

Maps of associated maximum flood durations, velocities, and wave heights were processed in a similar manner to that of the flood depths and extents in that they were gridded onto a common 2 m mesh and then combined. Data that fell outside the flood map extents were removed so that the footprints of all data layers are consistent.

2.3.1. Uncertainty

Uncertainties in the numerical model outputs and DEM measurements were combined with estimates of vertical land motion (VLM, Section 2.3.2) to produce spatially varying offsets that were added/subtracted to/from the modeled flood elevations to produce maps of flood potential; essentially spatially-characterizing the vertical uncertainty from the model. This was done for each of the 40 scenarios, yielding maximum and minimum potential flood extents (the upper and lower bounds of total uncertainty) for each flood projection. Models compared well to available observation stations within the SCB (Section 3), but model comparisons remain limited for the size and scope of the study area and may not address spatial variability of error. Therefore, a much larger value of ±50 cm was used to represent model uncertainty. This larger value was used in an effort to mitigate for the low number of available and tested storms compared to the geographic scope. Uncertainties associated with the baseline DEM were set at ±0.18 m, equivalent to the 95% confidence level for topographic lidar measurements in open terrain [90].

2.3.2. Vertical Land Motion

Spatially variable measurements of vertical land motion based on GPS data and statistical and physical tectonic models, largely attributed to tectonic movement of the San Andreas Fault System from Howell et al. [91], were also incorporated into uncertainty. For each SLR scenario, spatially variable, 1 km resolution rasters of uplift and subsidence values were derived from Howell et al. [91]. For consistency with long-term cliff/beach projections, uplift and subsidence rates were applied through year 2100 for all SLR amounts above 1.0 m, while earlier time frames (2044/2069/2088) were used for lower SLR scenarios, as described in Section 2.2.2.2. The corresponding amounts of uplift and subsidence were then added to or subtracted from the baseline DEM to account for uncertainty due to vertical land motion.

Maximum rates of uplift (0.4 mm/year) and subsidence (0.6 mm/year) within our study area equate to a maximum of 3.4 cm of uplift and 5.2 cm of subsidence for the 1 m SLR scenario. In the future, these rates could be refined using more spatially-resolved data sources (e.g., InSAR) to complement the GPS data which are fairly sparse along the Southern California coast. However, in general, even the highest rates of vertical motions that have been recorded in the San Francisco Bay Area [92] and more locally in the Santa Ynez mountains [93] rarely exceed more than 2 and 6 mm/year, respectively, and therefore are quite small relative to the expected rates of SLR by the middle and end of the 21st century.

3. Results—Model Validation

The model setup and simulation scheme were tested by comparing model outputs to observed water level variations due to astronomic tides and non-tidal residuals (storm surge and other anomalous water levels), wave heights, wave runup, and short-term morphologic change.

The root-mean-square-difference (*rmsd*) and *bias* were calculated,

$$rmsd = \left[\frac{\sum_{i=1}^{N} (obs_i - mdl_i)^2}{N} \right]^{1/2} \tag{3}$$

$$bias = \frac{1}{N} \sum_{i=1}^{N} (mdl_i - obs_i) \tag{4}$$

where *obs* is the observation data, *mdl* is the model data, *i* is the individual time point data, and *N* is the total number of time-points analyzed. The *rmsd* represents the standard deviation of the residuals (difference between the observed and modeled values), the *bias* describes the model's overall offset from observations. Regarding flood extents, an extreme paucity of region-scale spatially-varying flood data exist for model comparisons. Certain temporally-limited datasets have been painstakingly collected for specific locations [94], but comparisons of flood extents across the study area is infeasible at the current time.

The model's ability to replicate tidal variations was tested over a month-long time-period (October to November 2010) to capture full variations in spring and neap cycles. Modeled tidal variations were compared to NOAA predicted tides at the four tide-stations within the SCB: La Jolla (station ID: 941030), Los Angeles (station ID: 9410660), Santa Monica (station ID: 9410840), and Santa Barbara (station ID: 9411340). Comparisons between time-series of the modeled and predicted tides are very good at all four stations, being less than 6 cm for both the *rmsd* and *bias* (Figure 9).

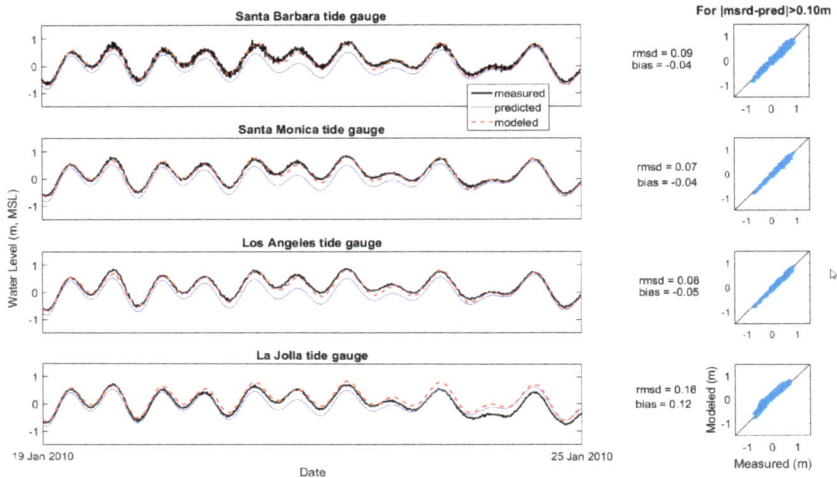

Figure 9. Comparison of modeled and measured water levels at the Santa Barbara, Santa Monica, Los Angeles, and La Jolla tide gauges during the January 2010 storm.

Wave model accuracy was tested against the January 2010 storm by comparing hindcast wave heights, periods, and directions to observed values at 18 buoys within the Southern California Bight (see Figure 8 for buoy locations; Figure 10 for observation-model comparisons; and Table ?? in Appendix B for full comparisons). In addition to *rms* values, non-dimensional Wilmott skill scores are used to aid in quantifying model skill of wave simulations [95]. See Appendix B for complete skill scores.

The model's ability to simulate wave heights across the area is generally good to great (Wilmott skill ranging 0.69 to 0.96, Table ??), and in conjunction with the *rmsd* values, shows that model performance increases with the finer Tier II grids. *Rmsd* values range 19–51 cm for the Tier II grids and 28–67 cm for the Tier I grids. Peak wave directions are quite good with *rmsd* values less than 3°. Peak wave periods are modeled with good to fair skill (0.48 < skill < 0.69). The lower peak wave period skill compared to wave height skill is likely a reflection of the "jumpy" nature of peak wave periods in multimodal regions such as SCB [25]. Skill scores of mean wave period are likely to show better performance, but are not listed here as it is the peak periods that are used for boundary conditions to Tier III XBeach models.

Wave runup was evaluated by running XBeach for a time-period of available runup measurements at Ocean Beach, just south of the Golden Gate near San Francisco in central California. Runup measurements were obtained during 3-h daylight intervals in May 2006 when offshore waves ranged between 1–2 m and peak wave periods up to 14 s (at NDBC buoy 46026) using a camera system [96]. The foreshore beach slope was mild with an average slope of 0.03. Computed *rms* values between the observed and modeled runup height for four separate 3-h periods ranged 10–16 cm [96].

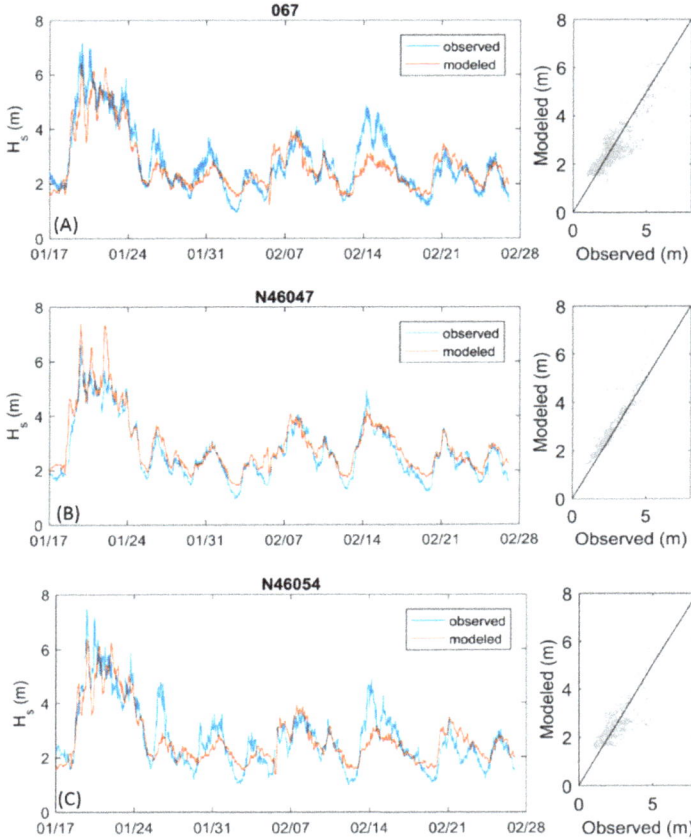

Figure 10. Time series and scatter comparisons of measured and modeled wave heights at three buoy locations within the SCB. See Appendix B for more information.

4. Discussion—Projected Hazards Case Study: Del Mar, California

Changes in flood extents and hazards between scenarios can be investigated to determine thresholds and hotspots of vulnerability with respect to SLR and storm intensity. To highlight the significance of dynamic storm contributions to flood risks, and the incorporation of these processes in CoSMoS, results for Del Mar, California, are showcased.

Looking across this community, the storm-related impacts of flooding are clearly visible (Figure 11) for all SLR scenarios. Considering SLR-related flooding only, 0.8 km² of Del Mar is impacted with SLR of 150 cm (Figure 12). Notably, however, flood extents from a 100-year storm for the same SLR impact more than twice the area than when considering SLR alone, underlining the importance of dynamically-driven water levels in coastal flooding. This is especially true for lower amounts of SLR, probable in the beginning to middle of the 21st century. With higher amounts of SLR, flood extents reach the extent of the lagoon basin and the surrounding high terrain, limiting further exposure. Taking a closer look at the local scale, identification of specific at-risk features and locations are clearly ascertained. Runup points, illustrating landward wave-driven wetted extents, show even more extensive effects inland. In contrast, CoSMoS flood extents signify presence of 1–2 cm of water for at least a minute, to demonstrate a more representative impact of flooding. Runup is sometimes used as the TWL at shore to drive flood illustrations in other vulnerability models [97] despite potentially limited volumes of seawater associated with the uppermost and intermittent swash. However, as wave-driven wetting and flooding projections are separated in CoSMoS results, users can either include or isolate impacts as appropriate for their site, planning needs, and level of risk tolerance.

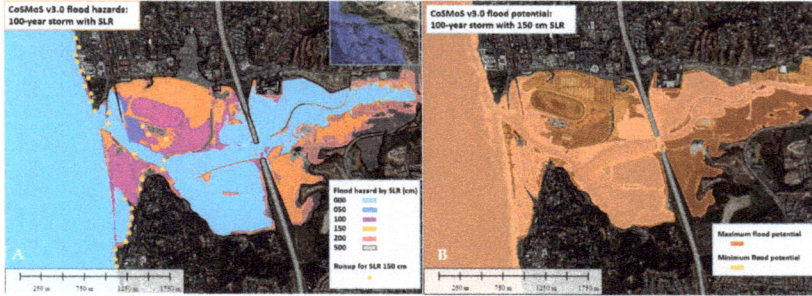

Figure 11. Projected flood hazards for Del Mar, California: (**A**) flood hazard projections for 100-year storm scenario and various SLR. Maximum runup generated along 100-m space transects, representing landward wave-driven potential wetting extent, for SLR 150 cm is shown in orange circles; and (**B**) flood potential for 100-year storm scenario and 150 cm SLR.

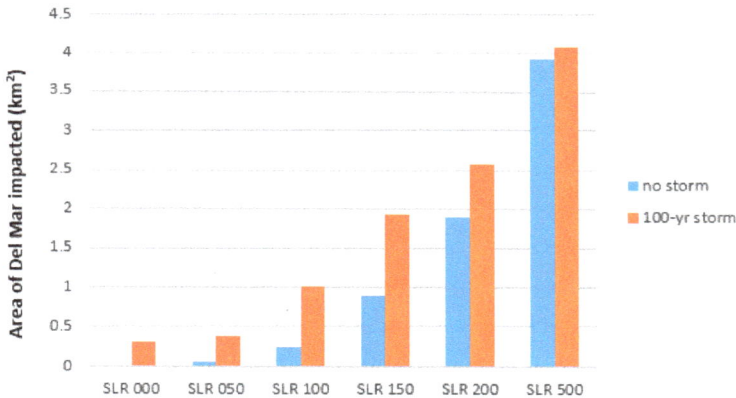

Figure 12. Area of projected flooding impacts for Del Mar for various SLR considering background conditions (no storm) and a 100-year storm.

Considering uncertainty illustrated in flood potential (Figure 11), marsh areas are particularly vulnerable. The low-elevation stretches of marsh show significant increases in vulnerability with maximum flood potential, or the upper range of model uncertainty and DEM accuracy. As elevation-data collection methods for the DEM showed the greatest uncertainty in regions of densely-vegetated parcels such as marsh areas, due to the inability of lidar at penetrating the thick plant matter [98], this uncertainty should be incorporated for any risk or adaptation study. For comprehensive analyses of risk, runup and flood potential (uncertainty) should be considered, in addition to explicit flood projections.

Looking at long-term shoreline change hazards around the lagoon entrance to Del Mar, existing beaches appear particularly vulnerable for conditions at or greater than 150 cm (Figure 13). Long-term projections indicate beaches retrograde landward to existing breakwater and urban infrastructure for all SLR scenarios above 150 cm. However, impacts may occur sooner, as locations for the SLR 100 cm show significant retrograding, coming to within less than 7 m of this urban boundary for much of the beach area shown. Similarly, cliff retreat for the bluffs north of the lagoon entrance show ~40 m of landward regression for the same SLR 150 cm scenario; greater erosion is seen for higher SLR scenarios. The subsequent impacts of this cliff erosion, as incorporated into long-term profile and DEM evolution, can be seen in the flooding projections, as flood extents progress inland along the bluff-backed coastline for each consecutive SLR scenario.

An extensive presentation and discussion of projected hazards across the SCB, to include examination of spatial variation in flood extents and contributions to flood hazards, is presented in Part 2 of this manuscript [17].

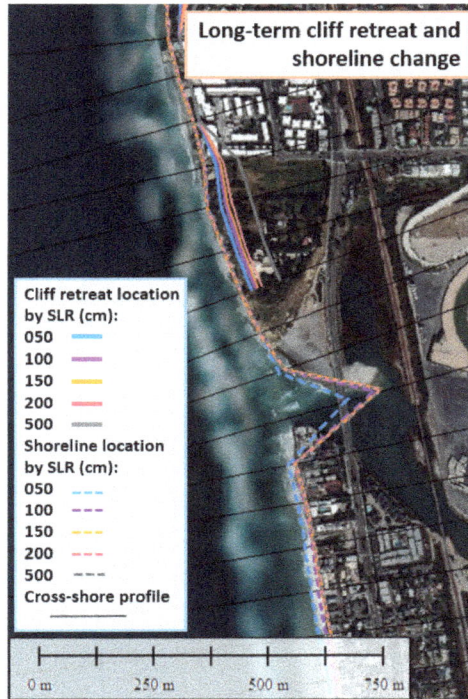

Figure 13. Long-term projected cliff retreat (solid colored lines) and sandy beach shoreline change (dashed lines) for various SLR at seaward lagoon entrance of Del Mar, California. In this projected management scenario, sandy beach erosion is limited to the urban boundary, and cliff retreat limited in locations with existing coastal armoring. Cross-shore transect/profile locations, at which long-term changes were derived, are shown by black lines. Cliff retreat projections are made only for transects with existing coastal cliffs/bluffs. Note that sandy beach projections converge for SLR 150–500 cm for a large portion of this area; all projections converge/show no change in locations where beach width is minimal (e.g., at the foot of the cliffs north of the lagoon mouth).

5. Conclusions

The overarching concept of CoSMoS is to leverage projections of global climate patterns over the 21st century from the most recent (CMIP5) Global Climate Models (GCMs). Coarse resolution GCM projections are downscaled to the local level and used as boundary conditions to sophisticated ocean modeling tools that simulate complex physics to accurately predict local coastal water levels and flooding for the full range of expected SLR (n = 10: 0–2 m in 0.25 m increments, and 5 m) and storm scenarios (n = 4: average daily/background conditions, annual, 20-year and 100-year). Resulting model projections include spatially explicit estimates of flood extent, depth, duration, uncertainty, water elevation, wave runup, maximum wave height, maximum current velocity, and long-term shoreline change and bluff retreat.

This latest version of CoSMoS has incorporated several additional contributions to coastal water levels and change. Fluvial discharge projections have been included in the hydrodynamic simulations, to capture effects of back-flooding at 37 river mouths in the SCB. Wind setup, locally-generated waves, and other wind-driven processes are inclusively simulated with dynamically downscaled SLPs and winds. Long-term morphodynamic change, derived from century-scale sandy beach and cliff retreat projections, is captured and used in hydrodynamic simulations of future conditions. By including these projections, driven by a range of SLR and wave conditions, a more comprehensive hazards projection is achieved.

Long-term shoreline change models do not include feedback between beach and cliff erosion, however. This feedback is important for complex sediment systems, such as in the study area and other portions of California. Additionally, the profile evolution scheme does not capture along-shore transport and full redistribution of sediment. Future studies will include effort to address feedback between the cliff and shore models, as well as improved profile evolution algorithms.

The inclusion of fluvial discharge is an important addition for river areas along the coast. Several studies suggest the joint probability of co-occurrence between coastally-driven water levels and river flood events are not directly proportional [99,100]. CoSMoS includes estimates of fluvial discharge based on the selected storms' SLP gradient intensity, not a direct hydrologic projection. In addition, the hydrographs used in the simulations are idealized and peak fluvial discharges are first-order estimates. The simple model employed is efficient and appropriate for inclusion in the prescribed wave-driven flood scenarios. However, the model system would benefit from realizations of discharges for a given storm using improved atmospheric relationships. Similarly, pluvial flooding, related to the intensity and duration of precipitation, can be an important contribution of flooding for many coastal urban areas. Pluvial components are not included, as it is currently beyond the scope of the modeling system; however, such events are likely attributed to atmospheric rivers in California, [101] and the co-occurrence of these phenomena with extreme coastal wave events is currently being investigated.

Additionally, future development of CoSMoS aims to better capture overland flow and flood projections in overtopping and spill-over situations. While several types of overland flow models exists, models are limited in: (1) ability resolve high-resolution flow over large, complex regions; (2) public/open-source access; or (3) interoperability with other parts of the CoSMoS framework. Feasible options for future CoSMoS developments involve the use of flexible-mesh hydrodynamic models and which may also account for infragravity components.

Although the limitations exist for the models and computations contributing to this storm modeling system, CoSMoS provides a comprehensive, physically-consistent, and explicit projection of coastal hazards for the 21st century. Hazard projections are adaptable and scalable, covering a large region with high-resolution projections for a wide range of SLR and storm conditions, to meet the needs of varied users and interests involved in community-scale planning. The thorough modeling approach presented here is designed to provide the necessary data to fully assess potential coastal risks with our changing climate, and make decisions to mitigate and act.

Author Contributions: All authors contributed to the paper and portions of the CoSMoS framework.

Acknowledgments: Funding for this project was provided from: California Natural Resources Agency for California's 4th Climate Assessment, *Climate Change Impacts for Southern California*; California Department of Fish & Wildlife, *Imperiled Southern California Coastal Plant Species: Assessment of Threats from Climate Change, Including Rising Sea Levels, Storms and Coastal Erosion*; California State Coastal Conservancy, *Southern California Climate Change Vulnerability Study*; City of Imperial Beach, *Imperial Beach Climate Change Vulnerability Study*; and USGS Coastal and Marine Geology Program. Authors would also like to acknowledge support through a cooperative agreement with Deltares. Data from CoSMoS is available online at http://dx.doi.org/10.5066/F7T151Q4.

Conflicts of Interest: The authors declare no conflict of interest.

Appendix A

Appendix A.1. WaveWatch III Setting Details

Bathymetry and shoreline positions were populated with the 2-min Naval Research Laboratory Digital Bathymetry Data Base (DBDB2) v3.0 and National Geophysical Data Center Global Self-Consistent Hierarchical High-Resolution Shoreline (GSHHS) [102]. Wave spectra were computed with 15° directional resolution and 25 frequency bands ranging non-linearly from 0.04 to 0.5 Hz. Wind-wave growth and whitecapping was modeled with the Tolman and Chalikov [103] source term package and nonlinear quadruplet wave interactions were computed with the Hasselmann et al. [104] formulation. Bulk wave parameter statistics (significant wave height, H_s; peak wave period, T_p; and peak wave direction, D_p) were saved at daily time-steps (integrated over 24 h) at each grid point and hourly at select points in deep water offshore of the continental shelf (see 42 and 56 for further details).

Appendix A.2. Tier I (Delft 3D FLOW-WAVE) Setting Details

Bathymetry was derived from the National Geophysical Data Center (NGDC) Coastal Relief Model [105]. Spatially varying astronomic tidal amplitudes and phases derived from the Oregon State University TOPEX/Poseidon global tide database [87] were applied along all open boundaries of the Tier I FLOW grid. Thirteen tidal constituents were represented: M2, S2, N2, K2, K1, O1, P1, Q1, MF, MM, M4, MS4, and MN4.

The SWAN model was run in a stationary mode, with a JONSWAP spectrum with peak enhancement factor of 3.3 at the open boundary forcing, 36 directional bins (i.e., 10° discretization), and 35 frequencies with logarithmic spacing from 0.0418 Hz to 1.00 Hz. Directional spreading used was 4°, and depth induced breaking was computed with the Battjes and Janssen [106] formulation and a breaking index of 0.73; whitecapping is described with the default Komen et al. [107] expression. Bottom friction is based on the JONSWAP formulation, with the friction coefficient set at 0.067 m^2/s [108].

Appendix A.3. Tier II (Delft 3D FLOW-WAVE) Setting Details

All model settings of the Tier II domains are identical to those used for Tier I runs, with the exception of the time-step (10 s) and threshold depth (1 cm) in the hydrodynamic FLOW models. The threshold depth is used within the model to assign a grid cell as either wet or dry. For the flooding and drying scheme, the bottom is assumed to be represented as a staircase of tiles centered around the grid cell water level points. If the total water level drops below 1 cm, then the grid cell is set to dry. The grid cell is again set to wet when the water level rises and the total water depth is greater than the threshold.

Model grid bathymetry and topography were generated using the 2-m resolution DEM (Section 2.2.2.1) [49] in the near- and onshore regions, and the 1/3 arc-second NOAA coastal relief model [105] seaward of the three nautical mile limit.

Appendix A.4. Tier III (XBeach) Setting Details

Sediment transport is computed in XBeach with the Soulsby-van Rijn [109] transport formula and bore averaged equilibrium sediment concentrations. A median grain diameter of 0.25 mm and sediment thickness of 2 m was assumed for all profile models. Bottom roughness is set to a uniform Chezy value of 65, horizontal background viscosity of 0.01 m^2/s, and a flooding and drying threshold depth of 1 cm, similar to Tier II. Profile sections that were initialized with steepness in excess of 32° (angle of repose of natural sand) are assumed to be hard structures or cliffs and set to be immobile (disabling erosions/accretion during the storm). All simulations are run with a morphological acceleration factor of 10 to speed up the morphological time scale relative to the hydrodynamic timescale and thus reduce computation time.

Appendix A.5. Fluvial Discharge Data Sources

Source discharge and draining area data were gathered from USGS gauging stations [110], USGS 12-digit and 8-digit hydrologic units (see [111] for a description of hydrologic units) and local water district maps. Future projections of SLP were obtained from the same downscaled atmospheric forcing fields that were used to identify and simulate the coastal storms. Historical atmospheric data fields were gathered from Scripps [14,45]. Peak discharge rates were calculated from records spanning at least 14 years (60-year mean record length).

Appendix B

Wave model accuracy was tested against the January 2010 storm by comparing hindcast wave heights, periods, and directions to observed values at 18 buoys within the Southern California Bight Wilmott skill scores range from 0 to 1, with 1 being perfect agreement between the model and observations. Skill scores between 0.8 and 1 are considered great; a score between 0.6 and 0.8 is considered good; and a score between 0.3 and 0.6 is fair. These are highlighted as green, yellow, and gray, respectively, where computed model skill and the collocation of the finest grid (either Tier I or named Tier II grid abbreviations as in Figure 8) corresponding to each buoy location are listed.

Table B1. Comparison of modeled and measured waves for the January 2010 validation storm. Grid refers to either Tier I or a Tier II grid abbreviation (Figure 8).

NDBC ID	CDIP ID	Lat (Degrees N)	Lon (Degrees W)	Grid	Significant Wave Height		Peak Wave Period		Peak Wave Direction	
					rms, Meters	Skill	rms, Seconds	Skill	rms, Degrees	Count
46086	-	32.49083	118.03472	Tier I	0.37	0.92	2.02	0.62	-	982
46069	-	33.67444	120.21167	Tier I	0.60	0.96	2.57	0.66	-	93
46054	-	34.26472	120.47694	Tier I	0.62	0.91	1.81	0.61	-	981
46053	-	34.25250	119.85333	gc, is	0.25	0.75	1.88	0.63	-	978
46025	-	33.74944	119.05278	Tier I	0.28	0.76	2.19	0.55	-	983
46221	28	33.85500	118.63400	mk	0.25	0.83	2.36	0.53	1.7	1966
46242	43	33.21980	117.43940	cb, ty	0.19	0.55	3.08	0.51	1.2	1614
46224	45	33.17778	117.47215	cb, ty	0.27	0.74	2.66	0.48	1.5	1944
46219	67	33.22480	119.88180	Tier I	0.67	0.79	1.82	0.69	2.2	1964
46215	76	35.20382	120.85931	Tier I	0.50	0.89	2.25	0.60	2.0	1968
46222	92	33.61791	118.31701	mk	0.27	0.88	2.09	0.58	2.3	1963
46231	93	32.74700	117.37000	sd	0.31	0.90	2.20	0.55	2.4	1968
46223	96	33.45800	117.76700	cb	0.25	0.79	2.30	0.52	1.9	1947
46225	100	32.93342	117.39083	cb	0.27	0.84	2.37	0.54	1.9	1968
46216	107	34.33300	119.80300	gc, is	0.29	0.88	1.79	0.54	2.0	1968
46217	111	34.16692	119.43465	gc, is	0.22	0.69	2.26	0.50	2.3	1941
46241	161	33.00300	117.29200	cb, ty	0.25	0.71	2.27	0.56	1.9	1968
46238	167	33.76000	119.55000	Tier I	0.51	0.89	1.78	0.62	2.3	1967

References

1. Merkens, J.L.; Reimann, L.; Hinkel, J.; Vafeidis, A.T. Gridded population projections for the coastal zone under the Shared Socioeconomic Pathways. *Glob. Planet. Chang.* **2016**, *145*, 57–66. [CrossRef]
2. Polade, S.D.; Gershunov, A.; Cayan, D.R.; Dettinger, M.D.; Pierce, D.W. Precipitation in a warming world: Assessing projected hydro-climate changes in California and other Mediterranean climate regions. *Sci. Rep.* **2017**, *7*, 10783. [CrossRef] [PubMed]
3. Diaz, D.B. Estimating global damages from sea level rise with the Coastal Impact and Adaptation Model (CIAM). *Clim. Chang.* **2016**, *137*, 143–156. [CrossRef]
4. Church, J.A.; Clark, P.U.; Cazenave, A.; Gregory, J.M.; Jevrejeva, S.; Levermann, A.; Merrifield, M.; Milne, G.A.; Nerem, R.S.; Nunn, R.D.; et al. Sea Level Change. In *Climate Change 2013: The Physical Science Basis. Contribution of Working Group I to the Fifth Assessment Report of the Intergovernmental Panel on Climate Change*; Stocker, T.F., Qin, D., Plattner, G.K., Tignor, M., Allen, S.K., Boschung, J., Nauels, A., Xia, Y., Bex, V., Midgley, P.M., Eds.; Cambridge University Press: Cambridge, UK; New York, NY, USA, 2013.
5. Ranasinghe, R. Assessing climate change impacts on open sandy coasts: A review. *Earth-Sci. Rev.* **2016**, *160*, 320–332. [CrossRef]
6. Harper, B.; Hardy, T.; Mason, L.; Fryar, R. Developments in storm tide modelling and risk assessment in the Australian region. *Nat. Hazards* **2009**, *51*, 225–238. [CrossRef]
7. Mousavi, M.; Irish, J.; Frey, A.; Olivera, F.; Edge, B. Global warming and hurricanes: The potential impact of hurricane intensification and sea level rise on coastal flooding. *Clim. Chang.* **2011**, *104*, 575–597. [CrossRef]
8. Graham, N.E.; Cayan, D.R.; Bromirski, P.D.; Flick, R.E. Multi-model projections of twenty-first century North Pacific winter wave climate under the IPCC A2 scenario. *Clim. Dyn.* **2013**, *40*, 1335–1360. [CrossRef]
9. Camus, P.; Menéndez, M.; Méndez, F.J.; Izaguirre, C.; Espejo, A.; Cánovas, V.; Perez, J.; Rueda, A.; Losada, I.J.; Medina, R. A weather-type statistical downscaling framework for ocean wave climate. *J. Geophys. Res. Oceans* **2014**, *119*, 7389–7405. [CrossRef]
10. Hoeke, R.K.; McInnes, K.L.; O'Grady, J.G. Wind and Wave Setup Contributions to Extreme Sea Levels at a Tropical High Island: A Stochastic Cyclone Simulation Study for Apia, Samoa. *J. Mar. Sci. Eng.* **2015**, *3*, 1117–1135. [CrossRef]
11. Barnard, P.L.; O'Reilly, B.; van Ormondt, M.; Elias, E.; Ruggiero, P.; Erikson, L.H.; Hapke, C.; Collins, B.D.; Guza, R.T.; Adams, P.N.; et al. *The Framework of a Coastal Hazards Model: A Tool for Predicting the Impact of Severe Storms*; Open-File Report, 2009-1073; U.S. Geological Survey: Reston, VA, USA, 2009; p. 19. Available online: http://pubs.usgs.gov/of/2009/1073/ (accessed on 2 February 2018).
12. Barnard, P.L.; van Ormondt, M.; Erikson, L.H.; Eshleman, J.; Hapke, C.; Ruggiero, P.; Adams, P.N.; Foxgrover, A.C. Development of the Coastal Storm Modeling System (CoSMoS) for predicting the impact of storms on high-energy, active-margin coasts. *Nat. Hazards* **2014**, *31*. [CrossRef]
13. Porter, K.; Wein, A.; Alpers, C.; Baez, A.; Barnard, P.L.; Carter, J.; Corsi, A.; Costner, J.; Cox, D.; Das, T.; et al. *Overview of the ARkStorm Scenario*; Open-File Report, 2010-1312; U.S. Geological Survey: Reston, VA, USA, 2010; 183p. Available online: https://pubs.usgs.gov/of/2010/1312/ (accessed on 2 February 2018).
14. Pierce, D.W.; Cayan, D.R.; Dehann, L. *Creating Climate Projections to Support the 4th California Climate Assessment*; University of California at San Diego, Scripps Institution of Oceanography: La Jolla, CA, USA, 2016.
15. O'Neill, A.C.; Erikson, L.H.; Barnard, P.L. Downscaling wind and wavefields for 21st century coastal flood hazard projections in a region of complex terrain. *Earth Space Sci.* **2017**, *4*, 314–334. [CrossRef]
16. Erikson, L.H.; O'Neill, A.C.; Barnard, P.L. Estimating fluvial discharges coincident with 21st century coastal storms modeled with CoSMoS. In Proceedings of the International Coastal Symposium (ICS), Busan, Korea, 13–18 May 2018; Shim, J.S., Chun, I., Lim, H.S., Eds.; Coastal Education and Research Foundation, Inc., J. of Coastal Research: Coconut Creek, FL, USA, 2018.
17. Erikson, L.H.; Barnard, P.L.; Finzi-Hart, J.; Hayden, M.; Jones, J.; Wood, N.; Fitzgibbon, M.; Foxgrover, A.C.; Limber, P.; Vitousek, S.; et al. Projected 21st century coastal flooding in the Southern California Bight. Part 2: Extreme events and variations in coastal response. *J. Mar. Sci. Eng.*. under review.
18. Christiansen, R.L.; Yeats, R.S.; Graham, S.A.; Niem, W.A.; Niem, A.R.; Snavely, P.D. Post-Laramide geology of the U.S. Cordilleran region. In *The Cordilleran Region: Conterminous U.S.: Geology of North America*; Burchfiel, B.C., Lipman, P.W., Zoback, M.L., Eds.; Geological Society of America: Boulder, CO, USA, 1992; Volume G-3, pp. 261–406.

19. Hogarth, L.; Babcock, J.; Driscoll, N.; Dantec, N.; Haas, J.; Inman, D.; Masters, P. Long-term tectonic control on Holocene shelf sedimentation offshore La Jolla, California. *Geology* **2007**, *35*, 275. [CrossRef]

20. O'Reilly, W.C.; Guza, R.T. A Comparison of Two Spectral Wave Models in the Southern California Bight. *Coast. Eng.* **1993**, *19*, 263–282. [CrossRef]

21. O'Reilly, W.C.; Guza, R.T.; Seymour, R.J. Wave prediction in the Santa Barbara Channel. In Proceedings of the 5th California Islands symposium, Santa Barbara, CA, USA, 29–31 March 1999.

22. Rogers, W.; Kaihatu, J.; Hsu, L.; Jensen, R.; Dykes, J.; Holland, K. Forecasting and hindcasting waves with the SWAN model in the southern California Bight. *Coast. Eng.* **2007**, *54*, 1–15. [CrossRef]

23. Adams, P.N.; Inman, D.L.; Graham, N.E. Southern California Deep-Water Wave Climate: Characterization and Application to Coastal Processes. *J. Coast. Res.* **2008**, *24*, 1022–1035. [CrossRef]

24. Crosby, S.C.; O'Reilly, W.C.; Guza, R.T. Modeling long period swell in southern California: Practical boundary conditions from buoy observations and global wave model predictions. *J. Atmos. Ocean. Technol.* **2016**, *33*, 1673–1690. [CrossRef]

25. Hegermiller, C.A.; Rueda, A.; Erikson, L.H.; Barnard, P.L.; Antolinez, J.A.A.; Mendez, F.J. Controls of multimodal wave conditions in a complex coastal setting. *Geophys. Res. Lett.* **2017**, *44*, 315–323. [CrossRef]

26. National Oceanic and Atmospheric Administration (NOAA). *Tides and Currents*; NOAA Center for Operational Oceanographic Products and Services: Silver Spring, MD, USA, 2017. Available online: http://tidesandcurrents.noaa.gov/ (accessed on 2 January 2018).

27. Scripps Institute of Oceanography (SIO). *Coastal Data Information Program (CDIP)*; Data from 2015; University of California at San Diego: La Jolla, CA, USA, 2015; Available online: https://cdip.ucsd.edu/ (accessed on 15 January 2016).

28. Flick, R.E. Comparison of California tides, storm surges, and mean sea level during the El Niño winters of 1982–1983 and 1997–1998. *Shore Beach* **1998**, *66*, 7–11.

29. Bromirski, P.D.; Flick, R.E.; Cayan, D.R. Storminess variability along the California coast: 1858–2000. *J. Clim.* **2003**, *16*, 982–993. [CrossRef]

30. Storlazzi, C.D.; Griggs, G.B. Influence of El Nino–Southern Oscillation (ENSO) events on the evolution of central California's shoreline. *Geol. Soc. Am. Bull.* **2000**, *112*, 236–249. [CrossRef]

31. Bromirski, P.D.; Cayan, D.R.; Flick, R.E. Wave spectral energy variability in the northeast Pacific. *J. Geophys. Res.* **2005**, *110*, C03005. [CrossRef]

32. Flick, R.E.; Murray, J.F.; Ewing, L.C. Trends in United States datum statistics and tide range. *J. Waterw. Port Coast. Ocean Eng.* **2003**, *129*, 155–164. [CrossRef]

33. Vermeer, M.; Rahmstorf, S. Global sea level linked to global temperature. *Proc. Natl. Acad. Sci. USA* **2009**, *106*, 21527–21532. [CrossRef] [PubMed]

34. National Research Council. *Sea-Level Rise for the Coasts of California, Oregon, and Washington: Past, Present, and Future*; The National Academies Press: Washington, DC, USA, 2012. [CrossRef]

35. Hinkel, J.; Jaeger, C.; Nicholls, R.J.; Lowe, J.; Renn, O.; Peijun, S. Sea-level rise scenarios and coastal risk management. *Nat. Clim. Chang.* **2015**, *5*, 188–190. [CrossRef]

36. Carson, M.; Kohl, A.; Stammer, D.; Slangen, A.; Katsman, C.; van de Wal, R.; Church, J.; White, N. Coastal sea level changes, observed and projected during 20th and 21st century. *Clim. Chang.* **2016**, *134*, 269–281. [CrossRef]

37. Cayan, D.R.; Kalansky, J.; Iacobellis, S.; Pierce, D. *Creating Probabilistic Sea Level Rise Projections to Support the 4th California Climate Assessment*; California Energy Commission: Sacramento, CA, USA, 2016; 16-IEPR-04, TN 211806. Available online: http://docketpublic.energy.ca.gov/PublicDocuments/16-IEPR-04/TN211806_20160614T101823_Creating_Probabilistic_Sea_Leve_Rise_Projections.pdf (accessed on 15 February 2018).

38. DeCanto, R.M.; Pollard, D. Contribution of Antarctica to past and future sea-level rise. *Nature* **2016**, *531*. [CrossRef] [PubMed]

39. Chen, X.; Zhang, X.; Church, J.; Watson, C.S.; King, M.A.; Monselesan, D.; Legresy, B.; Harig, C. The increasing rate of global mean sea-level rise during 1993-2014. *Nat. Clim. Chang.* **2017**, *7*, 492–495. [CrossRef]

40. Taylor, K.E.; Stouffer, R.J.; Meehl, G.A. An overview of CMIP5 and the experiment design. *Am. Meteorol. Soc.* **2012**, *93*. [CrossRef]

41. Tolman, H.L.; Balasubramaniyan, B.; Burroughs, L.D.; Chalikov, D.V.; Chao, Y.Y.; Chen, H.S.; Gerald, V.M. Development and Implementation of Wind-Generated Ocean Surface Wave Modelsat NCEP. *Weather Forecast.* **2002**, *17*, 311–333. [CrossRef]

42. Erikson, L.H.; Hegermiller, C.A.; Barnard, P.L.; Ruggiero, P.; van Ormondt, M. Projected wave conditions in the Eastern North Pacific under the influence of two CMIP5 climate scenarios. *Ocean Model.* **2015**, *96*, 171–185. [CrossRef]

43. Lesser, G.R.; Roelvink, J.A.; van Kester, J.A.; Stelling, G.S. Development and validation of a three-dimensional morphological model. *Coast. Eng.* **2004**, *51*, 883–915. [CrossRef]

44. Booij, N.; Ris, R.C.; Holthuijsen, L.H. A third-generation wave model for coastal regions. 1. Model description and validation. *J. Geophys. Res.* **1999**, *104*, 7649–7666. [CrossRef]

45. Pierce, D.W.; Cayan, D.R.; Thrasher, B. Statistical Downscaling Using Localized Constructed Analogs (LOCA). *J. Hydrometeorol.* **2014**, *15*, 2558–2585. [CrossRef]

46. Roelvink, D.; Reniers, A.; Van Dongeren, A.; van Thiel de Vries, J.; McCall, R.; Lescinski, J. Modelling storm impacts on beaches, dunes and barrier islands. *Coast. Eng.* **2009**, *56*, 1133–1152. [CrossRef]

47. Roelvink, D.; Reniers, A.; Van Dongeren, A.; van Thiel de Vries, J.; Lescinski, J.; McCall, R. *XBeach Model Description and Manual*; Report; Unesco-IHE Institute for Water Education, Deltares and Delft University of Tecnhology: Delft, The Netherlands, 2010.

48. Stockdon, H.F.; Holman, R.A.; Howd, P.A.; Sallenger, A.H. Empirical parameterization of setup, swash, and runup. *Coast. Eng.* **2006**, *53*, 573–588. [CrossRef]

49. Danielson, J.J.; Poppenga, S.K.; Brock, J.C.; Evans, G.A.; Tyler, D.J.; Gesch, D.B.; Thatcher, C.A.; Barras, J.A. Topobathymetric elevation model development using a new methodology—Coastal National Elevation Database. *J. Coast. Res.* **2016**, *76*, 75–89. [CrossRef]

50. Vitousek, S.; Barnard, P.L. A non-linear, implicit one-line model to predict long-term shoreline change. In Proceedings of the Coastal Sediments 2015, San Diego, CA, USA, 11–15 May 2015; Wang, P., Rosati, J.D., Cheng, J., Eds.; World Scientific: Hackensack, NJ, USA, 2015; p. 14. [CrossRef]

51. Vitousek, S.; Barnard, P.L.; Limber, P.; Erikson, L.; Cole, B. A model integrating longshore and cross-shore processes for predicting long-term shoreline response to climate change. *J. Geophys. Res. Earth Surface* **2017**, *122*, 782–806. [CrossRef]

52. Limber, P.W.; Barnard, P.L.; Hapke, C. Towards projecting the retreat of California's coastal cliffs during the 21st Century. In Proceedings of the Coastal Sediments 2015, San Diego, CA, USA, 11–15 May 2015; Wang, P., Rosati, J.D., Cheng, J., Eds.; World Scientific: Hackensack, NJ, USA, 2015; p. 14. [CrossRef]

53. Limber, P.W.; Barnard, P.L.; Vitousek, S.V.; Erikson, L.H. A model ensemble for projecting multi-decadal coastal cliff retreat during the 21st century. *J. Geophys. Res. Earth Surface.* under review.

54. Erikson, L.H.; O'Neill, A.; Barnard, P.L.; Vitousek, S.; Limber, P.W. Climate change-driven cliff and beach evolution at decadal to centennial time scales. In Proceedings of the 8th International Conference on Coastal Dynamics 2017, Helsingor, Denmark, 12–16 June 2017. Paper No. 210.

55. Kanamitsu, M.; Kanamaru, H. 57-year California Reanalysis Downscaling at 10km (CaRD10) Part 1: System Detail and Validation with Observations. *J. Clim.* **2007**, *20*, 5527–5552. [CrossRef]

56. Erikson, L.H.; Hegermiller, C.E.; Barnard, P.L.; Storlazzi, C.D. *Wave Projections for United States Mainland Coasts*; Pamphlet to Accompany Data Release; U.S. Geological Survey: Santa Cruz, CA, USA, 2017; 172p.

57. Arblaster, J.M.; Meehl, G.A.; Karoloy, D.J. Future climate change in the Southern Hemisphere: Competing effects of ozone and greenhouse gases. *Geophys. Res. Lett.* **2011**, *38*, L02701. [CrossRef]

58. Hemer, M.A.; Fan, Y.; Mori, N.; Semedo, A.; Wang, X.L. Projected changes in wave climate from a multi-model ensemble. *Nat. Clim. Chang.* **2013**, *3*, 471–476. [CrossRef]

59. Yin, J. A consistent poleward shift of the storm tracks in simulations of 21st century climate. *Geophys. Res. Lett.* **2005**, *32*, L18701. [CrossRef]

60. National Ocean and Atmospheric Administration. *California Merged Topobathy Dataset*; Dataset; NOAA: Silver Spring, MD, USA, 2013. Available online: https://catalog.data.gov/dataset/2013-noaa-coastal-california-topobathy-merge-project (accessed on 1 July 2015).

61. Hegermiller, C.A.; Antolinez, J.A.A.; Rueda, A.C.; Camus, P.; Perez, J.; Erikson, L.H.; Barnard, P.L.; Mendez, F.J. A multimodal wave spectrum-based approach for statistical downscaling of local wave climate. *J. Phys. Oceanogr.* **2016**, *47*. [CrossRef]

62. Hegermiller, C.A.; Erikson, L.H.; Barnard, P.L. *Nearshore Waves in Southern California: Hindcast, and Modeled Historical and 21st-Century Projected Time Series*; Data Release; U.S. Geological Survey: Santa Cruz, CA, USA, 2016.

63. U.S. Army Corp of Engineers (USACE). *Wave Information Studies (WIS) Wave Model*; Pacific Data; USACE Waterways Experiment Station: Vicksburg, MS, USA, 2014. Available online: http://wis.usace.army.mil (accessed on 1 September 2015).

64. Erikson, L.H.; Espejo, A.; Barnard, P.L.; Serafin, K.; Hegermiller, C.A.; O'Neill, A.C.; Ruggiero, P.; Limber, P. Identification of global climate model-driven storm events that result in similar coastal flood potentials along coherent shoreline sections. *Coast. Eng.*. under review.

65. Pfeffer, W.T.; Harper, J.T.; O'Neel, S. Kinematic constraints on glacier contributions to 21st-century sea-level rise. *Science* **2008**, *331*, 1340–1343. [CrossRef] [PubMed]

66. Horton, B.P.; Rahmstorf, S.; Engelhart, S.E.; Kemp, A.C. Expert assessment of sea-level rise by AD 2100 and'AD 2300. *Quat. Sci. Rev.* **2014**, *84*, 1–6. [CrossRef]

67. Ruggiero, P.; Komar, P.D.; McDougal, W.G.; Marra, J.J.; Beach, R.A. Wave runup, extreme water levels and erosion of properties backing beaches. *J. Coast. Res.* **2001**, *17*, 409–419.

68. Walkden, M.J.A.; Hall, J.W. A predictive mesoscale model of the erosion and profile development of soft rock shores. *Coast. Eng.* **2005**, *52*, 55–563. [CrossRef]

69. Walkden, M.J.A.; Dickson, M. Equilibrium erosion of soft rock shores with a shallow or absent beach under increased SLR. *Mar. Geol.* **2008**, *251*, 75–84. [CrossRef]

70. Trenhaile, A.S. Predicting the response of hard and soft rock coasts to changes in sea level and wave height. *Clim. Chang.* **2011**, *109*, 599–615. [CrossRef]

71. Hackney, C.; Darby, S.E.; Leyland, J. Modelling the response of soft cliffs to climate change: A statistical, process-response model using accumulated excess energy. *Geomorphology* **2013**, *187*, 108–121. [CrossRef]

72. California Coastal Commission (CCC). *Coastal Erosion Armoring Dataset*; CCC: Sacramento, CA, USA, 2005. Available online: https://catalog.data.gov/dataset/coastal-erosion-armoring-2005 (accessed on 1 July 2016).

73. California Coastal Records Project. Available online: http://www.californiacoastline.org/ (accessed on 15 January 2015).

74. Carpenter, N.E.; Stuiver, C.; Nicholls, R.; Powrie, W.; Walkden, M. Investigating the recession process of complex soft cliff coasts: An Isle of Wight case study. *Coast. Eng. Proc.* **2012**, *1*, 123. [CrossRef]

75. Yates, M.L.; Guza, R.T.; O'Reilly, W.C. Equilibrium shoreline response: Observations and modeling. *J. Geophys. Res.* **2009**, *114*, C09014. [CrossRef]

76. Castedo, R.; Murphy, W.; Lawrence, J.; Paredes, C. A new process–response coastal recession model of soft rock cliffs. *Geomorphology* **2012**, *177*, 128–143. [CrossRef]

77. Kline, S.W.; Adams, P.N.; Limber, P.W. The unsteady nature of sea cliff retreat due to mechanical abrasion, failure and comminution feedbacks. *Geomorphology* **2014**, *219*, 53–67. [CrossRef]

78. Young, A.P.; Guza, R.T.; Flick, R.E.; O'Reilly, W.C.; Gutierrez, R. Rain, waves, and short-term evolution of composite seacliffs in southern California. *Mar. Geol.* **2009**, *267*, 1–7. [CrossRef]

79. Anderson, T.R.; Fletcher, C.H.; Barbee, M.M.; Frazer, L.N.; Romine, B.M. Doubling of coastal erosion under rising sea level by mid-century in Hawaii. *Nat. Hazards* **2015**, *78*, 75–103. [CrossRef]

80. Long, J.W.; Plant, N.G. Extended Kalman filter framework for forecasting shoreline evolution. *Geophys. Res. Lett.* **2012**, *39*, L13603. [CrossRef]

81. Google earth V 7.1.8.3036. (2015–2016). In *Southern California Imagery 2015-2016, United States*; Eye alt. ground-level to 5000 ft., SIO, NOAA, U.S.; Navy, NGA, GEBCO; DigitalGlobe: Westminster, CO, USA, 2016; Available online: http://www.earth.google.com (accessed on 1 July 2016).

82. Horsburgh, K.L.; Wilson, C. Tide–surge interaction and its role in the distribution of surge residuals in the North Sea. *J. Geophys. Res.* **2007**, *112*, CO8003. [CrossRef]

83. Petroliagkis, T.I.; Voukouvalas, E.; Disperati, J.; Bildot, J. *Joint Probabilities of Storm Surge, Significant Wave Height and River Discharge Components of Coastal Flooding Events*; Technical Reports, EUR 27824 EN; Joint Research Center: Ispra, Italy, 2016. [CrossRef]

84. Warrick, J.A.; Farnsworth, K.L. Sources of sediment to the coastal waters of the Southern California Bight. In *Earth Science in the Urban Ocean—The Southern California Continental Borderland*; Lee, H.J., Normark, W.R., Eds.; Special Paper 454; Geological Society of America: Boulder, CO, USA, 2009; pp. 39–52.

85. Ghorbani, M.A.; Singh, V.P.; Sivakumar, B.; Kashani, M.H.; Atre, A.A.; Asadi, H. Probability distribution functions for unit hydrographs with optimization using genetic algorithm. *Appl. Water Sci.* **2017**, *7*, 663–676. [CrossRef]

86. Ris, R.C.; Booij, N.; Holthuijsen, L.H. A third-generation wave model for coastal regions: Part II—Verification. *J. Geophys. Res.* **1999**, *104*, 7667–7682. [CrossRef]
87. Egbert, G.D.; Bennett, A.F.; Foreman, M.G.G. TOPEX/POSEIDON tides estimated using a global inverse model. *J. Geophys. Res.* **1994**, *99*, 24821–24852. [CrossRef]
88. Storlazzi, C.D.; Griggs, G.B. The 1997–98 El Niño and erosion processes along the central coast of California. *Shore Beach* **1998**, *66*, 12–17.
89. Cayan, D.R.; Bromirski, P.D.; Hayhoe, K.; Tyree, M.; Dettinger, M.D.; Flick, R.E. Climate change projections of sea level extremes along the California coast. *Clim. Chang.* **2008**, *87* (Suppl. 1), 57. [CrossRef]
90. Dewberry. *USGS National Enhance Elevation Assessment*; Dewberry: Farifax, VA, USA, 29 March 2012; Available online: http://www.dewberry.com/docs/default-source/documents/neea_final-report_revised-3-29-12.pdf?sfvrsn=0 (accessed on 1 November 2017).
91. Howell, S.; Smith-Konter, B.; Frazer, N.; Tong, X.; Sandwell, D. The vertical fingerprint of earthquake cycle loading in southern California. *Nat. Geosci. Lett.* **2016**, *9*, 611–614. [CrossRef]
92. Bürgmann, R.; Hilley, G.; Ferretti, A.; Novali, F. Resolving vertical tectonics in the San Francisco Bay Area from permanent scatter InSAR and GPS analysis. *Geology* **2006**, *34*, 221–224. [CrossRef]
93. Wehmiller, J.F.; Sarna-Wojcicki, A.; Yerkes, R.F.; Lajoie, K.R. Anomalously high uplift rates along the Ventura—Santa Barbara Coast, California—Tectonic implications. *Tectonophysics* **1979**, *52*, 380. [CrossRef]
94. Gallien, T.W.; Barnard, P.L.; van Ormondt, M.; Foxgrover, A.C.; Sanders, B.F. A parcel-scale coastal flood forecasting prototype for a Southern California urbanized embayment. *J. Coast. Res.* **2012**, *29*, 642–656. [CrossRef]
95. Willmott, C.J. On the validation of models. *Phys. Geogr.* **1981**, *2*, 184–194.
96. Barnard, P.L.; Eshleman, J.L.; Erikson, L.H.; Hanes, D.M. *Coastal Processes Study at Ocean Beach, San Francisco, CA: Summary of Data Collection 2004–2006*; U.S. Geological Survey: Reston, VA, USA, 2007; Open-File Report 2007–1217. Available online: https://pubs.usgs.gov/of/2007/1217/of2007-1217.pdf (accessed on 2 February 2018).
97. Federal Emergency Management Agency (FEMA). *Guidance for Flood Risk Analysis and Mapping*; Guidance Doc 39; Coastal floodplain mapping; FEMA: Washington, DC, USA, 2015. Available online: https://www.fema.gov/media-library-data/1450470604373-131dbdfcb81af2cf67788650d08aef5e/Coastal_Floodplain_Mapping_Guidance_Nov_2015.pdf (accessed on 2 January 2018).
98. Foxgrover, A.C.; Finalyson, D.P.; Jaffe, B.E.; Takekawa, J.Y.; Thorne, K.M.; Spragens, K.A. *2010 Bathymetric Survey and Digital Elevation Model of Corte Madera Bay, California*; Open-File Report 2011-1217; U.S. Geological Survey: Reston, VA, USA, 2011. Available online: http://pubs.usgs.gov/of/2011/1217/ (accessed on 28 December 2017).
99. Zheng, F.; Westra, S.; Sisson, S.A. Quantifying the dependence between extreme rainfall land storm surge in the coastal zone. *J. Hydrol.* **2013**, *505*, 172–781. [CrossRef]
100. Zheng, F.; Westra, S.; Leonard, M.; Sisson, S.A. Modeling dependence between extreme rainfall and storm surge to estimate coastal flooding risk. *Water Resour. Res.* **2014**, *50*, 2050–2071. [CrossRef]
101. Konrad, C.P.; Dettinger, M.D. Flood runoff in relation to water vapor transport by atmospheric rivers over the Western United States, 1949–2015. *Geophys. Res. Lett.* **2017**, *44*. [CrossRef]
102. Wessel, P.; Smith, W.H.F. A Global Self-consistent, Hierarchical, High-resolution Shoreline Database. *J. Geophys. Res.* **1996**, *101*, 8741–8743. [CrossRef]
103. Tolman, H.L.; Chalikov, D. Source Terms in a Third-Generation Wind Wave Model. *J. Phys. Oceanogr.* **1996**, *26*, 2497–2518. [CrossRef]
104. Hasselmann, S.K.; Hasselmann, K.; Allender, J.H.; Barnett, T.P. Computations and parametrizations of the nonlinear energy transfer in a gravity-wave spectrum, Part II: Parameterizations of the nonlinear energy transfer for application in wave models. *J. Phys. Ocean.* **1995**, *15*, 1378–1391. [CrossRef]
105. National Geophysical Data Center (NGDC). *U.S. Coastal Relief Model—Southern California*; NGDC, NOAA: Boulder, CO, USA, 2003.
106. Battjes, J.A.; Janssen, J.P.F.M. Energy Loss and Set-up due to Breaking of Random Waves. In Proceedings of the 16th International Conference on Coastal Engineering, Hamburg, Germany, 27 August–3 September 1978; American Society of Civil Engineers: Reston, VA, USA, 1978; p. 19. [CrossRef]
107. Komen, G.J.; Cvaleri, L.; Donelan, M.; Hasselmann, K.; Janssen, P.A.E.M. *Dynamics and Modelling of Ocean Waves*; Cambridge University Press: Cambridge, UK, 1994; ISBN 0-521-47047-1.

108. Hasselmann, K.; Barnett, T.; Bouws, E.; Carlson, H.; Cartwright, D.; Enke, K.; Ewing, J.; Gienapp, H.; Hasselmann, D.; Krusemann, P.; et al. *Measurements of Wind-Wave Growth and Swell Decay During the Joint North Sea Wave Project (JONSWAP)*; Deutsches Hydrographisches Institut: Hamburg, Germany, 1973; p. 95.
109. Soulsby, R.L. *Dynamics of Marine Sands*; Thomas Telford: London, UK, 1997.
110. U.S. Geological Survey (USGS). *National Water Information System*; Water Data for the Nation; USGS: Reston, VA, USA, 2015. Available online: http://waterdata.usgs.gov/nwis/ (accessed on 1 October 2015).
111. Seaber, P.R.; Kapinos, F.P.; Knapp, G.L. *Hydrologic Unit Maps*; Water-Supply Paper 2294; US Geological Survey: Reston, VA, USA, 1987.

Journal of
Marine Science and Engineering

MDPI

Article

Projected 21st Century Coastal Flooding in the Southern California Bight. Part 2: Tools for Assessing Climate Change-Driven Coastal Hazards and Socio-Economic Impacts

Li Erikson [1,*], Patrick Barnard [1], Andrea O'Neill [1], Nathan Wood [2], Jeanne Jones [3], Juliette Finzi Hart [1], Sean Vitousek [4], Patrick Limber [1], Maya Hayden [5], Michael Fitzgibbon [5], Jessica Lovering [1] and Amy Foxgrover [1]

[1] U.S. Geological Survey, Pacific Costal and Marine Science Center, 2885 Mission Street, Santa Cruz, CA 95060, USA; pbarnard@usgs.gov (P.B.); aoneill@usgs.gov (A.O.); jfinzihart@usgs.gov (J.F.-H.); plimber@usgs.gov (P.L.); jlovering@usgs.gov (J.L.); afoxgrover@usgs.gov (A.F.)
[2] U.S. Geological Survey, Western Geographic Science Center, 2130 S.W. Fifth Avenue, Portland, OR 97201, USA; nwood@usgs.gov
[3] U.S. Geological Survey, Western Geographic Science Center, 345 Middlefield Road, Menlo Park, CA 94025, USA; jmjones@usgs.gov
[4] Civil and Materials Engineering, University of Illinois at Chicago, 2095 Engineering Research Facility, 842W. Taylor Street (M/C 246), Chicago, IL 60607-7023, USA; vitousek@uic.edu
[5] Point Blue Conservation Science, 3820 Cypress Drive #11, Petaluma, CA 94954, USA; mhayden@pointblue.org (M.H.); mfitzgibbon@pointblue.org (M.F.)
* Correspondence: lerikson@usgs.gov; Tel.: +1-831-460-7563

Received: 7 June 2018; Accepted: 26 June 2018; Published: 2 July 2018

Abstract: This paper is the second of two that describes the Coastal Storm Modeling System (CoSMoS) approach for quantifying physical hazards and socio-economic hazard exposure in coastal zones affected by sea-level rise and changing coastal storms. The modelling approach, presented in Part 1, downscales atmospheric global-scale projections to local scale coastal flood impacts by deterministically computing the combined hazards of sea-level rise, waves, storm surges, astronomic tides, fluvial discharges, and changes in shoreline positions. The method is demonstrated through an application to Southern California, United States, where the shoreline is a mix of bluffs, beaches, highly managed coastal communities, and infrastructure of high economic value. Results show that inclusion of 100-year projected coastal storms will increase flooding by 9–350% (an additional average 53.0 ± 16.0 km^2) in addition to a 25–500 cm sea-level rise. The greater flooding extents translate to a 55–110% increase in residential impact and a 40–90% increase in building replacement costs. To communicate hazards and ranges in socio-economic exposures to these hazards, a set of tools were collaboratively designed and tested with stakeholders and policy makers; these tools consist of two web-based mapping and analytic applications as well as virtual reality visualizations. To reach a larger audience and enhance usability of the data, outreach and engagement included workshop-style trainings for targeted end-users and innovative applications of the virtual reality visualizations.

Keywords: coastal hazards; sea-level rise; coastal storms; climate change; exposure; socio-economic vulnerability; data visualization

---

1. Introduction

Increases in sea-level rise (SLR), nuisance flooding, and changing storm patterns in coastal areas are raising awareness of the need to mitigate, plan, and consider alternatives in construction guidelines for the safety of future and planned construction and human health and safety [1–4]. Varied approaches have been developed to identify and map such coastal hazards for coastal planners and decision-makers [5–12]. However, few studies account for the combined effects of SLR and storm-driven coastal flooding on the local scale across vast geographic expanses; even fewer studies account for non-stationary changes in projected water levels and their resulting exposure hazards and socio-economic impacts [7,8,11,13–15]. To address this void, the Coastal Storm Modeling System (CoSMoS) was developed to provide planners, managers, policy-makers, and engineers with local-scale (approximately 10–100 m) data on probable future coastal exposure hazards across large geographic scales (approximately one hundred to several thousand kilometers) [13–15].

The third-generation CoSMoS model and its application in Southern California (USA), using a mid-emissions climate scenario (representative concentration pathway (RCP) 4.5), are presented in Part 1 [16] of this two-part manuscript. More extreme wave climate conditions are illustrated for California in the RCP 4.5 scenario [17] and, accordingly, it is used for detailed hazard simulations in CoSMoS [16]. The CoSMoS framework projects global changes, which are driven by Global Climate Models (GCMs) to local scales via a suite of regional and local scale models simulating coastal hazards in response to projections of 21st century waves, storm surges, anomalous variations in water levels, river discharge, tides, and SLR. A detailed discussion of the methodology, modeling framework, recent improvements, model validations/limitations, and an incorporation of uncertainty into coastal hazard projections can be found in Part 1 [16].

In this second part of the manuscript, results of the modeled hazards are presented and conjoined with land cover, population statistics, and socio-economic data to provide 21st century hazard-exposure estimates along the largely developed Southern California coastline, a region that thrives on tourism, software, automotive, ports, finance, and biomedical industries, contributing to more than 50% of California's Gross Domestic Product (GDP), ranked the fifth largest worldwide [18].

The *hazards* of interests in this study are coastal erosion and flooding. *Exposure* refers to the presence of various societal elements (e.g., people, buildings, resources, critical facilities, and infrastructure) that are in hazard zones, and therefore susceptible to damage or loss. *Vulnerability* describes the characteristics of individuals and assets as well as larger socioeconomic factors that influence the degree to which an individual, system, or community is susceptible to the damaging effects of a hazard. For example, although a large area of residential housing may be equally exposed to coastal flooding, the vulnerability of individual households will vary due to demographic characteristics of homeowners (which influence one's ability to prepare for and mitigate potential losses) and due to differences in the types of structures (which influence the ability to withstand impacts). In addition, two adjacent towns may have equal hazard exposure, but their overall vulnerability to flooding varies if the number of homes exposed to flooding represents 5% of available housing in one town and 95% of housing in the other town.

To communicate coastal hazards, exposures, and vulnerabilities as well as assess the socio-economic impacts, the data are made available via two publicly accessible web tools (Figure 1). "Our Coast, Our Future" [19] (OCOF; www.ourcoastourfuture.org) is a web application for data visualization, synthesis, and access to all output products from the CoSMoS model. The OCOF mapping interface provides coastal managers and the general public with a user-friendly means to visualize how various SLR scenarios alone and in combination with three different future return-period coastal storms are projected to flood or erode. Users can export summary tables and reports detailing changes in flood extent by scenario on a scale relevant to local planners. The Hazard Exposure Reporting and Analytics [20] (HERA; https://www.usgs.gov/apps/hera/) web tool translates the flooding hazards into community-based exposure statistics and quantifies populations, property, and critical infrastructure at risk in terms of exposure statistics and monetary values on a community level.

(a) Our Coast Our Future (OCOF) web-tool

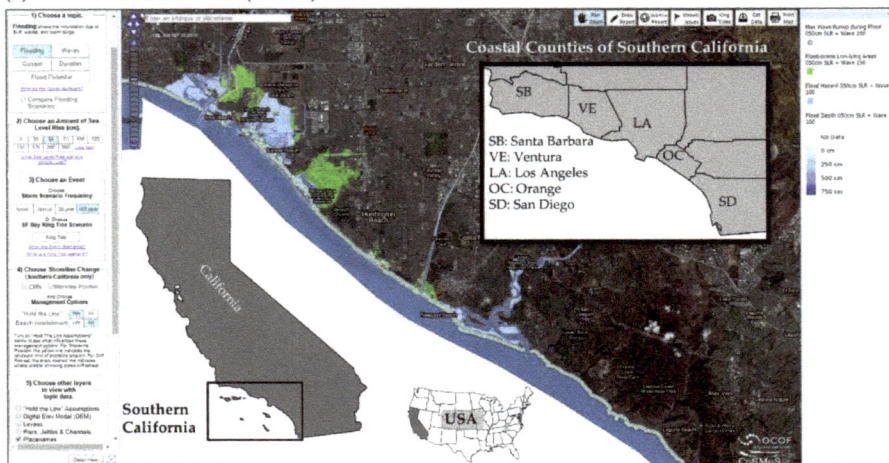

(b) Hazard Exposure Reporting and Analytics (HERA) web-tool

Figure 1. Web applications for visualization, synthesis, socio-economic analyses, and data download. (**a**) Screen-grab from the "Our Coast, Our Future" (OCOF) web tool showing coastal hazards associated with currents, wave heights, flood extents/depths, durations, and shoreline change for each of the modeled sea-level rise (SLR) and storm scenarios (www.ourcoastourfuture.org). Inset shows location of Southern California and the five coastal counties within the Southern California region; (**b**) Screen-grab from the Hazard Exposure and Analytics (HERA) web tool developed to aid in analysis of exposure and socio-economic statistics (https://www.usgs.gov/apps/hera/). Both applications contain a suite of tools and options for visualizing (maps and graphs) and synthesizing the model results. Example screen-grabs are for the Orange County (OC) area.

The scientific methods that underpin the projected hazards are based on state-of-the-art science that includes many of the latest developments and understandings of coastal processes (i.e., non-linear effects of currents and waves, reflection, refraction, and blocking of wave energy due to complex bathymetry; see Barnard et al. and O'Neill et al. [15,16]) making it difficult to communicate assumptions, limitations, as well as the strength and value of the data to non-technical and non-specific science-educated audiences. To address these concerns, both traditional and innovative stakeholder engagement, training, and outreach efforts have been tested. Although the translation of the science remains challenging, we have pinpointed several tactics that are likely to be useful in similar large-region, high-resolution studies elsewhere.

The aim of this paper is to (1) highlight the need to account for dynamic water levels in addition to static SLR for estimating future coastal flood hazards and vulnerabilities along high-energy coastal environments such as Southern California, USA and (2) to present means and introduce innovative approaches for conveying the hazards, socio-economic impacts, and underlying scientific basis to a broad audience including coastal managers, planners, and engineers. The remainder of this manuscript describes the data and methods used to quantify exposures and vulnerabilities along the Southern California coast. Results pertaining to erosion and flood hazards are presented with a particular emphasis on the added risk when storms are accounted for in addition to SLR. Results of the socio-economic impact analyses are then presented for a select set of assets and demographics rather than all available results to illustrate the use and applicability of stakeholder user-tools. The final sections present strategies and innovative outreach activities for dissemination of the information as well as a summary and discussion of findings.

2. Methods

The methods developed for simulating hazards with the latest generation of CoSMoS are outlined in Part 1 of this manuscript [16], which include presentation and discussion on models, selection of storm conditions, model validation/limitations, and uncertainty. The latest iteration of CoSMoS is implemented in Southern California, an active and complex tectonic region spanning over 500 km and five counties (Figure 1). The coastal landscape is generally characterized as beaches backed by semi-resistant bedrock sea-cliffs as well as coastally constrained estuaries and low-lying areas at the foot of coastal mountain ranges (see Part 1 for more details). CoSMoS-modeled hazards for Southern California include outputs of coastal erosion, wave heights, wave runup, total water levels, current speeds, flood extents, and flood depths/durations for 40 'scenarios' consisting of all combinations of 10 SLR elevations (0–200 cm SLR in 25 cm increments, plus 500 cm), three coastal storm intensities (annual, 20-year, and 100-year), and a no-storm condition [16,21,22]. Coastal erosion outputs included management scenarios involving beach nourishment and the existence and maintenance of hard structures (see Section 3.1.1). Low-lying flood-prone areas and uncertainties in flood extents and shoreline change are also generated as part of the model output.

Resulting model data were converted to static GeoTIFF rasters (flood depth, wave height, and current velocity), polygon shapefiles (flood extent, low-lying areas, and uncertainty) or point shapefiles (wave runup and shoreline change) and were processed for the OCOF cyberinfrastructure to display and provide exposure hazard map data. The cyberinfrastructure was built on the Open Source Geospatial Foundation stack of software. Simple raster tiles were first rendered from the GeoTIFF data layers; point and polygon layers were loaded into PostgreSQL (an open-source database; version 9+; available https://www.postgresql.org/)/PostGIS (an open-source, GIS-support software program; version 2+; available https://postgis.net/) database and piped through the GeoServer web service for rendering and display on the map (see Appendix A for a list of terms, acronyms, and software platforms). Initially, the data were provided for download as large zip files but these proved to be too cumbersome for many users; a re-organization of the data tiled by scenario, output product (e.g., flood depth/extent, duration, wave height), and individual counties has improved users' ability to access the data and increased overall user satisfaction.

Geospatial data summarizing various population, business, land cover, and infrastructure were used to estimate community exposure to a given flood hazard zone in HERA [20]. Residential populations were estimated using block-level population counts compiled from the 2010 US Census [23]. Demographic and economic factors, such as age, health, ethnicity, race, and health, and tenancy, can amplify an individual's sensitivity to hazards [24,25]; therefore, the 2010 block-level data were used to estimate demographic attributes related to these socio-economic indicators of sensitivity.

Business populations and regional trends of exposure were estimated in HERA using employee counts organized by North American Industry Classification System (NAICS) codes [23] at individual businesses using a georeferenced, proprietary employer database [26]. Business types based on NAICS codes were generalized in this analysis into five classes: (1) government and critical facilities; (2) manufacturing; (3) services; (4) natural resources; and (5) trade.

Land cover indicators include the amount and type of land in hazard zones based on 30-m-resolution data extracted from the 2011 National Land Cover Database (NLCD) [27]. The HERA application currently focuses on land classified in NLCD as either wetlands or developed.

Hazard exposure of critical facilities and infrastructure was estimated using the length of rail and road networks (infrastructure) and the number of schools, medical facilities, police stations, and fire stations (facilities). These facilities are considered critical because they provide public safety services or house vulnerable populations. Data sources for critical facilities and infrastructure include a wide array of county and federal sources [28].

For each variable, geographic information system (GIS) software was used to overlay data representing community boundaries, the community indicator, and a specified flood hazard zone. Two variables for each asset were estimated at the community level: (1) a total amount (or length, in miles, for road and rail networks) of an asset in a hazard zone and (2) a community percentage. For resident and employee populations in hazard zones, the community percentage reflects the exposed amount compared to the total amount within a community. For the business types, percentages reflect the number of businesses of a certain type divided by the total number of that business type in the community. For the demographic attributes, community percentages reflect the percentage of a specific demographic attribute relative to the total number of residents in the hazard zone, not the community total. Spatial analysis of vector data focused on determining if points (businesses and critical facilities), lines (roads and rails), or polygons (census blocks) were inside hazard zones. If census-block polygons overlapped hazard polygons, final population values were adjusted proportionately using the spatial ratio of each sliver within or outside of a hazard zone.

3. Results

3.1. Projected Hazards

3.1.1. Coastline Change

Shoreline management scenarios involving beach nourishment and the existence and maintenance of hard structures to limit erosion were simulated with the cliff recession and sandy shoreline change models (see Section 2.2 of Part 1) [16,29–31]. Two management scenarios were investigated for the cliff recession projections: (1) cliff recession unlimited by cliff armoring; and (2) cliff recession everywhere except where armoring exists (as of 2016). For the sandy shoreline projections, four management scenarios were simulated, representing all four combinations of: (A) no beach nourishment or continued rates of historical beach nourishment; and (B) the existence ("hold-the-line") or non-existence of hard structures that limit erosion. For the "hold-the-line" scenarios, scenario erosion was limited to an 180,000-point polyline digitized from aerial photos (Google Earth™, 2015/2016) that represents the division of beach and urban infrastructure.

Applied and averaged over 2156 transects, the cliff model projects that recession rates will increase by 62%, 92%, 150%, and 220% relative to historical rates for the 50, 93, 150, and 200 cm SLR scenarios, respectively [29]. The highest rates of increase are projected for Santa Barbara, Los Angeles, and southern Ventura counties. A total land loss of 7–82 m for the 25–200 cm SLRs is projected, but this assumes much variability along the coast and increasing uncertainty with the projected greater recession rates. Examples are shown for the 50 and 200 cm SLRs in Figure 2, where it can be seen that the greatest recession rates are in Los Angeles and San Diego counties.

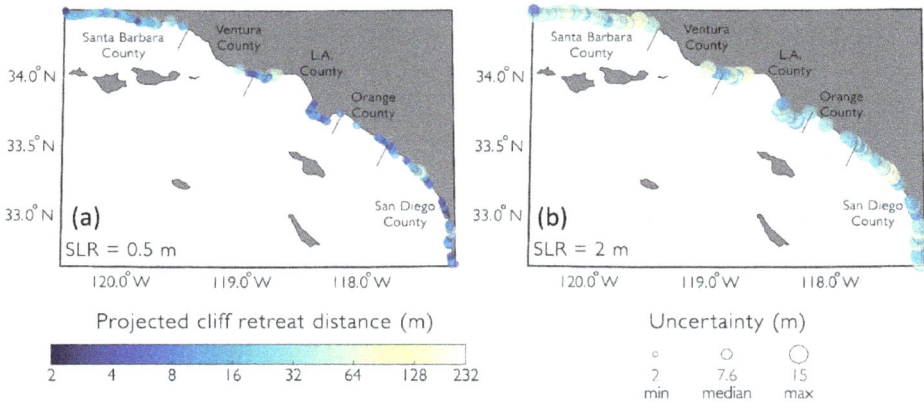

Figure 2. Map of cliff recession model results for two sea-level rise (SLR) scenarios: (**a**) 50 cm SLR and (**b**) 200 cm SLR. The colors represent cliff retreat distance, while the size of the marker represents the amount of uncertainty.

Results of the shoreline change model (sand and gravel beaches) projects average beach losses of 21–68 m for the 100–200 cm SLRs, depending on the management scenario [30] (Figure 3). The inclusion of the additional effect of seasonal[1] erosion (driven by larger-than-average wave conditions) increases this range from 32 m to 106 m of average beach loss by 2100 (Figure 3; see Vitousek et al. [30] for details). This amount of beach recession may result in 31–67% of beaches in Southern California being completely eroded to the landward limit of coastal infrastructure or cliffs by 2100. Furthermore, 25–65% of beaches may erode more than 5 m into existing coastal infrastructure and homes if allowed to migrate landward, unimpeded by seawalls and hardened structures.

The results also indicate that there is little overall difference between the nourished and unnourished management scenarios (scenarios 1 and 2): the continued nourishment scenarios only reduce the extent of erosion compared to the unnourished scenarios by 2–3 m on average (Figure 3). The model only projects any stability or significant decrease in erosion of the shoreline in local cases, where very large nourishments are assumed to continue.

3.1.2. Flood Hazards

Flood hazards, simulated with CoSMoS and evolved coastlines assuming management scenario 1 ("hold-the-line" and no nourishment, Figure 3), indicate that 10–380 km^2 of land along the Southern California coast will be permanently inundated with the 25–500 cm SLR, no storm, scenarios.

Inundation due to SLR alone is summarized for each of five coastal Southern California counties in Figure 4a. Ventura County is consistently the most vulnerable region for SLR scenarios above 1 m, where permanent inundation extents are estimated at 50 and 91 km^2 for the 200 and 500 cm SLR scenarios, respectively. Orange and San Diego Counties are projected to experience slightly less inundation, while Santa Barbara County is less vulnerable with a maximum flooded area of ~20 km^2 due largely to the high cliffs and bluffs that front this coastline. For the more near-term SLR estimate of 25 cm (approximately by the year 2030) [32], Los Angeles and San Diego will be the most affected, with an estimated 3–4 km^2 of inundation. The susceptibility of each county to the different states of

---

[1] average beach loss shorelines are based on a 1 January (mid-winter) time horizon, whereas the max seasonal erosion is based on the upper limit of the 95% confidence interval (2nd standard deviation) using all the model-projected positions.

SLR are due to variations in local topography and flood protection infrastructure that hinder or allow inland flow of ocean water.

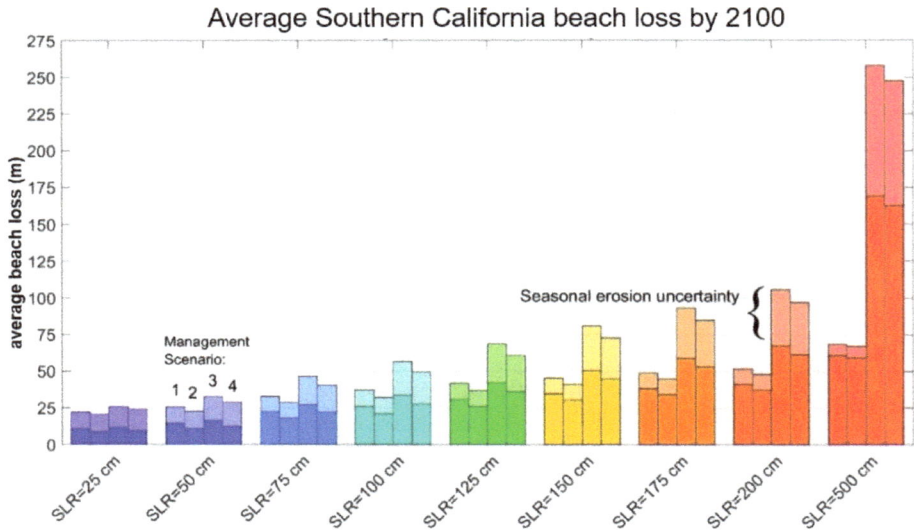

Figure 3. Average beach loss in Southern California derived from the shoreline change model under different management and sea-level rise (SLR) scenarios (see O'Neill et al. and Vitousek et al. [16,30] for model details). The four management scenarios are (1) "hold-the-line" (hard structure to limit erosion) with no continued nourishment; (2) "hold-the-line" with continued nourishment; (3) no structures/limit on erosion and no continued nourishment; and (4) no structures/limit on erosion and continued nourishment. Figure modified from Vitousek et al. [30].

Inclusion of storms (and the combined effects of associated waves, surge, and discharge) in the model simulations significantly increases the flood extents across all counties. For example, the 100-year coastal storm floods, on average, an additional 4.5 ± 2.0, 16.0 ± 5.5, 9.0 ± 6.0, 12.0 ± 8.0, and 11.5 ± 4.0 km^2 of land in Santa Barbara, Ventura, Los Angeles, Orange, and San Diego Counties, respectively (Figure 4b–f; Table 1). These extents predominantly include the coastally constrained estuaries and low-lying areas in each county as well as hundreds of kilometers of impacted cliff- and infrastructure-backed shoreline. Viewed from a perspective of percentages, the inclusion of storms represents substantial increases for the lower-end SLRs, whereas, for the higher SLRs, the percentages are smaller because the total areas inundated by SLR alone increases and dominates the flood signal. For example, flood extents increase >10-fold (from 1.2 to 15.5 km^2, an increase of 14.5 km^2) when the 100-year coastal storm in Ventura County is included for the low-end 25 cm SLR, but 'only' by 26% (from 49.5 km^2 to 62.5 km^2) for the 200 cm SLR, although the areal increase is similar (13.0 km^2 compared to 14.5 km^2). Taking the entirety of Southern California into consideration, percentages of increased flood-area due to storms decrease from ~350% for the 25 cm SLR scenario to 9% for the 500 cm SLR scenarios (last column in Table 1).

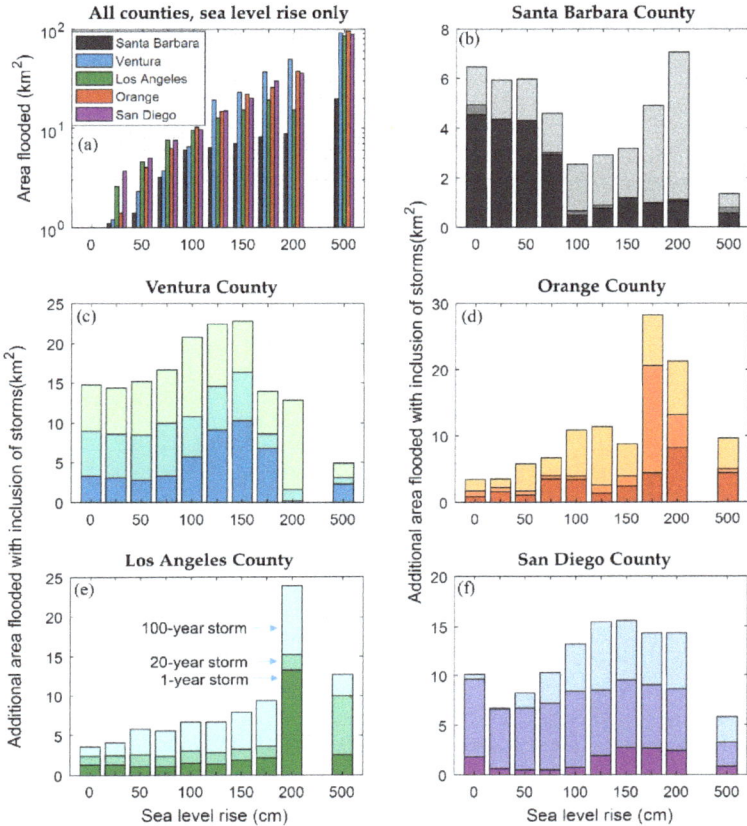

Figure 4. Areas projected to flood within each of the five coastal counties in Southern California due to sea-level rise (SLR) alone and in combination with coastal storms. (**a**) Total extents of flooded areas in each county for the SLR only scenario (no storm). Note the logarithmic vertical axis to capture the large differences between the 25 cm and 500 cm SLR scenarios; (**b**–**f**) Areas flooded beyond the SLR-only scenario, when the annual (darker color), 20-year, and 100-year (lightest color) costal storms are simulated in combination with SLR.

3.2. Projected Exposures and Vulnerabilities

Integration of modeled shoreline retreat and flood hazards with geospatial demographic and socio-economic data shows that 20,000–164,000 residents, 150–1330 km of road, and as much as 22 km^2 of agricultural land are at risk of being permanently flooded in Southern California with a 25–200 cm SLR. Building replacement values are estimated to be between $3.65 billion and $26.10 billion (2006 value, unadjusted for inflation). Inclusion of the 100-year coastal storm increases hazard exposures to 30,800–256,000 residents (a 55–110% increase), 340–2300 km of road (a 50–115% increase), and building replacement costs to $6.95–$38.25 billion (up by 42–91%).

Table 1. Area (km^2) flooded due to sea-level rise (SLR) alone and SLR in combination with the 100-year coastal storm (SLR only/SLR plus 100-year coastal storm). Bottom row shows increase in flood area from impact of 100-year coastal storm (average across all SLR values) for each county and across all counties.

SLR (cm)	County						All Counties	
							Increased Flooding	
	Santa Barbara	Ventura	Los Angeles	Orange	San Diego	Total	km^2	Percent
25	1.1/7.0	1.2/15.6	2.6/6.8	1.4/4.8	3.7/10.5	10/45	35	347%
50	1.4/7.4	2.3/17.5	4.6/10.4	4.0/10.4	5.0/13.2	17/58	41	237%
75	3.2/7.8	3.7/20.4	7.6/13.2	6.2/12.9	7.5/17.8	28/72	44	155%
100	6.0/8.6	6.5/27.2	9.5/16.3	10.2/21.1	9.6/22.8	42/96	54	129%
125	6.4/9.3	19.1/41.5	12.6/19.2	14.8/26.2	14.9/30.4	68/127	59	87%
150	7.0/10.1	23.0/45.8	15.2/23.1	21.9/30.7	20.1/35.7	87/145	58	67%
175	8.2/13.1	37.0/51.0	19.2/28.6	25.6/54.0	29.9/44.3	120/191	71	59%
200	8.8/15.9	49.4/62.4	15.2/39.0	37.5/58.8	35.8/50.1	147/226	79	54%
500	19.6/21.0	91.3/96.1	85.0/97.7	95.5/105.1	88.9/94.7	380/415	34	9%
Increase in flooding with 100-year storm (average across all SLRs) *								
km^2	4.5 ± 2.0	16.0 ± 5.5	9.0 ± 6.0	12 ± 8.0	11.5 ± 4.0		53.0 ± 16.0	
percent	157%	327%	84%	101%	99%			

* rounded to 0.5 km^2.

3.2.1. County Level

From a county perspective, residents within Los Angeles, Orange, and San Diego counties are at a greater risk to coastal flooding compared to the other two counties (Ventura and Santa Barbara) within coastal Southern California (Figure 5a). Overall, 80–90% of residents exposed to flooding for any combination of SLR and coastal storm in Southern California, reside within Los Angeles, Orange, or San Diego counties. More than 18,000 residents are at risk of being permanently flooded in these three counties for the lowest modeled 25 cm SLR and no coastal storm. Inclusion of the 100-year coastal storm increases residential flood exposure by 176% (from 5170 to 14,270 residents), 62% (from 5970 to 9650 residents), and 57% (from 7150 to 11,250 residents) in Los Angeles, Orange, and San Diego Counties, respectively.

Replacement costs of buildings (residential and commercial) are projected to be highest in the same three counties: Los Angeles, Orange, and San Diego. Together, these three counties will experience more than 85% of the cost burden associated with coastal flooding in Southern California (Figure 5b). Replacement values range from $0.95–$4.54 billion in Los Angeles, $4.28–$5.62 billion in San Diego, and $1.17–$6.62 billion in Orange counties for the 25–200 cm SLR. Including coastal storms increases the cost burden by 6–450%. In Los Angeles, Orange, and San Diego counties, replacement values approximately double from approximately $1 billion to $2 billion when storms are included with the 25 cm SLR.

3.2.2. Community Level

Identification of the most vulnerable communities within each of the counties was performed with the HERA tool, which provides graphs and maps of (1) magnitudes (e.g., total number of residents, length of road, or replacement values) and (2) percentages of an asset that fall within the hazard zone and within each community. For example, the number of residents and percent of total residents in Orange County communities that are projected to be affected by a 50 cm SLR exclusive and inclusive of the 100-year storm are shown in Figure 6. The bar plot indicates that residents within Newport, Huntington, and Seal Beach areas are most vulnerable and that the 100-year coastal storm floods additional residential areas within these communities. Similar plots and assessments are provided on the web-tool for infrastructure, developed and undeveloped land areas, and employees.

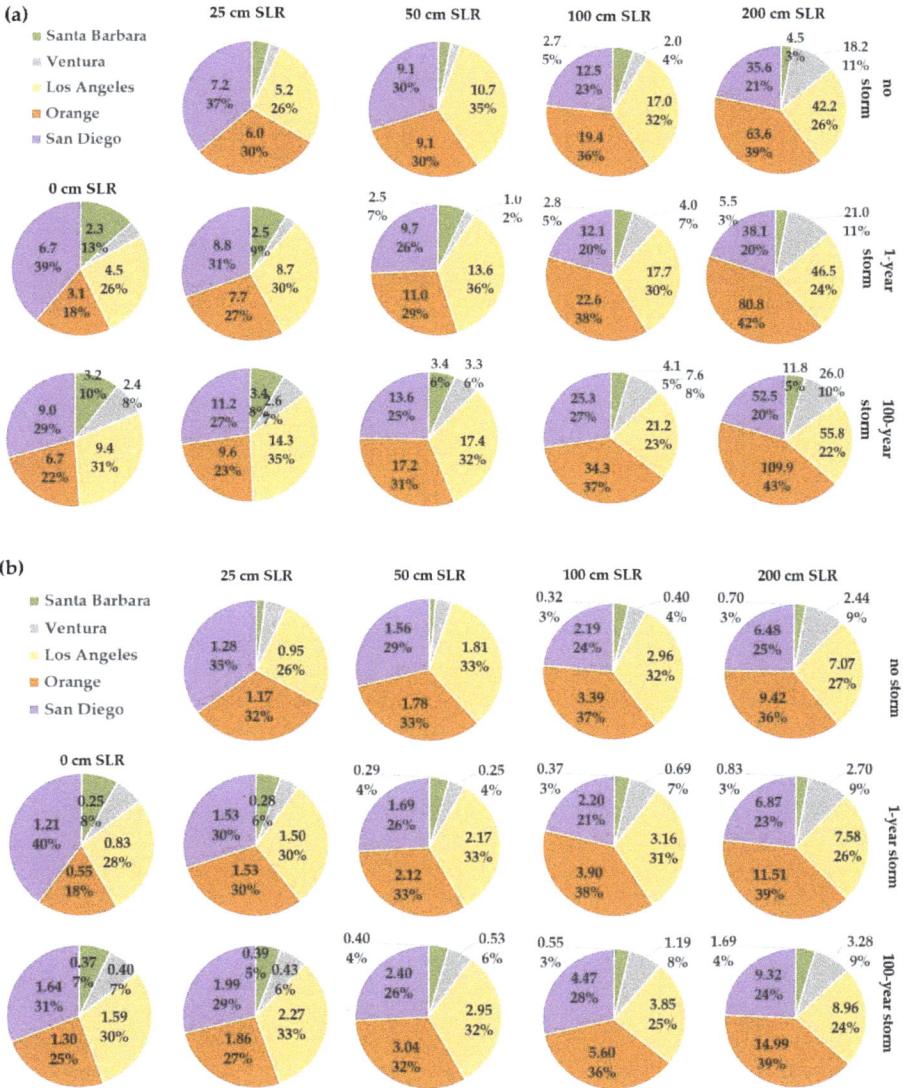

Figure 5. Example of socio-economic impacts resulting from flood hazards modeled for sea-level rise (SLR) scenarios of 25, 50, 100, and 200 cm without storms and in combination the annual and 100-year projected storms. (**a**) County residents (1000) and (**b**) building replacement values in billions of dollars. Pie charts show the proportion of residents affected and cost burdens within each of the five Southern California coastal counties (number of residents, cost of building replacements, and percent of totals with respect to impacts to all of Southern California). Results for the 75, 125, 125, 150, 175, and 500 cm SLR scenarios and 20-year return period storms are not shown but are commensurate with the trends shown.

Number of residents		Percent of residents	
50 cm SLR & no storm	50 cm SLR & 100-yr storm	50 cm SLR & no storm	50 cm SLR & 100-yr storm
Costa Mesa 12	Costa Mesa 12	Costa Mesa 0%	Costa Mesa 0%
Dana Point 92	Dana Point 121	Dana Point 0%	Dana Point 0%
Fountain Valle 0	Fountain Valle 0	Fountain Valle 0%	Fountain Valle 0%
Huntington Beach 1225	Huntington Beach 4754	Huntington Beach 1%	Huntington Beach 3%
Irvine 16	Irvine 33	Irvine 0%	Irvine 0%
Laguna Beach 190	Laguna Beach 230	Laguna Beach 1%	Laguna Beach 1%
Los Alamitos 0	Los Alamitos 0	Los Alamitos 0%	Los Alamitos 0%
Newport Beach 6818	Newport Beach 8860	Newport Beach 8%	Newport Beach 10%
San Clemente 101	San Clemente 151	San Clemente 0%	San Clemente 0%
San Juan Capistrano 0	San Juan Capistrano 0	San Juan Capistrano 0%	San Juan Capistrano 0%
Santa Ana 0	Santa Ana 0	Santa Ana 0%	Santa Ana 0%
Seal Beach 74	Seal Beach 2272	Seal Beach 0%	Seal Beach 9%
Uninc. Orange Cty 585	Uninc. Orange Cty 809	Uninc. Orange Cty 0%	Uninc. Orange Cty 1%
Westminster 0	Westminster 0	Westminster 0%	Westminster 0%

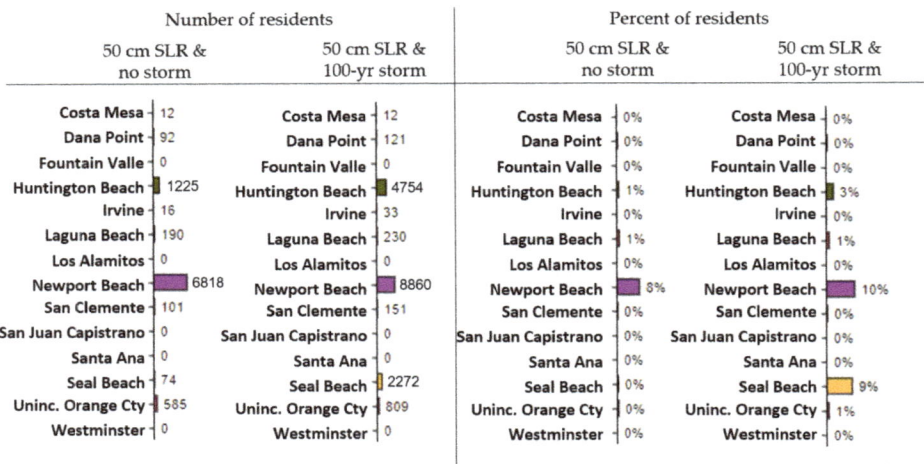

Figure 6. Example outputs (screen-grab: https://www.usgs.gov/apps/hera/) from the Hazard Exposure and Analytics (HERA) web tool comparing the number and percent of persons in each community that reside within the flood hazard zone for the 50 cm SLR with and without the projected 100-year coastal storm. Example is for Orange County.

In addition to comparing statistics between communities, HERA provides a means to evaluate exposures and vulnerabilities within each community. Magnitudes and percentages of affected demographics (over the age of 65, under the age of 5, ethnicity, race, renter versus owner residents, and head of household), land cover types (barren land, developed land, forest, pasture crops, shrub grass, and wetlands), infrastructure (highways, secondary streets and roads, railroads, critical facilities), economic assets (parcel values, building replacement values by occupancy class such as commercial, government, industrial, educational), and employee sectors (e.g., manufacturing, services, trade) are itemized and displayed spatially and summarized in graphs according to modeled hazards. An example of the spatial distribution and number of residents, including a breakdown of affected ethnicities, is shown in Figure 7b–d for Huntington Beach in Orange County. The data in Figure 7d indicates that this particular community hosts a fairly diverse population and that residents of all ethnicities will be affected by a 50 cm SLR; moreover, additional populations across all ethnicities will be exposed in case of the 100-year storm in combination with the 50 cm SLR. For the 50 cm SLR inclusive and exclusive of the 100-year storm, the greatest building replacement cost burden will be on the community's residents (Figure 7f). The lowest cost burden is expected to be for government structures, of which there is only one identified within the flood hazard zone (not shown). The data also indicate that government and education infrastructure are prone to damage by SLR and less by future coastal storms, as indicated by the equal or nearly equal cost-burden for the no storm and 100-year storm scenarios.

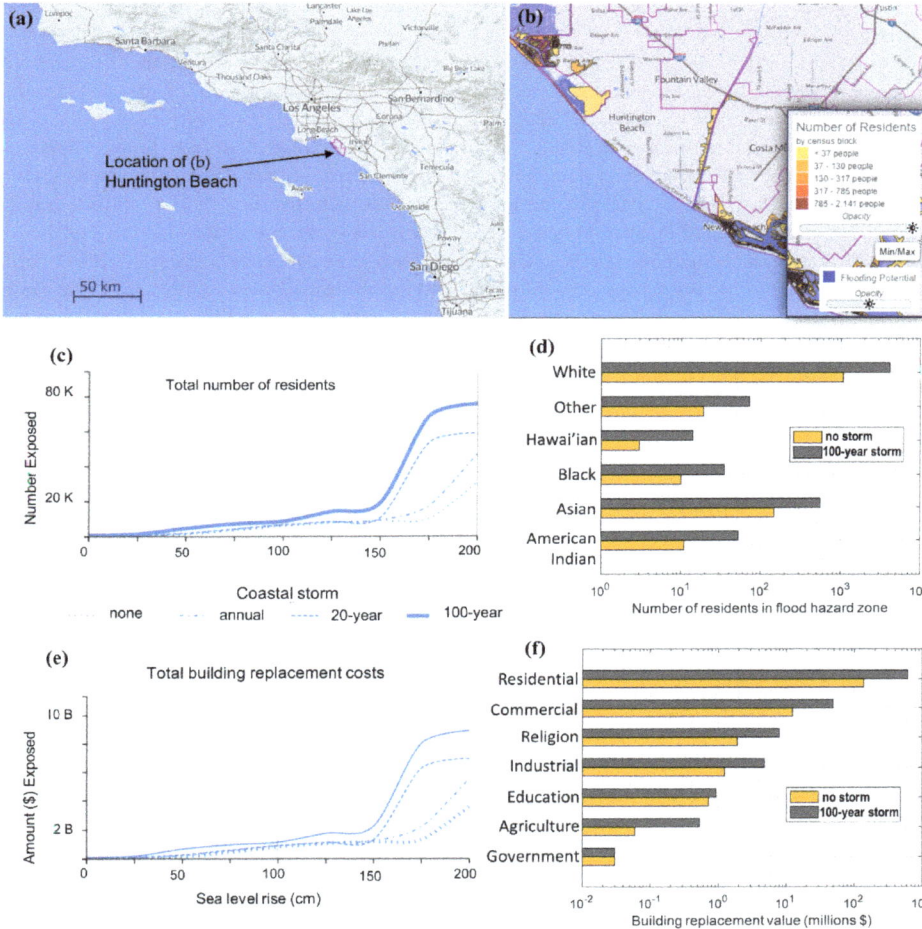

Figure 7. Example outputs from the Hazard Exposure and Analytics (HERA) web tool showing populations affected and building replacement costs within the Huntington Beach community in Orange County because of CoSMoS-modeled sea-level rise (SLR) and coastal storms. (**a**) Screen grab of HERA map showing Huntington Beach in Southern California; (**b**) Screen grab of spatial display in HERA showing the number of Huntington Beach residents exposed with 50 cm SLR in combination with a 100-year coastal storm (same as Figure 1b); (**c**) Total number of community residents exposed for all combinations of SLR (up through the 200 cm SLR) and coastal storms; (**d**) The number of community residents, grouped by ethnicity, exposed to the 50 cm SLR inclusive and exclusive of the 100-year storm; (**e**) Total building replacement costs within the community for all combinations of SLR (up through the 200 cm SLR) and coastal storms; (**f**) Building replacement costs, grouped by occupancy class, exposed to the 50 cm SLR inclusive and exclusive of the 100-year storm.

4. Stakeholder Engagement and Outreach

Outreach and engagement with planners, engineers, emergency managers, and environmental scientists from coastal cities, counties, utilities, state agencies, non-governmental organizations, and the private sector were conducted both in advance of and following the release of model results.

Outreach was designed and delivered in collaboration with established and trusted regional partners or networks to ensure local relevance. Workshops were held prior to the development of web tools to gain an understanding of what type of data, formats, and displays might be most suitable for end-users. Once model results were complete and incorporated into the web tools, high-level trainings and demonstrations of the OCOF and HERA web tools were conducted. To bolster interest and use, trainings were tailored to the specific needs and interests of local stakeholders and immediate access to the data in their respective areas was provided. For instance, the San Diego County workshop included a panel highlighting local projects engaged in SLR planning as well as a separate session that focused on model details, assumptions, and limitations for more technical end-users. For the Los Angeles County workshop, community planning exercises were conducted where attendees could view the SLR and erosion projections on paper maps and brainstorm adaptation ideas. Over the course of the Southern California project, we participated in eight workshops over three years, reaching over 500 participants across all Southern California counties.

In addition to the more traditional outreach and engagement strategies, virtual reality (VR) and 360-degree 3-D videos were used for communicating future coastal flood hazards. For instance, in the city of Santa Monica, California, projected flood extents were used to create virtual images of the beach under different states of SLR, both with and without the effects of coastal storms (see mobileowl.co/samo; Figure 8a–c). Possible adaptation strategies were also presented in some locales. Residents and visitors were guided through a series of images in a virtual-reality viewer that showed the beach

(1)   in its present state (Figure 8a),
(2)   flooded due to a 100-year storm modeled with CoSMoS,
(3)   in the future with 2 m of SLR (Figure 8b),
(4)   in the future with 2 m of SLR and the 100-year coastal storm, and potentially
(5)   modified with a possible adaptation strategy (Figure 8c).

As users moved through the images, they were asked a series of questions on demographics and levels of concern as the flood hazards increase. These images were available via an in situ platform, nicknamed the "owl" (by Owlized™, see Figure 8a inset), that was placed on Santa Monica Pier from November 2016 to January 2017; the "owls" were used to support Santa Monica's local coastal planning and outreach efforts. The augmented images within the viewer provided visceral opportunities to visualize the complex scientific information used by the Santa Monica community in its planning. The visualizations are still available via the online mobile viewer (mobileowl.co/samo) and continue to be viewed by interested stakeholders.

A second innovative technology and approach using VR allows users to visualize how coastal areas may be effected everyday under future SLR conditions. A video system developed in partnership with Google™ and consisting of 16 GoPro™ cameras (the GoPro Odyssey™) was used to film 360-degree videos that show beaches during the highest ('King') tides of the year (Figure 8d–f). The videos were uploaded to YouTube and, using VR headsets, residents can explore and observe their local beach during these extreme conditions, visualizing how it might appear under normal tide conditions in the coming century if sea level continues to rise. Both sets of VR visualizations ("owl" and King tide videos) have been used for education and outreach purposes to help make complex scientific information more accessible. They can both be accessed via mobile platforms as well as via home computers with sophisticated VR headsets (such as any Google™ VR viewer).

Figure 8. Images showing stills from virtual reality simulations of sea-level rise (SLR) potentials. (**a–c**) Example images (screen-grabs: mobileowl.co/samo) from the virtual reality (VR) viewer shown in (**a**). (**a**) Present day conditions and the 1920's shoreline position at Santa Monica Beach in Los Angeles County, California (CA). Inset photo shows use of the viewfinder for observing and scanning the VR images; (**b**) VR image of CoSMoS-modeled shoreline position with a 2 m SLR; (**c**) VR image of a possible adaptation option employing vegetated dunes; (**d–f**) Example images (screen-grabs: https://www.youtube.com/watch?v=FQl93W469vl) from the 360-degree 3-D viewer and video. Videos of annual high tides are filmed to serve as proxies for future sea levels under everyday normal tide conditions; the videos are viewed in 3-D by users using any VR headset. Inset photos in (**d**) show the 16 GoPro™ camera system used to capture the 360-degree view.

5. Discussion and Summary of Findings

The overarching concept of CoSMoS is to leverage projections of global climate patterns over the 21st century from recent Global Climate Models (GCMs). Coarse resolution GCM projections are downscaled to the local level and used as boundary conditions to sophisticated ocean modeling tools that simulate complex physics to accurately predict local coastal water levels and flooding for the full range of expected SLR ($n = 10$: 0–2 m in 0.25 m increments and 5 m) and storm scenarios ($n = 4$: average daily/background conditions, annual, 20-year, and 100-year). Resulting model projections include spatially explicit estimates of flood extent, depth, duration, uncertainty, water elevation, wave run-up, maximum wave height, maximum current velocity, and long-term shoreline change and cliff retreat.

The model system produces coastal hazard projections suitable to aid local climate adaptation planning. The results are provided to the public via two heavily vetted and user-tested web tools, one that presents the hazards in a map-style interface, ("Our Coast, Our Future" (OCOF): www.ourcoastourfuture.org) and a second one that integrates the hazards with geospatial demographic and socio-economic data to provide information on exposures and vulnerabilities (Hazard Exposure Reporting and Analytics (HERA): https://www.usgs.gov/apps/hera/). Both tools have the dual benefit of providing user-friendly web tools that allow the interested public to explore complex scientific information, similar to how they would use a web browser to explore a local map, as well as providing robust scientific information that can be used by municipal, county, and statewide planners and managers in their coastal adaptation and local hazard mitigation plans. Additionally, the web applications can be used by diverse audiences for multiple purposes. For those coastal communities that do not have access to geographical information system (GIS) specialists, OCOF and HERA allow these communities to explore the full suite of SLR and storm hazards and impacts to incorporate into their planning efforts. For communities with access to technical GIS capabilities, web applications are used as a public outreach tool that allows interested residents and community members to explore the scientific information to supplement their own understanding. Thus, they play an important education/outreach function supporting local coastal adaptation planning.

Using a continuous time series of nearshore wave conditions as well as storm surge and sea level anomaly levels in combination with SLR, the coastline change models developed for this study indicate a spatial average of ~25–40 m of beach erosion and ~10–85 m cliff retreat in Southern California [30,31] (rounded to the nearest 5 m). This amount of shoreline retreat would completely erode as much as 67% of the beaches in Southern California [30]. Lower SLR scenarios result in less but not insignificant erosion; for example, 50 cm of SLR results in an average of 15 m of beach loss and 10 m of cliff retreat.

Flood hazards modeled with the deterministic, dynamic CoSMoS model and including projected coastline change, show that, across Southern California, 100 to 200 cm of SLR will inundate 40–150 km^2 of land. It also shows that the effects of a 100-year coastal storm would flood an additional 54% (150–230 km^2) to 129% (40–340 km^2) of land area. More near-term projections of 25 cm SLR (by approximately the year 2030) [32] are estimated to permanently flood ~10 km^2, with an intermittent flood extent increasing the area affected by nearly 350% (10–45 km^2) when the 100-year storm is also taken into consideration. The results demonstrate that, if sea level continues to rise, many areas will be impacted by flooding in both the long- and short-term, and that storm conditions, combined with even small amounts of SLR expected within just a few decades, will substantially increase the exposure hazard.

Translated to socio-economic impacts, 25–200 cm of SLR places ~20,000–164,000 residents at risk of being permanently flooded along the Southern California shores. Building replacement values are estimated to be between $3.64 billion and $26.10 billion (2006 value, unadjusted for inflation). Accounting for the 100-year storm exposes an additional 56–109% of residents and increases building replacement costs by 46% to $38.2 billion, thus highlighting the importance of including storms in vulnerability assessments.

For actual implementation of hazard mitigation or climate adaptation actions, quantified projections of impacts at the community level are invaluable. The HERA tool provides analytics summarized in the form of maps and graphs for evaluation of vulnerable areas, populations, infrastructure, and economic sectors as well as the ability to download all the data for off-line in-depth local-scale analysis and planning. Using this tool, social vulnerability at the community level can be evaluated according to relative distribution of income, race, age, access to resources, viability of critical infrastructure (e.g., hospitals, roads), building replacement costs, and diversity of economic assets [24,33].

To communicate the availability, uses, and implications of the modeled hazards, exposures, and vulnerabilities, numerous workshops and outreach activities, tailored to the specific needs and interests of local stakeholders and delivered through existing, trusted networks and partnerships have been and continue to be held in regions where the model system has been applied. At each workshop, an overview of the CoSMoS model and regional results are provided, demonstrations and hands-on trainings of the web tools are conducted, and access to data specific to the region is highlighted and discussed. In addition to the more traditional outreach and engagement strategies, virtual reality technology and activities are being developed and applied to reach greater audiences and to better communicate future coastal flood hazards. CoSMoS, and the associated web tools OCOF and HERA, are used for local and state-level coastal planning as well as hazard mitigation planning by approximately 30 coastal cities and counties in California. It is also utilized by many of the state agencies, such as the California Coastal Commission, California Department of Emergency Services and the California Department of Transportation, nongovernmental organizations, and regional-scale collaborations. Although both the modeling methods and outreach activities continue to evolve, the success of providing useful information for coastal planners, engineers, and other stakeholders is underpinned by close relationships and ties with regional partners and local stakeholders who help envision, build, and develop effective products and tools for the critical end-user.

Author Contributions: All authors contributed writing to the paper. P.B. conceived the modeling system and coupling to web-tools. L.E. led model development, implemented parts of the model, and compiled and led the writing of the paper. A.O. developed and implemented SoCal CoSMoS; J.L. assisted with model development and application. A.F. led the GIS components of CoSMoS. S.V. and P.L. developed and implemented the coastal change models. J.F.-H. led the virtual reality and stakeholder outreach activities including innovative technology developments. N.W. and J.J. developed the HERA tool. M.F. led the development of the "Our Coast, Our Future" tool. M.H. led outreach and updates of the "Our Coast, Our Future" tool.

Funding: Funding for this project was provided from the California Natural Resources Agency for California's 4th Climate Assessment, Climate Change Impacts for Southern California; California Department of Fish & Wildlife, Imperiled Southern California Coastal Plant Species: Assessment of Threats from Climate Change, Including Rising Sea Levels, Storms and Coastal Erosion; California State Coastal Conservancy, Southern California Climate Change Vulnerability Study; City of Imperial Beach, Imperial Beach Climate Change Vulnerability Study; and USGS Coastal and Marine Geology Program.

Acknowledgments: Authors would like to acknowledge support through a cooperative agreement with Deltares. Authors would also like to thank key partners in outreach efforts: USC Sea Grant, Tijuana River Estuarine Research Reserve, San Diego Climate Collaborative, California Coastal Commission, City of Carpinteria, California Sea Grant, State of California Coastal Conservancy, County of Ventura, County and City of Santa Barbara, City of Santa Monica, California Ocean Protection Council, Santa Monica Bay Restoration Commission, Los Angeles Regional Collaborative, Climate Science Alliance, UC Irvine Sustainability Initiative, Southern California Coastal Water Research Project, ESA, TerraCosta, and Climate Access. Data from CoSMoS is available online at http://dx.doi.org/10.5066/F7T151Q4. Any use of trade, firm, or product names is for descriptive purposes only and does not imply endorsement by the U.S. Government.

Conflicts of Interest: The authors declare no conflict of interest.

Appendix

List of terms, acronyms, and software platforms used in the text

3-D	Three-dimensional
360-degree	Image collection and display in a continuous 360-degree arc
CoSMoS	Coastal Storm Modeling System
GCM	Global Climate Model
GIS	Geographic Information System
GeoTIFF	Common raster file format with georeferencing information
Google	Technology company with applicable expertise in internet, mapping, and image sharing services; Mountain View, CA, USA
Google Earth	Program that renders a three-dimensional view of Earth and imagery; version 6+; Google: Mountain View, CA, USA
GoPro	A commercial camera company; San Mateo, CA, USA
HERA	Hazard Exposure Reporting and Analytics
NAICS	North American Industry Classification System
NLCD	National Land Cover Database
LA	Los Angeles County

OC	Orange County
OCOF	"Our Coast, Our Future"
Open Source Geospatial Foundation	Non-profit organization that supports open-source GIS formats and technology; https://www.osgeo.org/
"owl"	Virtual reality viewer used to communicate 21st century flood impacts
Owlized	A commercial virtual-reality company; San Francisco, CA, USA
PostGIS	An open-source, GIS-support software program; version 2+; available https://postgis.net/
PostgreSQL	An open-source relational database; version 9+; available https://www.postgresql.org/
SB	Santa Barbara County
SD	San Diego County
SLR	Sea-level Rise
RCP	Representative Concentration Pathway
VE	Ventura County
VR	Virtual Reality

References

1. Wahl, T.; Brown, S.; Haigh, I.; Nilsen, J. Coastal sea levels, impacts, and adaptation. *J. Mar. Sci. Eng.* **2018**, *6*, 19. [CrossRef]

2. Mentaschi, L.; Vousdoukas Michalis, I.; Voukouvalas, E.; Dosio, A.; Feyen, L. Global changes of extreme coastal wave energy fluxes triggered by intensified teleconnection patterns. *Geophys. Res. Lett.* **2017**, *44*, 2416–2426. [CrossRef]

3. Hauer, M.E.; Evans, J.M.; Mishra, D.R. Millions projected to be at risk from seal-level rise in the continental united states. *Nat. Clim. Chang.* **2016**, *6*, 691–695. [CrossRef]

4. Sweet, W.V.; Park, J. From the extreme to the mean: Acceleration and tipping points of coastal inundation from sea level rise. *Earth Future* **2014**, *2*, 579–600. [CrossRef]

5. Purvis, M.J.; Bates, P.D.; Hayes, C.M. A probabilistic methodology to estimate future coastal flood risk due to sea level rise. *Coast. Eng.* **2008**, *55*, 1062–1073. [CrossRef]

6. Van Dongeren, A.; Ciavola, P.; Viavattene, C.; de Kleermaker, S.; Martinez, G.; Ferreira, O.; Costa, C.; McCall, R. RISK-KIT: Resilience-Increasing Strategies for Coasts-toolKit. *J. Coast. Res.* **2014**, *70* (Suppl. S1), 366–371. [CrossRef]

7. Prime, T.; Brown, J.M.; Plater, A.J. Physical and economic impacts of sea-level rise and low probability flooding events on coastal communities. *PLoS ONE* **2015**, *10*, e0117030. [CrossRef] [PubMed]

8. Knight, P.J.; Prime, T.; Brown, J.M.; Morrissey, K.; Plater, A.J. Application of flood risk modelling in a web-based geospatial decision support tool for coastal adaptation to climate change. *Nat. Hazards Earth Syst. Sci.* **2015**, *15*, 1457–1471. [CrossRef]

9. Sánchez-Arcilla, A.; García-León, M.; Gracia, V.; Devoy, R.; Stanica, A.; Gault, J. Managing coastal environments under climate change: Pathways to adaptation. *Sci. Total Environ.* **2016**, *572*, 1336–1352. [CrossRef] [PubMed]

10. Calil, J.; Reguero, B.G.; Zamora, A.R.; Losada, I.J.; Mendez, F.J. Comparative coastal risk index (CCRI): A multidisciplinary risk index for Latin America and the Caribbean. *PLoS ONE* **2017**, *12*. [CrossRef] [PubMed]

11. Brown, J.M.; Morrissey, K.; Knight, P.; Prime, T.D.; Almeida, L.P.; Masselink, G.; Bird, C.O.; Dodds, D.; Plater, A.J. A coastal vulnerability assessment for planning climate resilient infrastructure. *Ocean Coast. Manag.* **2018**, *163*, 101–112. [CrossRef]

12. Christie, E.K.; Spencer, T.; Owen, D.; McIvor, A.L.; Moller, I.; Viavattene, C. Regional coastal flood risk assessment for a tidally dominant, natural coastal setting: North Norfolk, southern North Sea. *Coast. Eng.* **2018**, *134*, 177–190. [CrossRef]

13. Barnard, P.L.; O'Reilly, B.; van Ormondt, M.; Elias, E.; Ruggiero, P.; Erikson, L.H.; Hapke, C.; Collins, B.D.; Guza, R.T.; Adams, P.N.; et al. *The Framework of a Coastal Hazards Model: A Tool for Predicting the Impact of Severe Storms*; Open-File Report 2009-1073; U.S. Geological Survey: Reston, VA, USA, 2009; 19p. Available online: http://pubs.usgs.gov/of/2009/1073/ (accessed on 2 February 2018).

14. Barnard, P.L.; van Ormondt, M.; Erikson, L.H.; Eshleman, J.; Hapke, C.; Ruggiero, P.; Adams, P.N.; Foxgrover, A.C. Development of the coastal storm modeling system (cosmos) for predicting the impact of storms on high-energy, active-margin coasts. *Nat. Hazards* **2014**, *74*, 1095–1125. [CrossRef]

15. Barnard, P.L.; Erikson, L.H.; Fitzgibbon, M.; Foxgrover, A.; Finzi Hart, J.; Limber, P.; O'Neill, A.C.; van Ormondt, M.; Vitousek, S.; Wood, N.; et al. Dynamic flood modeling essential to assess the coastal impacts of climate change. *Sci. Adv.* **2018**. in review.

16. O'Neill, A.; Erikson, L.; Barnard, P.; Limber, P.; Vitousek, S.; Warrick, J.; Foxgrover, A.; Lovering, J. Projected 21st century coastal flooding in the Southern California Bight. Part 1: Development of the third generation cosmos model. *J. Mar. Sci. Eng.* **2018**, *6*, 59. [CrossRef]

17. Erikson, L.H.; Hegermiller, C.A.; Barnard, P.L.; Ruggiero, P.; van Ormondt, M. Projected wave conditions in the Eastern North Pacific under the influence of two CMIP5 climate scenarios. *Ocean Model.* **2015**, *96*, 171–185. [CrossRef]

18. Bureau of Economic Analysis (BEA). *Gross Domestic Prodcut (GDP)*; Dataset; United States Department of Commerce, BEA: Suitland, MD, USA, 2017. Available online: https://www.bea.gov/ (accessed on 15 January 2017).

19. Point Blue Conservation Science and US Geological Survey. *Our Coast, Our Future*; Web Application; Point Blue Conservation Science: Petaluma, CA, USA, 2018; Available online: www.ourcoastourfuture.org (accessed on 1 May 2018).

20. Jones, J.M.; Henry, K.; Wood, N.; Ng, P.; Jamieson, M. HERA: A dynamic web application for visualizing community exposure to flood hazards based on storm and sea level rise scenarios. *Comput. Geosci.* **2017**, *109*, 124–133. [CrossRef]

21. Erikson, L.H.; Barnard, P.L.; O'Neill, A.C.; Vitousek, S.; Limber, P.; Foxgrover, A.C.; Herdman, L.H.; Warrick, J. *Cosmos 3.0 Phase 2 Southern California Bight: Summary of Data and Methods; Summary of Methods to Accompany Data Release*; U.S. Geological Survey: Santa Cruz, CA, USA, 2017. [CrossRef]

22. Erikson, L.H.; Barnard, P.L.; O'Neill, A.; Limber, P.; Vitousek, S.; Finzi Hart, J.; Hayden, M.; Jones, J.; Wood, N.; Fitzgibbon, M.; et al. *Assessing and Communicating the Impacts of Climate Change on the Southern California Coast California's Fourth Climate Change Assessment*; Report; California Natural Resources Agency: Sacramento, CA, USA, 2018; Volume CRNA-CCC4A-2018, p. 75.

23. United States Census Bureau. *Block Census Data*; Dataset; U.S. Department of Commerce, United States Census Bureau: Suitland, MD, USA, 2012. Available online: https://www.census.gov/data.html (accessed on 1 October 2012).

24. Rufat, S.; Tate, E.; Burton, C.G.; Maroof, A.S. Social vulnerability to floods: Review of case studies and implications for measurement. *Int. J. Disaster Risk Reduct.* **2015**, *14*, 470–486. [CrossRef]

25. Wood, N.; Ratliff, J.; Peters, J. *Community Exposure to Tsunami Hazards in California*; Scientific Investigations Report, 2012-5222; U.S. Geological Survey: Reston, VA, USA, 2012.

26. Infogroup. *Employer Database*; Online Government Dataset; Infogroup: Papillion, NE, USA, 2012; Available online: http://www.referenceusagov.com (accessed on 20 October 2012).

27. Homer, C.G.; Dewitz, J.A.; Yang, L.; Jin, S.; Danielson, P.; Xian, G.; Coulston, J.; Herold, N.D.; Wickham, J.D.; Megown, K. Completion of the 2011 national land cover database for the conterminous united states-representing a decade of land cover change information. *Photogramm. Eng. Remote Sens.* **2015**, *81*, 345–354.

28. Jones, J.M.; Wood, N.; Ng, P.; Henry, K.; Jones, J.L.; Peters, J.; Jamieson, M. *Community Exposure in California to Coastal Flooding Hazards Enhanced by Climate Change, Reference Year 2010*; Dataset; U.S. Geological Survey: Reston, VA, USA, 2016; Volume 2018.

29. Limber, P.W.; Barnard, P.L.; Hapke, C. Towards projecting the retreat of California's coastal cliffs during the 21st Century. In Proceedings of the Coastal Sediments 2015, San Diego, CA, USA, 11–15 May 2015; Wang, P., Rosati, J.D., Cheng, J., Eds.; World Scientific: Hackensack, NJ, USA, 2015; p. 14. [CrossRef]

30. Vitousek, S.; Barnard, P.L.; Limber, P.W.; Erikson, L.; Cole, B. A model integrating longshore and cross-shore processes for predicting long-term shoreline response to climate change. *J. Geophys. Res. Earth Surf.* **2017**, *122*, 782–806. [CrossRef]

31. Limber, P.W.; Barnard, P.L.; Vitousek, S.; Erikson, L.H. A model ensemble for projecting multi-decadal coastal cliff retreat during the 21st century. *J. Geophys. Res. Earth Surf.* **2018**. [CrossRef]

32. Cayan, D.R.; Kalansky, J.; Iacobellis, S.; Pierce, D. *Creating Probabilistic Sea Level Rise Projections to Support the 4th California Climate Assessment*; 16-IEPR-04, TN 211806; California Energy Commission: Sacramento, CA, USA, 2016. Available online: http://docketpublic.energy.ca.gov/PublicDocuments/16-IEPR-04/TN211806_20160614T101823_Creating_Probabilistic_Sea_Leve_Rise_Projections.pdf (accessed on 15 February 2018).
33. Fielding, J.L. Inequalities in exposure and awareness of flood risk in England and wales. *Disasters* **2011**, *36*, 477–494. [CrossRef] [PubMed]

Journal of
*Marine Science
and Engineering*

MDPI

Review

Quantifying Economic Value of Coastal Ecosystem Services: A Review

Seyedabdolhossein Mehvar [1,2,*], Tatiana Filatova [3,4], Ali Dastgheib [2], Erik de Ruyter van Steveninck [2] and Roshanka Ranasinghe [1,2,5,6]

[1] Department of Water Engineering and Management, University of Twente, P.O. Box 217, 7500 AE Enschede, The Netherlands; r.ranasinghe@un-ihe.org
[2] Department of Water Science and Engineering, IHE Delft Institute for Water Education, P.O. Box 3015, 2601 DA Delft, The Netherlands; a.dastgheib@un-ihe.org (A.D.); e.deruijtervansteveninck@un-ihe.org (E.d.R.v.S.)
[3] Department of Governance and Technology for Sustainability, University of Twente, P.O. Box 217, 7500 AE Enschede, The Netherlands; t.filatova@utwente.nl
[4] School of Systems, Management and Leadership, Faculty of Engineering and Information Technology, University of Technology Sydney, Sydney, NSW 2007, Australia
[5] Harbour, Coastal and Offshore Engineering, Deltares, P.O. Box 177, 2600 MH Delft, The Netherlands
[6] School of Civil Engineering, University of Queensland, St. Lucia, Brisbane, QLD 4072, Australia
* Correspondence: s.mehvar@utwente.nl; Tel.: +31-068-442-6320

Received: 10 November 2017; Accepted: 29 December 2017; Published: 9 January 2018

Abstract: The complexity of quantifying ecosystem services in monetary terms has long been a challenging issue for economists and ecologists. Many case specific valuation studies have been carried out in various parts of the World. Yet, a coherent review on the valuation of coastal ecosystem services (CES), which systematically describes fundamental concepts, analyzes reported applications, and addresses the issue of climate change (CC) impacts on the monetary value of CES is still lacking. Here, we take a step towards addressing this knowledge gap by pursuing a coherent review that aims to provide policy makers and researchers in multidisciplinary teams with a summary of the state-of-the-art and a guideline on the process of economic valuation of CES and potential changes in these values due to CC impacts. The article highlights the main concepts of CES valuation studies and offers a systematic analysis of the best practices by analyzing two global scale and 30 selected local and regional case studies, in which different CES have been valued. Our analysis shows that coral reefs and mangroves are among the most frequently valued ecosystems, while sea-grass beds are the least considered ones. Currently, tourism and recreation services as well as storm protection are two of the most considered services representing higher estimated value than other CES. In terms of the valuation techniques used, avoided damage, replacement and substitute cost method as well as stated preference method are among the most commonly used valuation techniques. Following the above analysis, we propose a methodological framework that provides step-wise guidance and better insight into the linkages between climate change impacts and the monetary value of CES. This highlights two main types of CC impacts on CES: one being the climate regulation services of coastal ecosystems, and the other being the monetary value of services, which is subject to substantial uncertainty. Finally, a systematic four-step approach is proposed to effectively monetize potential CC driven variations in the value of CES.

Keywords: coastal ecosystems; ecosystem services; economic valuation; climate change

---

1. Introduction

For centuries people have lived in coastal zones (CZ) and benefited from the ecosystem services that these areas provide. CZ have always been popular due to their accessibility to resources, in particular due to the abundant supply of subsistence resources, and recreational and cultural activities [1]. While coastal areas cover only 4% of the earth's total land area and are equivalent to only 11% of the World's ocean area [2], they host one third of the World's population and are twice as densely populated as inland areas [3]. The population density grows in the CZ annually due to migration driven by global demographic and socio-economic changes [1]. Growing population and accompanying infrastructure build-up provoke agglomeration economies that attract even more people and capital to the CZ, which has resulted in 15 out of the 20 present-day megacities of the World being located in low elevation CZ [4].

Worldwide, the economies of coastal communities and their resilience highly depend on the ecosystem services that CZ provide. It is well known that coastal ecosystems undergo major changes triggered by direct and indirect drivers. Direct drivers include natural forcing drivers such as coastal hazards (e.g., flooding, erosion), the probability and severity of which is expected to increase with climate change (CC). CC impacts which are primarily due to the anthropogenic greenhouse gas effect can alter the atmospheric composition and thereby change the intricate dynamics of the marine area resulting in variations in coastal ecosystems. The potential CC impacts relevant to coastal ecosystems include variations in mean sea level (i.e., sea level rise), wave conditions, storm surge [5], ocean circulation, ocean acidification (due to higher levels of CO_2), water temperature and changes in precipitation [6].

However, there is very little known about how CC may affect the value of ecosystem services in the CZ. Other examples of direct drivers of ecosystem change include land conversion, which changes the local land use and land cover [7]. The two most important indirect drivers of the environmental changes in coastal areas are population growth and economic development. Both direct and indirect drivers lead to a loss of coastal ecosystem services (CES), which will damage ecosystems and will undermine further development in CZ. Ironically, it is often the monetary valuation of this loss in CES that attracts the attention of policy makers and stakeholders.

Interest in ecosystem services in both research and policy-making communities has grown rapidly [8]. Many studies have estimated the value of ecosystem services for different wetland types, most of which have been limited to a particular local-scale case study (e.g., [9–13]).

On a larger spatial scale, Chaikumbung et al. [14] reviewed 1432 valuation studies of wetlands worldwide with a goal to provide a meta-regression analysis of their economic value and factors that influence it. In addition, Rao et al. [15] estimated the global value of CES for specific coastal ecosystems ranging from 0.4–1998 US $/ha/year in 2003 corresponding to 0.5–2530 US $/ha/year in 2013. Other studies have indicated that CZ and the oceans together contribute more than 60% of the total economic value of the biosphere [16,17]. A more recent study [18] indicates that global land use has changed between 1997 and 2011 resulting in an ecosystem services loss of between US $4 and US $20 trillion per year, implying that CES may have experienced a proportional loss. However, studies estimating the monetary effects of CC impacts on CES are scarce. Such an endeavor often requires a multidisciplinary effort—even more than in traditional ecosystem valuation exercises that do not consider CC effects.

Despite the above mentioned local scale and global studies, a coherent review on the valuation of CES with a systematic description of fundamental concepts, key reported applications, and potential CC impacts on the monetary value of CES has not been undertaken to date. This review article takes a step towards addressing this large knowledge gap and is aimed at assisting researchers and policy makers in multidisciplinary fields to gain a better appreciation of the economic value of CES and potential CC impacts on CES.

Specifically, here we discuss a number of salient questions that one has to consider when seeking to estimate the value of certain coastal ecosystems. In particular, (1) What type of wetlands and ecosystems are being assessed? (2) What type of ecosystem services, goods and values need to be considered? (3) Which drivers of ecosystem change are applicable to the study? (4) What kind of valuation methods should be used for valuing a particular ecosystem service? (5) What type of data are needed given the limitations and costly process of its collection? (6) How CES have been valued in previous global and local case studies, and what are the highest and lowest valued services and the most frequently used valuation methods therein? (7) How important are CC impacts on CES, and to what extent can they potentially affect the value of services? (8) What are the gaps and challenges in assessing potential CC-driven changes in the value of CES? These questions are sequentially addressed in the subsequent sections of this article, following the sequential structure shown in Figure 1.

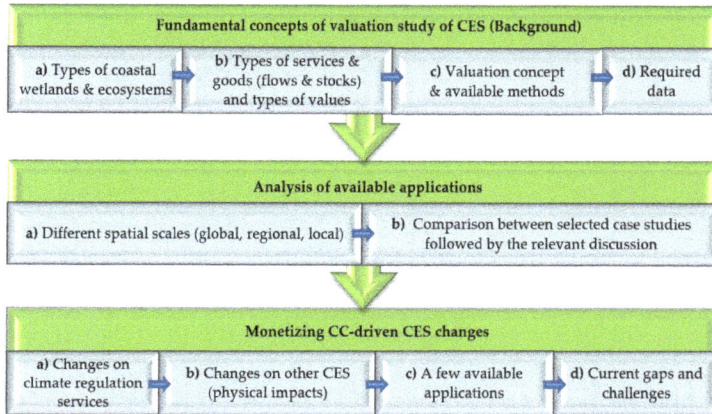

Figure 1. Schematic depiction of the sequential structure of this article.

As shown in Figure 1, this article first presents a background on coastal wetlands, ecosystems and their services/goods (Section 2.1), followed by a summary description of the concepts underlying ecosystem services valuation studies, current economic valuation methods and required data for conducting such studies (Section 2.2). This is followed by Section 3 which presents and analyses 30 selected local and regional-scale valuation studies of CES and two global-scale cases where they are clustered based on type of ecosystem to highlight the current status of valuation studies of CES. Section 4 presents the important and less known issue of monetizing climate change impacts on CES. Here, the link between CC impacts and monetary value of CES is discussed via a methodological framework.

2. Background

2.1. Coastal Wetlands, Ecosystems, Services and Goods

Wetlands are classified into three types: inland, coastal and marine, and human-made wetlands [3]. In general, the coastal and marine environment can extend up to 100 km inland and up to 50 m water depth in the ocean [19]. Coastal and marine wetlands include estuaries, lagoons, coastal peatlands and beaches. Nearshore areas with a maximum depth of 6 m at low tide are also considered part of coastal and marine wetlands as defined by the Ramsar Convention in 1971 [3,20]. Associated with coastal and marine wetlands (referred in this article as coastal wetlands), coastal ecosystems include mangrove forests, coral reefs, sea-grass beds, marshes, beach and dune systems as well as pelagic systems. In general, ecosystem services are defined as the immaterial services that are of benefit to humans with a monetary value which are generated by wetlands [7]. Thus, ecosystem services are the benefits (sometimes referred to as flows of the benefits) that people obtain from ecosystems, while ecosystem goods consist of food provision (such as fish, fiber), and raw materials (such as wood), sometimes also called stocks of natural ecosystems. From an ecological point of view, stocks and flows refer to the structural components and environmental functions respectively [21].

Tinch and Mathieu [22] stated that the ecosystem services framework focuses on the flows of valuable goods and services that are provided by the stock of natural resources. According to this study, these two terms should be differentiated since flow values are the ones that can be derived over a defined time interval, while stock values are the net present value sum of all flow values that may be derived from an ecosystem over a future period.

A variety of benefits can be explicitly classified as ecosystem services such as use and non-use values including existence and bequest values [3,20,23]. Table 1 indicates the classification of the main coastal and marine ecosystem services modified from [20,23].

Table 1. Values provided by coastal and marine ecosystem services.

Use Values		Non-Use Values
Direct Values	**Indirect Values**	**Existence and Bequest Values**
Food, fiber and raw materials provision	Flood control	Cultural heritage and spiritual benefits
Transport	Storm protection, wave attenuation	Resources for future generations
Water supply	CC impacts mitigation	Biodiversity
Recreation and tourism	Contaminant storage, detoxification	
Wild resources	Shoreline stabilization/erosion control	
Genetic material	Nursery and habitat for fishes and other marine species	
Educational opportunity	Nutrient retention and cycling	
Aesthetic	Regulation of water flow, water filtration	
Art	Source of food for sea organisms	
	Climate regulation, primary productivity as Oxygen production and CO_2 absorption, Carbon sequestration etc.	

Direct use values refer to the ecosystem services that can be directly used and associated with human well-being. Indirect use values include services that provide benefits outside the ecosystem. These latter values refer to ecosystem services with values that can be only measured indirectly, since they are only derived from supporting and protecting activities that have directly measurable values [24].

It should be noted that some of the cultural services (referred to as non-use values in Table 1) can also be included in other typologies of ecosystem services [25]. For example, recreation and tourism services offer non-consumptive values such as the enjoyment of recreational and cultural amenities (e.g., wildlife, bird watching and water sports) [26]. Recreational services can also be classified as a direct use value [27], which is how they are considered in this review (see Table 1).

Non-use or passive use values represent the value of ecosystem services which exist even if they are not used. These include existence and bequest values which refer to the public awareness of ecosystem services that exist and will persist for future generations to enjoy. Table 2 provides an overview of the coastal ecosystems and some of their attributed use-value services modified from [3].

Table 2. Overview of coastal ecosystems and their attributed use-value services.

Coastal Ecosystem	Direct Use Value	Indirect Use Value
Mangrove forests	Raw material (wood production), aesthetic, educational opportunities, artistic value	CC impact mitigation, storm protection and wave attenuation, shoreline stabilization and erosion control, flood control, nursery and habitat for fishes and other marine species, regulation of water flow and filtration, carbon sequestration, oxygen production and CO_2 absorption, contaminant storage and detoxification

Table 2. *Cont.*

Coastal Ecosystem	Direct Use Value	Indirect Use Value
Coral reefs	Aesthetic, recreation and tourism (snorkeling), educational opportunities, artistic value, raw material for building, jewelry and aquarium trade	Nursery and habitat for fishes and other marine species, wave attenuation and shoreline stabilization, nitrogen fixation
Sea-grass beds	Aesthetic, contribution to recreation and tourism (snorkeling)	Nursery and habitat for fishes and other marine species, source of food for sea organisms, shoreline stabilization and erosion control, primary productivity as oxygen production and CO_2 absorption, water filtration
Beach and dune systems	Recreation and tourism, fiber and raw material (wood source) provided by the dune vegetation, aesthetic value, artistic value	Flood control, erosion control, nursery for some marine species (turtles)
Pelagic systems	Food source, aesthetic value, tourism services, artistic value	Source of food for sea organisms, nursery and habitat for fishes and other marine species

2.2. Valuation of Ecosystem Services

In principle, economic valuation of ecosystem services is based on "people preference" and their choices. Therefore, it is quantified by the highest monetary value that a person is willing to pay in order to obtain the benefit of that particular service. The "willingness to accept" approach determines how much someone is willing to give up for a change in obtaining a certain ecosystem good or service [3]. Thus, the key outcome of valuation studies is to illustrate the importance of a healthy ecosystem for socio-economic prosperity and to monetize the gains that one may achieve or lose due to a change in ecosystem services [28].

The value of ecosystem services can be measured in three different forms [22]: (1) Total economic value (TEV) that refers to the value of a particular ecosystem service over the entire area covered by an ecosystem during a defined time period; (2) average value of an ecosystem service per unit, which is often indicated for a unit of area or time; (3) marginal value which is the additional value gained or lost by an incremental change in a provision of a particular service.

The valuation starts from estimating a TEV of an ecosystem, which is in fact a sum of consumer surplus and producer surplus. This is done by applying different valuation techniques. For example, in the case of tourism, producer surplus is the direct or indirect benefit from the local ecosystems for the tourism sector by considering the revenue made from tourists minus the costs of providing these services to them [29]. In addition, consumer surplus conveys the maximum amount that tourists are willing to pay for visiting the specific recreational area.

2.2.1. Valuation Methods

There are different ways of classifying economic methods used for valuing ecosystem services and goods: Revealed preference methods, stated preference methods, market price, and benefit transfer method. Table 3 shows an overview of these techniques including their attributed CES and goods adapted from [2,16,24,30–32].

Table 3. Overview of the valuation methods and their attributed coastal ecosystem services and goods.

Valuation Method		Description	Coastal Ecosystem Services and Goods
Revealed preference methods (use-value)	Production-based (net factor income)	Often used to value the ecosystem services that contribute to the production of commercially marketed goods	Regulating services such as oxygen production, CO_2 absorption, nitrogen fixation and carbon storage, providing fish nurseries, water purification, coastal protection
	Hedonic pricing	Commonly used to value the environmental services contributing to amenities. Property's price often represents the amenity value of ecosystems	Tourism and recreation, aesthetic, improving air quality
	Travel cost	Basically considers the travel costs paid by tourists and visitors to the environmental value of a recreation site	Tourism and recreation, recreational fishery and water sports
	Damage avoided cost, replacement cost	Based on either the cost that people are willing to pay to avoid damages or lost services, the cost of replacing services or the cost paid for substitute services providing the same functions and benefits	Buffering CC impacts such as wave attenuation, providing coastal protection against storms and erosion, flood impact reduction, water purification, carbon storage
Stated preference methods (both use and non-use value)	Contingent valuation (CVM)	The most applied method for both use and non-use values, based on surveys asking people their willingness to pay (WTP) to obtain an ecosystem service	Tourism and recreation, recreational fishery and water sports, aesthetic value, cultural and spiritual value, art value, educational value
	Contingent choice (CCM)	WTP is stated based on choices between different hypothetical scenarios of ecosystem conditions	
Market price		Often used for the ecosystem products that are explicitly traded in the market	Fiber, wood and sea food provision, raw material for building, and aquarium
Benefit transfer		It transfers available data from previous valuation studies for a similar application	Mostly applied for gross value of coastal ecosystems associated with recreation

2.2.2. Required Data

The data required for valuation of ecosystem services are collected by different means. Primary data are obtained via field observations and surveys, participatory approaches and stakeholder involvement which is often in the form of questionnaires and interviews. This is costly and time consuming, but the flexibility of the approach allows researchers to collect data on specific averting behaviors, attitudes, perceptions, and on the prices of averting behaviors stated by people [33]. This type of data is mostly obtained when using stated preference methods. For example, for valuation of recreation service, information is obtained from beach visitors stating their willingness to pay to obtain the benefit of recreational services of coastal ecosystems.

In absence of such primary data, one may use secondary data that are obtained from existing sources, such as global and national databases or available literature. Using secondary data is becoming more common given that in many studies the time and budget constraints apply. A widely used method in valuation studies relying on secondary data is the benefit transfer method, which derives information and estimates of values from previous studies [33]. Richardson et al. [34] described a coherent analysis of the benefit transfer method summarizing advancements, databases and analysis tools provided to simplify the application of this method.

3. Analysis of the Available Valuation Studies—Selected Sample

Valuation of CES can be done at different spatial scales, ranging from local and regional to global scale. In general, the spatial variability of ecological services might be of importance, affecting the net ecological benefits that they provide [35–37]. For example, Barbier et al. [36] showed that the magnitude of wave attenuation by coral reefs, salt marshes, sea-grass beds, and sand dunes varies spatially across ecosystems. This geographical influence was also observed by de Groot et al. [38].

3.1. Local and Regional Scale Applications

In this review, 30 local and regional valuation studies have been selected to represent a sample of current valuation studies. This sample has been chosen by searching the scientific database of google scholar considering applications in which the value of coastal ecosystem services has been estimated with a reported value. These selected references have been considered in such way to be able to distinguish them based on specific characteristics such as ecosystems and services considered, valuation methods used and estimated value. Here we have clustered these studies based on the type of coastal ecosystems that have been valued and Tables 4–7 show the ecosystem services provided, valuation methods used, and estimated results for each case.

Table 4. Selected applications of valuation of coral reef ecosystem services.

Reference	Valuation Method/s	Ecosystem Service/Good	Estimated Value
[39]	Stated preference	Tourism and recreation (marine national park in Seychelles)	US $88,000 (whole area)
[40]	Hedonic property price	Aesthetic (Indian ocean)	US $174 (per hectare)
[41]	Travel cost, stated preference	Recreation (Andaman sea of Thailand)	US $205.41 million (per year)
[29]	Production-based, avoided damage cost, travel cost, stated preference	Fishery, tourism, biodiversity, amenity, coastal protection (Guam)	US $141 million (per year)
[42]	Market price, net factor income, stated preference	Recreational and commercial fishing (Caribbean Netherlands, Bonaire)	US $400,000 and US $700,000 (per year)
[43]	Avoided damage cost	Protection to coastal erosion (Sri Lanka)	US $160–172,000 (per km of reef, per year)
[44]	Avoided damage cost	Habitat support for fisheries (Caribbean sea)	US $95–140 million (projected by 2015)
[44]	Avoided damage cost	Tourism (Caribbean sea)	US $300 million (projected by 2015)
[45]	Avoided damage cost	Coastal protection by wave dissipation (Bonaire Island, Caribbean, Netherlands)	US $33,000–70,000 (within 10 years–beyond 10 years)
[46]	Avoided damage cost	Coastal protection (Tobago, St. Lucia, (Caribbean)	US $18–33 million, US $28–50 million (annual values)
[44]	Replacement, substitute cost	Coastal protection in Caribbean coastline	US $750 million– 2.2 billion (annually)

The economic value of ecosystem services provided by coral reefs: Table 4 shows studies in which coral reef services have been valued at local and regional spatial scales. These services vary from recreational and tourism services to fishery, erosion control and coastal protection services. Depending on the service, different valuation methods have been used. Geographically, the selected case studies are mostly located in the Caribbean region highlighting this area as one of the important and large habitats of coral reefs. According to Table 4, coral reef services are valued at different estimations ranging from US $33,000–70,000 [45] to a very high value between US $750 million and US $2.2 billion [44]. This Table also shows that compared to other services, the coastal protection service provided by coral reefs is among the highest value estimated for this type of ecosystems.

Table 5. Selected applications of valuation of mangrove ecosystem services.

Reference	Valuation Methods	Ecosystem Service/Good	Estimated Value
[13]	Market price, replacement cost	Fishery, timber, carbon sequestration and storm protection (Vietnam)	US $3000 (per hectare, per year)
[47]	Avoided damage cost	Coastal protection, wood, habitat support for fishery (Thailand)	US $10,158–12,392 (per hectare)
[48]	Avoided damage cost	Storm (wind) protection (Odisha region, India)	US $177 (per hectare) (1999 price level)
[49]	Benefit transfer (from 48 selected studies)	Fisheries, fuel wood, coastal protection, water quality (Southeast Asia)	(mean) US $4185 (per hectare, per year) (2007 price level)
[12]	Replacement cost	Nutrient retention value (India)	US $232 (per hectare)

The economic value of ecosystem services provided by mangroves: Table 5 shows selected applications of CES valuations for mangroves indicating a range of estimations which is mostly considered per hectare of the study area. According to Table 5, fishery and storm protection are among the most frequently valued services reflecting the importance of these two services provided by mangroves. With respect to the estimated values, depending on the number of services considered, low values of US $177/ha [48] and US $232/ha [12] were estimated in the coastal area in India, while higher values between US $10,158/ha and US $12,392/ha were estimated in Thailand [47].

The economic value of other coastal ecosystems: Table 6 shows 7 selected case studies in which the value of other coastal ecosystems such as marshes, beaches and pelagic systems have been estimated using a variety of methods. These studies highlight the difference between types of values affecting estimated results. For example, Emerton and Kekulandala [11] estimated the total value of flood control services provided by marshes in Sri Lanka at USD 5 million per year, while Bell [50] estimated average values of USD 6471 and USD 981 respectively with respect to habitat support for fishery provided per acre of marsh on the East and West coasts of Florida in 1984 dollars. Among these selected applications, a high value of USD 23–44 billion was estimated by Molnar et al. [51] for food provision service of the marine area (referred to as pelagic system in this review) in British Colombia in 2004.

Table 6. Selected applications of valuations of other coastal ecosystems.

Reference	Ecosystem	Valuation Methods	Ecosystem Service/Good	Estimated Value
[11]	Marsh	Avoided damage cost	Flood attenuation (Colombo, Sri Lanka)	US $5 million (per year)
[50]	Marsh	Production-based	Habitat support for fisheries (Florida coast)	a. US $6471 (East) b. US $981 (West) (per acre)
[10]	Beach and dune system	Stated preference	Tourism (San Andres Island, Colombia)	US $997,468 (annual consumer surplus)
[52]	Pelagic system	Avoided damage cost, market price	Food provision (fish) (UK)	£513 million (in 2004)
[51]	Pelagic system	Benefit transfer (literature data)	Aesthetic and recreation (British Colombia)	US $23-44 billion (per year)
[53]	Pelagic system	Travel cost	Recreation (Baltic Sea)	€15 billion (total annual)
[9]	Pelagic system	Stated preference	Food provision (fish) (coast of Southeast Alaska)	US $248–313 Mean value for single-day private boat fishing trips

The economic value of combined coastal ecosystems: There are some available valuation studies in which coastal ecosystems have been valued as a whole system. These cases, which are shown in Table 7, indicate that one may value a combination of services [54] and goods grouped as coastal nature or a combination of two ecosystems such as coral reefs and mangroves as presented by Cooper et al. [54], and van Beukering and Wolfs [55]. Most of the selected cases in Table 7 were derived from studies that approximated the value of coastal nature in Caribbean Islands. Among the used methods, the net factor income has been applied for estimating research and artistic value of coastal nature as presented by van Beukering and Wolfs [55]. Selected applications show that direct-use value of the coastal environment was mostly considered in these studies indicating the highest value for providing erosion protection service at US $231–347 million in Belize in 2007 [54]. On the contrary, art value of the coastal environment was estimated at a relatively low annual value of US $290,000 in Bonaire Island [55]. Notably, all estimated values of services provided by the coastal environment are presented in annual value.

Table 7. Selected applications of valuation of combined coastal ecosystems.

Reference	Ecosystem	Valuation Methods	Ecosystem Service/Good	Estimated Value
[55]	Coastal nature	Net factor income	Research opportunity (Bonaire Island, Caribbean)	US $1,240,000–1,485,000 (in 2011)
[55]	Coastal nature	Net factor income	Pharmaceutic (Bonaire Island, Caribbean)	US $688,788 (annual)
[55]	Coastal nature	Net factor income	Art (Bonaire Island, Caribbean)	US $460,000 (annual)
[42]	Coastal nature	Stated preference	Tourism (Bonaire Island, Caribbean)	US $50 million (annual)
[56]	Coastal nature	Hedonic property price	Amenity (analysis of 1 million housing transactions) from 1996 to 2008 (UK)	£3700 (moving the bottom 1% postcode to the best 1% postcode (per year)
[54]	Coral reef and mangrove	Net factor income, avoided damage cost	a. Tourism b. Fisheries c. Erosion protection (Belize, Caribbean)	a. US $150–196 mil. b. US $14–16 mil. c. US $231–347 mil. (in 2007)
[55]	Coral reef and mangrove	Market price	Carbon sequestration (Bonaire Island, Caribbean)	US $290,000 (annual)

3.2. Global Scale Applications

The few reported global scale ecosystem service valuation studies have used the benefit transfer method, where global assessments have been derived based on the results of different local and regional case studies. For example, Costanza et al. [18] estimated the value of global ecosystem services at US $125 trillion per year (assuming updated unit values and changes to biome areas) and US $145 trillion per year (assuming only unit values changed) both in US $2007.

In contrast, the global valuation of coastal ecosystems done by de Groot et al. [38] estimated the total monetary value of a bundle of ecosystem services. Table 8 indicates the results of this study in standardized units (Int. $/ha/yr–2007 price level) for the coastal area categorized as open ocean, coral reefs, coastal systems and coastal wetlands. According to de Groot et al. [38], the open ocean (referred to as *pelagic system* in our review) represents the largest area of the marine ecosystem including deep sea (water and sea floor below 200 m). The coastal systems studies include several distinct ecosystems such as sea-grass fields, shallow seas of continental shelves, rocky shores and beaches, which are found in the terrestrial near-shore as well as the intertidal zones—i.e., until the 200 m depth contour. Moreover, de Groot et al. [38] separately studied coral reefs and coastal wetlands (mangroves and tidal marshes) because of the important and unique ecosystem services these systems provide.

Table 8. Global valuation of coastal ecosystem services (total monetary value per biome).

Coastal Wetlands/Ecosystems	No. of Estimates	Total of Service Mean Value	Total of Median Value	Total of Minimum Value	Total of Maximum Value
Open ocean	14	491	135	85	1664
Coastal systems	28	28,917	26,760	26,167	42,063
Coastal wetlands	139	193,845	12,163	300	887,828
Coral reefs	94	352,915	197,900	36,794	2,129,122

Source: de Groot et al. [38] (values in Int. $/ha/year, 2007 price level).

3.3. Discussion

The local and regional applications reviewed in this article indicate that *tourism and recreation* as well as storm protection services are among the most commonly valued services. This is in agreement with the conclusions made by [10,41,45]. In addition, these two types of services are often valued higher than other services. For example, Cooper et al. [54] conducted a valuation study in which tourism and erosion protection services of coral reefs and mangroves were valued at US $150–196 million and US $231–347 million, while providing fishery habitat value was estimated much lower at US $14–16 million. On the contrary, very little appears to be known about the value of cultural services of coastal ecosystems such as aesthetic and artistic values. Van Beukering and Wolfs [55] presented one of the few examples of valuing these less considered services in Caribbean Islands with an associated call for conducting similar studies in other coastal areas.

With respect to valuation methods used in the selected applications, avoided damage, replacement, and substitute cost methods are the most frequently used techniques for valuing storm/flood protection service provided by mangroves. In addition, stated preference and production-based methods (net factor income) are the second and third commonly used methods for the valuation of services, respectively.

The selected 30 studies also highlighted one of the main limitations in valuation of CES presenting a mostly incomplete measure of the ecosystem's value. The reason for this incomplete estimation might be the complexity of covering and valuing all the services provided by ecosystems in a particular area. Therefore, there are often missing services in the valuation studies. Data scarcity in some coastal areas together with the time consuming and high cost associated with data collection are also other important factors that discourage investigators from considering and valuing all services provided by a particular coastal ecosystem.

Tables 4–7 also illustrate that the range of estimated values quantitatively vary due to the many inconsistencies and irregularities in their characteristics. These discrepancies might be in temporal scale of studies, the way that data has been collected, number of services valued, type of the estimated values, location of the case studies, probability of the hazard occurrence and importance of the hinterlands (relevant for the replacement, substitute and avoided damage cost method) and other factors that make the estimated results not easily comparable. For instance, the high estimated value for the food provision service provided by marine areas in the study of Molnar et al. [51] may not be directly comparable with the results of Boero and Briand [52] for the same coastal ecosystem, due to the fact that location, type of valuation method and the services considered are different in these two studies. Also, in some of the selected applications, estimated value is presented per year (total value), while in some others (e.g., [44]), the marginal values have been projected for the future.

Apart from the mentioned irregularities, in some methods such as stated preference, socio demographic data such as age, level of education, mean salary etc. may affect the WTP stated by the visitors. This can affects the estimated value and consequently results in a totally different estimation for a certain CES. Therefore, all the aforementioned factors, discrepancies and inconsistencies add another challenge in comparing the results of different valuation studies.

4. Coastal Ecosystems and Climate Change Impacts: Monetizing Changes in the Value of Ecosystem Services

As mentioned in Section 1, the consequences of coastal hazards as direct drivers of change, not only affect the inhabitants of CZ, but also pose a considerable threat to the coastal environment. A remarkable proportion of coastal ecosystems is already under threat, with 50% of marshes, 35% of mangroves, 30% of coral reefs and 29% of the known global coverage of sea-grasses either lost or degraded already [24]. Climate change is likely to have a considerable impact on the threat levels faced by CES in future.

The World's coastlines are shaped by mean sea level, wave conditions, storm surge, and river flows, while CC driven variations in these forcing factors pose a considerable effect on the coastal area [5]. As a result, CC will significantly affect the direct or indirect benefits that humans obtain from these ecosystems [57]. Thus, CC will substantially alter or eliminate certain ecosystem services in the future. To better understand the impacts of CC on CES and to develop effective adaptation measures where possible, it is essential to improve our knowledge on the links between CC and ecosystem services and the corresponding economic impacts [58].

Academic literature discusses two types of links between CC and CES; climate regulation service provided by CES and CC impacts on CES (see Figure 2).

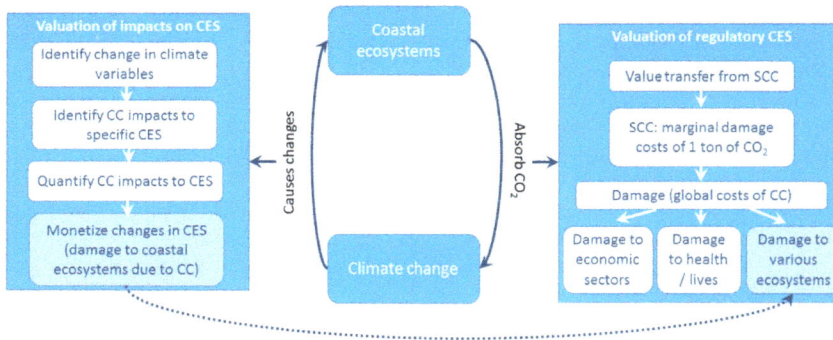

Figure 2. Links between climate change (CC), coastal ecosystem services (CES) and their valuation.

4.1. CC Link with Climate Regulation Service

Coastal ecosystems provide many services as indicated in Table 1; one of them being climate regulation through their ability to absorb CO_2 (right hand side of Figure 2). Blue carbon (CO_2) sequestered by mangroves, sea-grass beds, tidal marshes and vegetation present in other coastal ecosystems [59] attracts considerable attention in CC mitigation discussions. For example, Jerath et al. [60] performed a valuation study of regulatory climate service of mangrove forests of the Everglades National Park in Florida, USA.

The benefits of this regulatory service are compared to the average abatement costs of carbon that the mangroves provide. The valuation exercise of the regulatory climate service in this case is rather simple. Since CC has adverse effects globally, the common practice is to apply the value transfer method based on the social costs of carbon (SCC) to derive a single value of 1 ton of carbon absorbed. SCC indicates the marginal damage costs of 1 ton of CO_2 emitted, based on the global estimates of damages caused by CC [6].

Global CC induced damage in general comprises the monetary assessments of climate-induced damage to various economic sectors, health and human lives as well as ecosystems across the entire planet using global Integrated Assessment Models [61]. However, the monetary damage to various ecosystems including coastal are very rudimental [62,63]. In the Florida study, [60] used the IPCC value of SCC equal to $36/tCO_2$ in US $2007 price level, which translated into US $2015 per ton of carbon amounts to $167/tCO_2$. Furthermore, the total carbon absorption of coastal mangroves was potentially estimated and multiplied on this monetization rate. While this connection between CES and CC is an important research direction, it is the other link between CC and CES that is most relevant to this article.

4.2. CC Driven Changes on CES

CC will result in changes in temperature and hydrology, which will alter ecosystems. As coastal ecosystems evolve under CC, the benefits that they provide in the form of ecosystem services are likely to decline [57] and therefore, impact the socio-economic wellbeing of people. Any attempt to monetize CC driven variations in CES (left hand side of Figure 2) should necessarily follow a number of steps, as summarized below.

First, the results of global climate models have to be downscaled to a case study area to identify likely changes in climate variables and resulting physical changes in the coastal environment under different CC scenarios. Potentially, different CC mitigation [64,65] or adaptation [66] scenarios can be considered. When exploring CC impacts on the East coast of England, Turner et al. [65] used 17 climate model patterns till 2080 and a range of local weather data to estimate the key climate variables for coastal ecosystems. They included increases in average monthly and average monthly maximum temperatures, changes in monthly precipitation patterns, sea level rise (SLR) and changes in extreme events which are expected to be more frequent and severe in the future. CC-driven changes in temperature significantly influence the coastal and marine areas as they alter ocean conditions such as water temperature and biogeochemistry [67–69]. As a result, oceans become warmer and more stratified. In addition, higher levels of CO_2 lead to higher acidity of oceans [70]. Both will influence ocean flora and fauna. Increase in precipitation may cause growing nutrient fluxes due to heavy runoff from land leading to eutrophication risks in coastal wetlands [69].

Potential CC physical impacts on coasts—SLR and extreme events in particular—lead to episodic coastal inundation and permanent submersion of low lying land, episodic storm erosion of beaches and dunes, episodic formation and closure of small tidal inlets, and/or chronic coastal recession [5]. Coastal recession will result in less opportunity for beach recreational use. In general, less attractive beaches for recreation might be due to any of the CC driven impacts on sandy coasts listed in [5].

Secondly, these CC driven physical impacts need to be linked to specific ecosystem services that are currently provided by coastal wetlands in a particular location. This should ideally start with a qualitative assessment of changes in the provision of CES. In particular, it is necessary to assess whether changes in temperature, precipitation and other climatic variables will affect different use and non-use values of services provided by coastal ecosystems. Each of these two groups of CES should be specified and translated into measurable physical units. For instance, CC driven changes in temperature may be linked to changes in regulation services (categorized as indirect use value in this paper) measured through eutrophication, changes to food provision services measured in fishing stock, and changes to cultural and recreational ecosystem services measured through impacts on tourism and visitor numbers [65]. Similarly, SLR can be linked to regulation services measured through a need for flood protection to avoid land inundation [66] or to food provision services measured through increasing salinity of coastal soils [65].

Sumaila et al. [69] discussed the potential CC driven changes on food (fish) provision aspect of CES through change in primary fish productivity [71] caused by changes in physiology [72]. These CC driven changes on fish production affect economics of fisheries through a change in fishing costs due to adaptation to CC, a change in the relation between fish supply and ex-vessel revenue ultimately leading to changes in the price of fish and gross revenue. CC may also alter fish abundance by shifting their habitat and spatial distribution through changes in salinity, water temperature, vertical mixing rate, wind driven circulation etc. [68].

It should also be noted that the impacts of CC might alter CES differently depending on the ecosystem type and provided services. For example, changes in sea temperature may not always negatively affect the health condition of mangroves [73], since mangrove swamps can expand due to increases in temperature. Thus, it is important to note that ecosystems are not always damaged due to climate change impacts and in fact these ecosystems can extend under a warmer climate. Changes in precipitation patterns caused by CC may also positively affect the growth of mangroves and their areal extent [74,75]. However, decreases in the value of some CES may have a positive impact on other ecosystem services such as educational and research value, where damage due to CC impacts on coastal wetlands and their ecosystems might attract scientists and researchers to allocate higher academic budgets for undertaking research into these destructive impacts.

It should be noted that in quantification of CC-driven ecosystem changes, adaptation of some ecosystems to the physical changes can positively affect the services they provide. Moreover, it is very important to consider whether there is room for coastal ecosystems to migrate inland (mostly applicable for mangroves). For example, if SLR-induced inundation occurs causing inland salt intrusion, and if a landward migration is possible, then a favored habitat could be created for mangroves farther inland, resulting in inland expansion of such ecosystems. Therefore, these CC-driven changes can result in providing more mangrove services, and adding value rather than losing value.

Thirdly, the qualitative trends in CES provision need to be specified in quantitative terms. This often involves domain-specific modelling and data, for example, on fisheries or on land submersion. While research on the previous two steps is rather extensive, only a few studies have performed detailed quantitative analysis of physical impacts of CC on CES. One exception is the work of Cheung et al. [76] who estimated that ocean acidification in the North Atlantic will result in the reduction of fish growth, leading to a 20–30% decrease in harvests.

Finally, the CC driven variations in CES have to be monetized. Until recently, monetary valuation of potential losses of ecosystem services under different CC scenarios (left hand side of Figure 2) have been scarce leading to rudimental assessments of CC damages to ecosystems [63], which are part of global SCC (right hand side of Figure 2). Currently there is a fast growing body of literature with individual case studies as well as attempts for large-scale valuations. As the first estimate of the potential cost of adapting the World's fishing sector to climate change, Sumaila and Cheung [77] estimated that globally, the fishing sector may suffer from a $17 to $41 billion annual loss in landed value. Kragt et al. [78] presented another example in the Great Barrier Reef Marine Park in Australia, where reef trips by divers and snorkelers could decrease by 80% given a hypothetical decrease in coral and fish biodiversity, corresponding to a loss of AUD 103 million per year in tourism revenues.

Kuhfuss et al. [66] presented an example of a thorough valuation of CC driven changes in CES in coastal wetlands in France. Given a 1 m SLR scenario by 2100, regional coastal wetlands are expected to gain additional territory due to a retreat of agricultural and urban areas, and thus result in a value ranging between 10,790,000 to 16,188,000 of 2010 €. Fanning [79] also estimated that the annual value of CC driven impacts on ecosystem services at a coastal lagoon in Uruguay is US $178,487–290,540 and US $300,000 for regulatory (indirect use), and provisional (direct use) combined with cultural ecosystem services, respectively. Large-scale valuations have also indicated that annual damage to CES in Europe due to CC driven SLR could be about €2.9 billion by 2050 [80].

5. Concluding Remarks

This article aims to provide a coherent review on the valuation of coastal ecosystem services by systematically describing the main valuation concepts, and addressing the issue of climate change impacts on the monetary value of CES. To this end, it offers a systematic overview of the state-of-the-art and a CES assessment guideline for practitioners and researchers in interdisciplinary teams. To achieve this objective, firstly we present a summary of coastal wetlands, ecosystems and the services they provide, drivers of ecosystem change, valuation methods, and required data for performing valuation studies.

Secondly, we analyze two global scale and 30 local and regional scale case studies of coastal ecosystem valuation. This analysis has resulted in the following main observations: Valuation studies reviewed in this article consider coral reefs and mangroves as the most important coastal ecosystems, while sea-grass beds are the least investigated coastal ecosystems in terms of the monetary value provided by their services; With respect to ecosystem services provided, tourism and recreation as well as storm protection are the most common CES that have been monetized. These two services are often valued higher than other services provided by coastal ecosystems. On the contrary, cultural services such as aesthetic and artistic values have been hardly valued so far, indicating the necessity for further research regarding these services; With respect to the valuation methods, avoided damage, replacement and substitute cost methods are the most commonly used techniques for valuing storm or flood protection services. In addition, stated preference and production-based methods (net factor income) are also frequently used for valuation of many different ecosystem services; The most common limitation of the reviewed studies is the incomplete measure of ecosystem value they provided. The reason for this incomplete estimation can be associated with the complexity of considering and valuing all the services provided by ecosystems as well as data scarcity in the study areas.

Finally, this article addresses the important but poorly understood aspect of how foreshadowed climate change may affect coastal ecosystem services. Here, we present a framework that illustrates the two different ways in which CC and CES are linked. The first link represents how CC affects the climate regulation service of the coastal ecosystems. The second link, represents how CC may tangibly affect CES. Furthermore, the monetization of CC driven variations in the value of CES, which is subject to substantial uncertainty, is identified as a major challenge. To address this challenge, we propose a systematic approach involving the following steps: (1) Identification of changes in climate variables and resulting physical impacts on coastal ecosystem; (2) qualitative determination of the effects that CC-driven physical impacts identified in (1) may have on CES; (3) translation of the qualitative trends of CC driven impacts into quantitative estimates; and (4) monetization of quantitative CC impacts on CES. The fourth step represents the main contemporary knowledge gap in CES valuation studies due to uncertainties in quantifying future physical climate impacts on CES, and changes in WTP due to shifts across socio-demographic groups. Hence, quantifying potential CC driven losses or gains in CES value is an important future research direction that will ultimately enable much needed quantitative assessments of climate change associated environmental risk in coastal areas.

Acknowledgments: Authors are thankful for the financial support from the AXA Research Fund and the Ministry I&M–IHE Delft cooperation program. RR is supported by the AXA Research fund and the Deltares Strategic Research Programme 'Coastal and Offshore Engineering'.

Author Contributions: S.M. and A.D. conceived the overall approach and the main conceptual design of the article. S.M. collected the applications of valuation studies and has contributed to the analysis of the results, and wrote most parts of the article. T.F. helped with the conceptual design of the article, has contributed to the interpretation of the results, and has drafted parts of the manuscript and helped with critically revising the content of the text. E.d.R.v.S. has contributed to the development of this study and helped mostly with the ecological-related issues in this study. R.R. helped with the conceptual design of the article and provided the final and critical revision of the content of this review.

References

1. Neumann, B.; Vafeidis, A.T.; Zimmermann, J.; Nicholls, R.J. Future coastal population growth and exposure to sea-level rise and coastal flooding-a global assessment. *PLoS ONE* **2015**, *10*, e0118571. [CrossRef] [PubMed]
2. Barbier, E.B. Valuing ecosystem services for coastal wetland protection and restoration: Progress and challenges. *Resources* **2013**, *2*, 213–230. [CrossRef]
3. Assessment, M.E. *Ecosystems and Human Well-Being: Wetlands and Water Synthesis*; World Resources Institute: Washington, DC, USA, 2005; Volume 5.

4. Bierbaum, R.M.; Fay, M.; Ross-Larson, B. *World Development Report 2010: Development and Climate Change*; World Bank Group: Washington, DC, USA, 2009.

5. Ranasinghe, R. Assessing climate change impacts on open sandy coasts: A review. *Earth Sci. Rev.* **2016**, *160*, 320–332. [CrossRef]

6. Pachauri, R.K.; Allen, M.R.; Barros, V.R.; Broome, J.; Cramer, W.; Christ, R.; Church, J.A.; Clarke, L.; Dahe, Q.; Dasgupta, P. *Climate Change 2014: Synthesis Report. Contribution of Working Groups I, II and III to the Fifth Assessment Report of the Intergovernmental Panel on Climate Change*; IPCC: Geneva, Switzerland, 2014; p. 151. ISBN 978-92-9169-143-2.

7. Leemans, R.; de Groot, R.S. *Millennium Ecosystem Assessment: Ecosystems and Human Well-Being: A Framework for Assessment*; Island Press: Washington, DC, USA, 2003.

8. Braat, L.C.; de Groot, R. The ecosystem services agenda: Bridging the worlds of natural science and economics, conservation and development, and public and private policy. *Ecosyst. Serv.* **2012**, *1*, 4–15. [CrossRef]

9. Lew, D.K.; Larson, D.M. Is a fish in hand worth two in the sea? Evidence from a stated preference study. *Fish. Res.* **2014**, *157*, 124–135. [CrossRef]

10. Castaño-Isaza, J.; Newball, R.; Roach, B.; Lau, W.W. Valuing beaches to develop payment for ecosystem services schemes in colombia's seaflower marine protected area. *Ecosyst. Serv.* **2015**, *11*, 22–31. [CrossRef]

11. Emerton, L.; Kekulandala, L. *Assessment of the Economic Value of Muthurajawela Wetland*; Occasional Papers of IUCN Sri Lanka; IUCN-World Conservation uNion, Sri Lanka Country Office: Colombo, Sri Lanka, 2003; Volume 4.

12. Hussain, S.A.; Badola, R. Valuing mangrove ecosystem services: Linking nutrient retention function of mangrove forests to enhanced agroecosystem production. *Wetl. Ecol. Manag.* **2008**, *16*, 441–450. [CrossRef]

13. Vo, T.Q.; Künzer, C.; Oppelt, N. How remote sensing supports mangrove ecosystem service valuation: A case study in ca mau province, Vietnam. *Ecosyst. Serv.* **2015**, *14*, 67–75.

14. Chaikumbung, M.; Doucouliagos, H.; Scarborough, H. The economic value of wetlands in developing countries: A meta-regression analysis. *Ecol. Econ.* **2016**, *124*, 164–174. [CrossRef]

15. Rao, N.S.; Ghermandi, A.; Portela, R.; Wang, X. Global values of coastal ecosystem services: A spatial economic analysis of shoreline protection values. *Ecosyst. Serv.* **2015**, *11*, 95–105. [CrossRef]

16. Costanza, R.; d'Arge, R.; de Groot, R.; Faber, S.; Grasso, M.; Hannon, B.; Limburg, K.; Naeem, S.; O'neill, R.V.; Paruelo, J. The value of the world's ecosystem services and natural capital. *Nature* **1997**, *387*, 253–260. [CrossRef]

17. Martínez, M.; Intralawan, A.; Vázquez, G.; Pérez-Maqueo, O.; Sutton, P.; Landgrave, R. The coasts of our world: Ecological, economic and social importance. *Ecol. Econ.* **2007**, *63*, 254–272. [CrossRef]

18. Costanza, R.; de Groot, R.; Sutton, P.; van der Ploeg, S.; Anderson, S.J.; Kubiszewski, I.; Farber, S.; Turner, R.K. Changes in the global value of ecosystem services. *Glob. Environ. Chang.* **2014**, *26*, 152–158. [CrossRef]

19. Hassan, R.; Scholes, R.; Ash, N. *Ecosystems and Human Well-Being: Current State and Trends, vol 1. Findings of the Condition and Trends Working Group of the Millennium Ecosystem Assessment*; Island Press: Washington, DC, USA, 2005.

20. Barbier, E.B.; Acreman, M.; Knowler, D. *Economic Valuation of Wetlands: A Guide for Policy Makers and Planners*; Ramsar Convention Bureau: Gland, Switzerland, 1997.

21. Barbier, E.B. Valuing environmental functions: Tropical wetlands. *Land Econ.* **1994**, *70*, 155–173. [CrossRef]

22. Tinch, R.; Mathieu, L. *Marine and Coastal Ecosystem Services: Valuation Methods and Their Practical Application*; Biodiversity Series; UNEP-WCMC: Cambridge, UK, 2011.

23. Cesar, H.S. Coral reefs: Their functions, threats and economic value. In *Collected Essays on the Economics of Coral Reefs*; Cesar, H.S.J., Ed.; CORDIO, Department of Biology and Environmental Sciences, Kalmar University: Kalmar, Sweden, 2000; pp. 14–39.

24. Barbier, E.B.; Hacker, S.D.; Kennedy, C.; Koch, E.W.; Stier, A.C.; Silliman, B.R. The value of estuarine and coastal ecosystem services. *Ecol. Monogr.* **2011**, *81*, 169–193. [CrossRef]

25. Dluzewska, A. Cultural ecosystem services-framework, theories and practices. *Probl. Sustain. Dev.* **2016**, *12*, 101–110.

26. Chen, N.; Li, H.; Wang, L. A gis-based approach for mapping direct use value of ecosystem services at a county scale: Management implications. *Ecol. Econ.* **2009**, *68*, 2768–2776. [CrossRef]

27. Hein, L.; Van Koppen, K.; de Groot, R.S.; Van Ierland, E.C. Spatial scales, stakeholders and the valuation of ecosystem services. *Ecol. Econ.* **2006**, *57*, 209–228. [CrossRef]
28. Sukhdev, P.; Wittmer, H.; Miller, D. The economics of ecosystems and biodiversity (teeb): Challenges and responses. In *Nature in the Balance: The Economics of Biodiversity*; Oxford University Press: Oxford, UK, 2014; pp. 135–152.
29. Van Beukering, P.; Haider, W.; Longland, M.; Cesar, H.; Sablan, J.; Shjegstad, S.; Beardmore, B.; Liu, Y.; Garces, G.O. The economic value of guam's coral reefs. *Univ. Guam Mar. Lab. Tech. Rep.* **2007**, *116*, 102.
30. Ecosystem Valuation. Available online: http://www.ecosystemvaluation.org/ (accessed on 24 January 2017).
31. Bateman, I.J.; Mace, G.M.; Fezzi, C.; Atkinson, G.; Turner, K. Economic analysis for ecosystem service assessments. *Environ. Resour. Econ.* **2011**, *48*, 177–218. [CrossRef]
32. Russi, D.; ten Brink, P.; Farmer, A.; Badura, T.; Coates, D.; Förster, J.; Kumar, R.; Davidson, N. *The Economics of Ecosystems and Biodiversity for Water and Wetlands*; IEEP: London, UK; Brussels, Belgium, 2013.
33. Champ, P.A.; Boyle, K.J.; Brown, T.C. *The Economics of Non-Market Goods and Resources, a Primer on Nonmarket Valuation Second Edition*; Springer: Dordrecht, The Netherlands, 2017; Volume 13.
34. Richardson, L.; Loomis, J.; Kroeger, T.; Casey, F. The role of benefit transfer in ecosystem service valuation. *Ecol. Econ.* **2015**, *115*, 51–58. [CrossRef]
35. Aburto-Oropeza, O.; Ezcurra, E.; Danemann, G.; Valdez, V.; Murray, J.; Sala, E. Mangroves in the gulf of california increase fishery yields. *Proc. Natl. Acad. Sci. USA* **2008**, *105*, 10456–10459. [CrossRef] [PubMed]
36. Barbier, E.B.; Koch, E.W.; Silliman, B.R.; Hacker, S.D.; Wolanski, E.; Primavera, J.; Granek, E.F.; Polasky, S.; Aswani, S.; Cramer, L.A. Coastal ecosystem-based management with nonlinear ecological functions and values. *Science* **2008**, *319*, 321–323. [CrossRef] [PubMed]
37. Koch, E.W.; Barbier, E.B.; Silliman, B.R.; Reed, D.J.; Perillo, G.M.; Hacker, S.D.; Granek, E.F.; Primavera, J.H.; Muthiga, N.; Polasky, S. Non-linearity in ecosystem services: Temporal and spatial variability in coastal protection. *Front. Ecol. Environ.* **2009**, *7*, 29–37. [CrossRef]
38. De Groot, R.; Brander, L.; Van Der Ploeg, S.; Costanza, R.; Bernard, F.; Braat, L.; Christie, M.; Crossman, N.; Ghermandi, A.; Hein, L. Global estimates of the value of ecosystems and their services in monetary units. *Ecosyst. Serv.* **2012**, *1*, 50–61. [CrossRef]
39. Mathieu, L.F.; Langford, I.H.; Kenyon, W. Valuing marine parks in a developing country: A case study of the seychelles. *Environ. Dev. Econ.* **2003**, *8*, 373–390. [CrossRef]
40. Wilkinson, C.; Lindén, O.; Cesar, H.; Hodgson, G.; Rubens, J.; Strong, A.E. Ecological and socioeconomic impacts of 1998 coral mortality in the indian ocean: An enso impact and a warning of future change? *Ambio* **1999**, *28*, 188–196.
41. Seenprachawong, U. An economic analysis of coral reefs in the andaman sea of thailand. In *Marine and Coastal Ecosystem Valuation, Institutions, and Policy in Southeast Asia*; Springer: Singapor, 2016; pp. 31–45.
42. Schep, S.; van Beukering, P.; Brander, L.; Wolfs, E. *The Tourism Value of Nature on Bonaire Using Choice Modelling and Value Mapping*; IVM Institute for Environmental Studies: Amsterdam, The Netherlands, 2013.
43. Berg, H.; Öhman, M.C.; Troëng, S.; Lindén, O. Environmental economics of coral reef destruction in sri lanka. *Ambio* **1998**, *27*, 627–634.
44. Burke, L.; Maidens, J. *Reefs at Risk in the Caribbean*; World Resources Institute: Washington, DC, USA, 2004.
45. Van Zanten, B.; van Beukering, P. *Coastal Protection Services of Coral Reefs in Bonaire, Economic Values and Spatial Maps*; IVM Institute for Environmental Studies: Amsterdam, The Netherlands, 2012.
46. Burke, L.; Greenhalgh, S.; Prager, D.; Cooper, E. *Coastal Capital: Economic Valuation of Coral Reefs in Tobago and st. Lucia*; World Resources Institute (WRI): Washington, DC, USA, 2008.
47. Barbier, E.B. Valuing ecosystem services as productive inputs. *Econ. Policy* **2007**, *22*, 178–229. [CrossRef]
48. Das, S.; Crépin, A.-S. Mangroves can provide protection against wind damage during storms. *Estuar. Coast. Shelf Sci.* **2013**, *134*, 98–107. [CrossRef]
49. Brander, L.M.; Wagtendonk, A.J.; Hussain, S.S.; McVittie, A.; Verburg, P.H.; de Groot, R.S.; van der Ploeg, S. Ecosystem service values for mangroves in southeast asia: A meta-analysis and value transfer application. *Ecosyst. Serv.* **2012**, *1*, 62–69. [CrossRef]
50. Bell, F.W. The economic valuation of saltwater marsh supporting marine recreational fishing in the southeastern united states. *Ecol. Econ.* **1997**, *21*, 243–254. [CrossRef]

51. Molnar, M.; Kocian, M.; Batker, D. *Nearshore Natural Capital Valuation. Valuing the Aquatic Benefits of British Columbia's Lower Mainland*; A report; David Suzuki Foundation: Vancouver, BC, Canada, 2012; Earth Economics: Tacoma, WA, USA, 2012; p. 103.

52. Boero, F.; Briand, F. *Price and Value, Alternatives to Biodiversity Conservation (in the Seas)*; CIESM: Monaco, 2008.

53. Czajkowski, M.; Ahtiainen, H.; Artell, J.; Budziński, W.; Hasler, B.; Hasselström, L.; Meyerhoff, J.; Nõmmann, T.; Semeniene, D.; Söderqvist, T. Valuing the commons: An international study on the recreational benefits of the baltic sea. *J. Environ. Manag.* **2015**, *156*, 209–217. [CrossRef] [PubMed]

54. Cooper, E.; Burke, L.; Bood, N. *Coastal Capital: Belize-The Economic Contribution of Belize's Coral Reefs and Mangroves*; WRI Working Paper; World Resource Institute: Washington, DC, USA, 2009; p. 53.

55. Van Beukering, P.; Wolfs, E. *Essays on Economic Values of Nature of Bonaire. A Desk Study*; IVM Report (W12-14); Institute for Environmental Studies, VU University Amsterdam: Amsterdam, The Netherlands, 2012.

56. Gibbons, S.; Mourato, S.; Resende, G.M. The amenity value of english nature: A hedonic price approach. *Environ. Resour. Econ.* **2014**, *57*, 175–196. [CrossRef]

57. Grimm, N.B.; Groffman, P.; Staudinger, M.; Tallis, H. Climate change impacts on ecosystems and ecosystem services in the united states: Process and prospects for sustained assessment. *Clim. Chang.* **2016**, *135*, 97–109. [CrossRef]

58. Shaw, M.R.; Pendleton, L.; Cameron, D.R.; Morris, B.; Bachelet, D.; Klausmeyer, K.; MacKenzie, J.; Conklin, D.R.; Bratman, G.N.; Lenihan, J. The impact of climate change on california's ecosystem services. *Clim. Chang.* **2011**, *109*, 465–484. [CrossRef]

59. Vierros, M. Communities and blue carbon: The role of traditional management systems in providing benefits for carbon storage, biodiversity conservation and livelihoods. *Clim. Chang.* **2017**, *140*, 89–100. [CrossRef]

60. Jerath, M.; Bhat, M.; Rivera-Monroy, V.H.; Castañeda-Moya, E.; Simard, M.; Twilley, R.R. The role of economic, policy, and ecological factors in estimating the value of carbon stocks in everglades mangrove forests, south florida, USA. *Environ. Sci. Policy* **2016**, *66*, 160–169. [CrossRef]

61. Greenstone, M.; Kopits, E.; Wolverton, A. Developing a social cost of carbon for US regulatory analysis: A methodology and interpretation. *Rev. Environ. Econ. Policy* **2013**, *7*, 23–46. [CrossRef]

62. Hallegatte, S.; Mach, K.J. Make climate-change assessments more relevant. *Nature* **2016**, *534*, 613–615. [CrossRef] [PubMed]

63. Burke, M.; Craxton, M.; Kolstad, C.; Onda, C.; Allcott, H.; Baker, E.; Barrage, L.; Carson, R.; Gillingham, K.; Graff-Zivin, J. Opportunities for advances in climate change economics. *Science* **2016**, *352*, 292–293. [CrossRef] [PubMed]

64. Lane, D.; Jones, R.; Mills, D.; Wobus, C.; Ready, R.C.; Buddemeier, R.W.; English, E.; Martinich, J.; Shouse, K.; Hosterman, H. Climate change impacts on freshwater fish, coral reefs, and related ecosystem services in the united states. *Clim. Chang.* **2015**, *131*, 143–157. [CrossRef]

65. Turner, R.K.; Palmieri, M.G.; Luisetti, T. Lessons from the construction of a climate change adaptation plan: A broads wetland case study. *Integr. Environ. Assess. Manag.* **2016**, *12*, 719–725. [CrossRef] [PubMed]

66. Kuhfuss, L.; Rey-Valette, H.; Sourisseau, E.; Heurtefeux, H.; Rufray, X. Evaluating the impacts of sea level rise on coastal wetlands in languedoc-roussillon, france. *Environ. Sci. Policy* **2016**, *59*, 26–34. [CrossRef]

67. Daw, T.; Adger, W.N.; Brown, K.; Badjeck, M.-C. Climate change and capture fisheries: Potential impacts, adaptation and mitigation. Climate change implications for fisheries and aquaculture: Overview of current scientific knowledge. *FAO Fish. Aquac. Tech. Pap.* **2009**, *530*, 107–150.

68. Cochrane, K.; De Young, C.; Soto, D.; Bahri, T. Climate change implications for fisheries and aquaculture. *FAO Fish. Aquac. Tech. Pap.* **2009**, *530*, 212.

69. Sumaila, U.R.; Cheung, W.W.; Lam, V.W.; Pauly, D.; Herrick, S. Climate change impacts on the biophysics and economics of world fisheries. *Nat. Clim. Chang.* **2011**, *1*, 449–456. [CrossRef]

70. Mohanty, B.; Sharma, A.; Sahoo, J.; Mohanty, S. *Climate Change: Impacts on Fisheries and Aquaculture*; INTECH Open Access Publisher: Rijeka, Croatia, 2010.

71. MAB (Multi-Agency Brief). *Fisheries and Aquaculture in a Changing Climate*; FAO: Rome, Italy, 2009; p. 6.

72. Portner, H.O.; Knust, R. Climate change affects marine fishes through the oxygen limitation of thermal tolerance. *Science* **2007**, *315*, 95–97. [CrossRef] [PubMed]

73. McLeod, E.; Salm, R.V. *Managing Mangroves for Resilience to Climate Change*; World Conservation Union (IUCN): Gland, Switzerland, 2006.

74. Field, C.D. Impact of Expected Climate Change on Mangroves. In *Asia-Pacific Symposium on Mangrove Ecosystems*; Springer: Dordrecht, The Netherlands, 1995; pp. 75–81.

75. Snedaker, S.C. Mangroves and climate change in the florida and caribbean region: Scenarios and hypotheses. In *Asia-Pacific Symposium on Mangrove Ecosystems*; Springer: Dordrecht, The Netherlands, 1995; pp. 43–49.

76. Cheung, W.W.; Dunne, J.; Sarmiento, J.L.; Pauly, D. Integrating ecophysiology and plankton dynamics into projected maximum fisheries catch potential under climate change in the northeast atlantic. *ICES J. Mar. Sci.* **2011**, *68*, 1008–1018. [CrossRef]

77. Sumaila, U.R.; Cheung, W.W. *Cost of Adapting Fisheries to Climate Change*; World Bank Discussion Paper; World Bank: Washington, DC, USA, 2010.

78. Kragt, M.E.; Roebeling, P.C.; Ruijs, A. Effects of great barrier reef degradation on recreational reef-trip demand: A contingent behaviour approach. *Aust. J. Agric. Resour. Econ.* **2009**, *53*, 213–229. [CrossRef]

79. Fanning, A.L. Towards valuing climate change impacts on the ecosystem services of a uruguayan coastal lagoon. In *International Perspectives on Climate Change*; Springer: Cham, Switzerland, 2014; pp. 61–77.

80. Roebeling, P.; Costa, L.; Magalhães-Filho, L.; Tekken, V. Ecosystem service value losses from coastal erosion in europe: Historical trends and future projections. *J. Coast. Conserv.* **2013**, *17*, 389–395. [CrossRef]

Journal of
Marine Science and Engineering

MDPI

Article

Probabilistic Assessment of Overtopping of Sea Dikes with Foreshores including Infragravity Waves and Morphological Changes: Westkapelle Case Study

Patrick Oosterlo [1,*], Robert Timothy McCall [2], Vincent Vuik [1,3], Bas Hofland [1,2], Jentsje Wouter van der Meer [4,5] and Sebastiaan Nicolaas Jonkman [1]

[1]   Faculty of Civil Engineering and Geosciences, Delft University of Technology, P.O. Box 5048, 2600 GA Delft, The Netherlands; V.Vuik@tudelft.nl (V.V.); B.Hofland@tudelft.nl (B.H.); S.N.Jonkman@tudelft.nl (S.N.J.)
[2]   Deltares, Boussinesqweg 1, 2629 HV Delft, The Netherlands; Robert.McCall@deltares.nl
[3]   HKV Consultants, P.O. Box 2120, 8203 AC Lelystad, The Netherlands
[4]   Van der Meer Consulting BV, P.O. Box 11, 8490 AA Akkrum, The Netherlands; jm@vandermeerconsulting.nl
[5]   IHE Delft, P.O. Box 3015, 2601 DA Delft, The Netherlands
*   Correspondence: P.Oosterlo@tudelft.nl

Received: 19 March 2018; Accepted: 23 April 2018; Published: 1 May 2018

Abstract: Shallow foreshores in front of coastal dikes can reduce the probability of dike failure due to wave overtopping. A probabilistic model framework is presented, which is capable of including complex hydrodynamics like infragravity waves, and morphological changes of a sandy foreshore during severe storms in the calculations of the probability of dike failure due to wave overtopping. The method is applied to a test case based on the Westkapelle sea defence in The Netherlands, a hybrid defence consisting of a dike with a sandy foreshore. The model framework consists of the process-based hydrological and morphological model XBeach, probabilistic overtopping equations (EurOtop) and the level III fully probabilistic method ADIS. By using the fully probabilistic level III method ADIS, the number of simulations necessary is greatly reduced, which allows for the use of more advanced and detailed hydro- and morphodynamic models. The framework is able to compute the probability of failure with up to 15 stochastic variables and is able to describe feasible physical processes. Furthermore, the framework is completely modular, which means that any model or equation can be plugged into the framework, whenever updated models with improved representation of the physics or increases in computational power become available. The model framework as described in this paper, includes more physical processes and stochastic variables in the determination of the probability of dike failure due to wave overtopping, compared to the currently used methods in The Netherlands. For the here considered case, the complex hydrodynamics like infragravity waves and wave set-up need to be included in the calculations, because they appeared to have a large influence on the probability of failure. Morphological changes of the foreshore during a severe storm appeared to have less influence on the probability of failure for this case. It is recommended to apply the framework to other cases as well, to determine if the effects of complex hydrodynamics as infragravity waves and morphological changes on the probability of sea dike failure due to wave overtopping as found in this paper hold for other cases as well. Furthermore, it is recommended to investigate broader use of the method, e.g., for safety assessment, reliability analysis and design.

Keywords: probabilistic modelling; foreshore; infragravity waves; wave overtopping; morphological changes; safety assessment

1. Introduction

In recent years, Building with Nature solutions (soft solutions, such as sand nourishments) are increasingly considered as an alternative in flood protection, see e.g., Van Slobbe et al. [1]. These soft solutions are considered as a supplement to the usual hard measures (dikes, coastal structures). Three types of soft defences can be distinguished [2]. Here, one of these types is considered: a hybrid defence, more specifically a dike with a shallow sandy foreshore. This dike-foreshore system is a combination of a traditional dike and a nourished foreshore, consisting of a body of sand, possibly covered with vegetation. The foreshore reduces the loads on the dike. The dike remains part of the defence. A hybrid defence really is a combination in which both parts are of importance in the flood protection. Constructing a sandy foreshore in front of a dike can be a method to decrease the failure probability. However, the uncertainty in the morphological development of these foreshores leads to uncertainty with respect to their contribution in protection against flooding. The morphological stability of a foreshore during extreme conditions is not well-known. Furthermore, the influence of infragravity waves generated on the foreshore on the wave overtopping and failure probability is largely unknown. Increasing interest in soft solutions of dike managers, as well as local inhabitants (getting a nice beach), makes it interesting to investigate the contribution of foreshores to flood protection further.

The risk management and optimization of coastal structures have gained much attention the last years. In order to optimize a design, the uncertainties need to be taken into account as detailed as possible, by using probabilistic approaches. However, currently these approaches are often deterministic or semi-probabilistic. For example, the currently used methods in the Netherlands to determine the probability of dike failure due to wave overtopping only include a few stochastic variables and use a model to determine the wave hydrodynamics that does not include infragravity waves or morphological changes. By using a full-probabilistic approach, the risks could potentially be further reduced and the designs further optimized, see e.g., Wainwright et al. [3].

During a storm, several processes occur at a dike-foreshore system. Wave energy is dissipated on the foreshore by wave breaking, which causes a shift in the wave spectrum from the peak frequency to the lower frequencies due to the generation of infragravity waves, see e.g., Symonds et al.; Longuet-Higgins & Stewart [4,5]. The breaking waves also lead to erosion of the foreshore. Finally, if water levels and wave heights are large enough, wave overtopping may occur over the dike. Hence, the generation of infragravity waves, morphological changes of the foreshore and wave overtopping are important aspects in the design of dike-foreshore systems.

Infragravity waves are generated by groups of wind induced waves, and are sometimes called long waves or surfbeat in the surf zone. Infragravity waves are known to be important on dissipative beaches and for run-up [5]. Typical infragravity wave periods range from 25–30 s to 250–300 s [6]. As previous studies on dike-foreshore systems [4,7] indicated that wave energy dissipation causes a shift in the wave spectrum from the peak frequency to the lower frequencies (due to generation of infragravity waves), instead of the wave peak period, the $T_{m-1,0}$ wave period was recommended as a measure in these situations to calculate the wave run-up or wave overtopping [8–10]. This period is defined as $T_{m-1,0} = m_{-1}/m_0$, with m_k being the kth moment of the wave spectrum. The period is commonly used in coastal structure design [11], and the use of the $T_{m-1,0}$ wave period is incorporated in manuals like the EurOtop manual [12]. Furthermore, it was found, that on high foreshores where intense breaking occurs, the wave set-up becomes important for the wave run-up and overtopping as well [13].

Roeber & Bricker [14] showed that during Typhoon Haiyan, at the town of Hernani, the Philippines, an infragravity wave steepened into a tsunami-like wave that caused excessive damage and casualties. The fringing reef present near the town protected the community from moderate storms, but increased flooding during this extreme event. Typical for these reef environments is a very short wave-breaking zone over a steep reef face, which releases the bound infragravity waves with little energy loss. This indicates that it can be important to include the infragravity waves in the design of dike-foreshore systems.

For this paper, the failure mechanism wave overtopping was chosen, because it is one of the most important failure modes of dike-foreshore systems. The overtopping discharge depends on the wave conditions at the toe, slope geometry, and the existing freeboard. The most commonly used equations to describe wave overtopping are the ones from EurOtop [12].

Suzuki et al. [7] concluded that beach nourishments do not always reduce the wave overtopping discharge, due to set-up and generation of bores and infragravity waves. Van Gent & Giarrusso [15] found that for foreshores characterized by a bar, trough and terrace in front of a dike, the contribution of low frequency energy to the total wave energy could reach 30%. It has to be noted that most of the previous studies did not include the wave directional spreading, because they were based on physical flume model tests. However, the wave directional spreading can have a large influence on the generation of infragravity waves and the wave period [11].

Bakker & Vrijling; Madsen et al. [16,17] give extensive descriptions of many of the different available probabilistic calculation methods. The Joint Committee on Structural Safety [18] proposed a level-classification of probabilistic calculation methods for the reliability of an element (now ISO 2394:2015). Here, level I methods do not calculate failure probabilities, but use partial safety factors. Level II methods linearize the limit state function in the design point and approximate the probability distribution of each variable by a standard normal distribution. Level III methods or fully probabilistic methods calculate the probability of failure by considering the probability density functions of all variables.

Previous research considering probabilistic methods was often focused on different processes and failure mechanisms than wave overtopping (e.g., [19–22]). Furthermore, the previous research mostly considered only a few stochastic variables, with simple 1D empirical or numerical models and probabilistic methods like Monte Carlo Simulation, which require a large number of simulations (e.g., [19–23]).

Overall, probabilistic methods have been applied to coastal dikes and dunes, but not yet to the issues studied in this paper. Hence, a method which includes the influence of complex hydrodynamics such as infragravity waves, and morphological changes of a sandy foreshore on the probability of dike failure due to wave overtopping does not yet exist. Therefore, the main goal of this paper is to develop a framework to calculate the probability of sea dike failure due to wave overtopping, taking into account the morphological changes of the foreshore and complex hydrodynamics like infragravity waves. Furthermore, the goal is to determine the influence of using a process-based model with complex hydrodynamics versus using a calculation rule, and to determine the influence of morphological changes of the foreshore on the probability of failure due to wave overtopping for the here considered case.

First, the case study location is described (Section 2.1), then the choice and validation of the model framework are discussed (Section 2.2). Next, the input of the probabilistic calculations is briefly described (Section 2.3). After that, the results are given (Section 3), followed by a discussion (Section 4). The paper ends with several conclusions (Section 5).

2. Materials and Methods

2.1. Case Study Description

The configuration that is considered in this paper is roughly based on the hybrid sea defence at Westkapelle, on the coast of Walcheren in The Netherlands, see Figure 1. The measured bathymetrical transect 1832 from 2009 is shown in Figure 2. The average crest height of this dike is 12.6 m+NAP (Normaal Amsterdams Peil, Dutch ordnance level). This means that it is a very large dike, as extreme surge levels considered are around 5 m+NAP [24], and therefore the crest freeboard under design conditions is still more than 7 m. The average slope is around 1:8, and the revetment of the outer slope consists of asphalt above a berm and rubble stone penetrated with mastic asphalt below this berm. The berm is located around 7.5 m+NAP. In 2002, research indicated that (e.g., due to climate change) the design conditions should be increased [25]. The expected wave overtopping discharge during the new design conditions exceeded the acceptable level of 0.1 $ls^{-1}m^{-1}$ (return period of 1:4000 years). As the landward slope of the dike consists of bare sand, a larger discharge than 0.1 $ls^{-1}m^{-1}$ could not be accepted [12]. In 2008, it was decided to strengthen the dike with a foreshore with a volume of 2.5 million cubic meter of sand in front of the dike. The initial height and width of the foreshore were approximately 4 m+NAP and 75 m respectively, with an average foreshore slope of approximately 1:17. During normal conditions, this gives a dry beach in front of the dike.

Figure 1. The Netherlands, with the coast of Walcheren indicated (**left**). The Westkapelle sea defence, with the bathymetrical transects indicated (**right**). Transect 1832 is shown in red.

The profile used in the calculations is loosely based on the 2009 bathymetric survey transect 1832 at the Westkapelle sea defence, see Figure 2. Instead of the local crest height at transect 1832, the mean crest height of the Westkapelle sea defence is used, being 12.6 m+NAP. Furthermore, the bathymetry is simplified to two line sections based on the average slope (a horizontal beach and a foreshore slope), to be able to easily compare a situation without and with a foreshore. This simplification of the profile means that the volume of sand might not be completely equal to the actual foreshore at Westkapelle. A schematization of the Westkapelle sea defence is justified for this study, because the goal of the paper is to determine a framework which can include the influence of the complex hydrodynamics and cross-shore morphological changes during severe storms on the probability of failure.

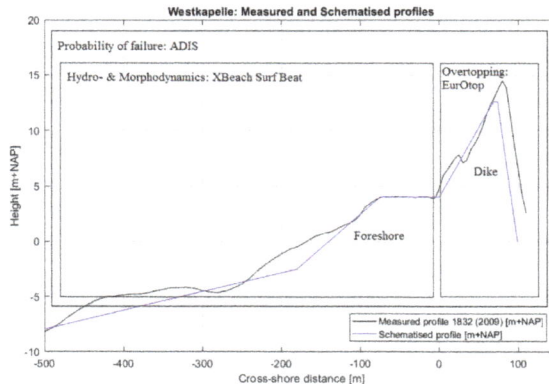

Figure 2. Measured 2009 bathymetric transect 1832 of the Westkapelle dike-foreshore system (black) and the profile used for the calculations (blue), with the models that were used in the model framework and the processes which they model. Note that only the section from −500 m to 150 m is shown.

2.2. Set-Up and Validation of Model Framework

The set-up of the framework is explained in more detail in this subsection. Because a single model that includes all the different relevant processes does not exist, a model framework is developed, in which different models are combined. In this study we use a combination of XBeach in surfbeat mode, to cope with the morphodynamic and hydraulic behavior, the EurOtop formulae on wave overtopping, and the probabilistic method Adaptive Directional Importance Sampling (ADIS). It combines accuracy in the modelling of processes with computational efficiency. The morphological changes that are considered here are the dominant short term storm morphological changes. The present paper builds on the work of Den Bieman et al. [21], who first combined ADIS and XBeach, to perform a dune erosion safety assessment.

2.2.1. Hydro- and Morphodynamics: XBeach and $h/2$-Assumption

XBeach is chosen to model the hydrodynamics and morphological changes. In surfbeat mode, XBeach can include the short wave height transformation, the wave set-up, the infragravity wave transformation and the morphological changes. It is mainly chosen for its inclusion of the infragravity waves and morphological changes, and its proven skill. XBeach is a nearshore numerical model used to assess the coastal response during storm conditions, and has been extensively calibrated and validated [26,27] (also refer to www.xbeach.org). XBeach is a coupled phase-averaged spectral wave model for sea swell and non-linear shallow water model for infragravity waves. It combines stochastic and deterministic approaches. XBeach solves the phase-averaged wave action equation. XBeach uses a form of the dissipation formula for wave breaking according to Roelvink [28] and a roller energy balance, which is coupled to the wave action balance, by Reniers et al. [29]. For the low frequency and mean flows, the shallow water equations are used. XBeach models sediment transport by a depth-averaged advection diffusion equation by Galappatti & Vreugdenhil [30]. The equilibrium sediment concentration is calculated by the sediment transport formula of Soulsby & Van Rijn [31].

A commonly used rule of thumb for gentle foreshore slopes is the simple $h/2$-assumption, where the wave height H_{m0} is half the water depth h. The assumption does not include the complex hydrodynamics such as infragravity waves or wave set-up, nor does it include morphological changes. Also, changes in wave period are not calculated by the model. The $h/2$-assumption was compared with XBeach 1D without the infragravity wave forcing for several situations. The models showed good agreement. The $h/2$-assumption calculation rule is used in this paper, to compare to the results of XBeach surfbeat, which does include both the complex hydrodynamics like infragravity waves, and morphological changes of the foreshore.

Even though XBeach has already been extensively validated, an additional validation was performed specifically for the case considered here, in Oosterlo [32]. The XBeach hydrodynamics were validated for this study by using the flume experiments of Van Gent [9] on which the EurOtop wave overtopping formula for $\xi_{m-1,0} > 7$ was based. XBeach showed good skill in predicting the H_{m0} and $T_{m-1,0}$. For a more detailed description of the validation, refer to Oosterlo [32].

In the modelling framework, the incoming waves are separated from the total wave signal as computed by XBeach by the method of Guza et al. [33]. For each hour of storm data, during which the boundary conditions are regarded as being constant, the wave spectra are calculated. From these spectra the wave height (H_{m0}) and wave period ($T_{m-1,0}$) at the dike toe are determined, which are used as input for the overtopping calculations.

2.2.2. Wave Overtopping: EurOtop

Because XBeach cannot be used to calculate small overtopping discharges as the ones considered in this paper [34], the EurOtop (first edition) formulae [35] are chosen to model the wave overtopping discharge, because they are the most commonly and extensively used wave overtopping formulae, and because they form a set of equations that can be used for all types of wave breaking conditions and slopes. The EurOtop [35] equations were based on a large set of measured data from different model tests in small and large scale as well as in wave flumes and wave basins. Furthermore, the EurOtop [35] formulae are the only formulae that include the infragravity waves in any way, by using the $T_{m-1,0}$ wave period. In the recent 2016 second edition [12], some equations have been updated. For sloping structures, where the freeboard is at least half the wave height, the differences between the first edition EurOtop [35] formulae and the new ones is quite small. For that reason one may also continue to use the first edition EurOtop [35] formulae in such a situation [12]. Because the previous is always the case for the situations considered in this paper, the equations of the first edition of EurOtop [35] are still used here. A brief description of the used EurOtop [35] equations is given in Appendix A.

In the modelling framework, the resulting wave parameters at the toe of the dike are entered into the EurOtop [35] equations, together with the freeboard, main wave direction, slope of the dike (average slope = 1:8) and roughness of the slope (for asphalt γ_f = 1.0 [12]), with which the wave overtopping discharge is calculated for each hour of the storm. The wave overtopping discharges during the storm are then compared and the largest one is used further in the ADIS calculation of the limit state function (see next subsection).

2.2.3. Probabilistic Calculation Method: Adaptive Directional Importance Sampling

Using limit states, it is possible to define a reliability function or limit state function. The general form of a limit state function is, e.g., Madsen et al. [17]):

$$Z = R - S \tag{1}$$

in which R indicates the strength parameters and S the load parameters. In the limit state, the limit state function Z is equal to 0. Failure occurs if $Z \leq 0$. The goal of the probabilistic calculations is to find the probability of failure and the corresponding design point. The design point is the point on the limit state function ($Z = 0$) with the largest probability density [17].

Adaptive Directional Importance Sampling (ADIS), a level III or fully probabilistic method, is chosen as the probabilistic calculation method, because ADIS can be used for small probabilities of failure, medium numbers of random variables, non-normal distributed variables, non-linear limit state functions, discontinuously shaped failure spaces, statistically dependent and independent stochastic variables and limit state functions, and multiple limit state functions [36]. But, most importantly, ADIS greatly reduces the number of calculations necessary, with respect to crude Monte Carlo Simulation. This means that more stochastic variables, more complex limit state functions and more detailed (hydrodynamic and morphological) models can be included, which include more physical processes, and can handle more complex cross-shore profiles and storm situations. Den Bieman et al. [21] used ADIS in combination with XBeach to perform a dune erosion safety assessment. The ADIS implementation in the OpenEarth toolbox [37] is used here.

ADIS is based on directional sampling, where directions are randomly sampled from standard normal space. With a chosen random direction, a line search algorithm is used, which searches the limit state $Z = 0$ in that direction, see Figure 3. When the limit state is reached or when the limit state cannot be found, a next random direction is chosen.

ADIS uses an adaptive response surface (ARS), a (simpler) function which approximates the limit state function and reduces the computational effort required. The response surface is useful, because it has a simple analytical solution, as opposed to the computationally more demanding XBeach computation. ADIS uses a second order polynomial as response surface. The second order polynomial is able to mimic the actual limit state function close enough in all but the most strongly non-linear of limit states [21]. If the root-mean-squared error between the adaptive response surface and the limit state function is small enough, the response surface is accepted and used instead of the limit state function, which reduces the computation time. The adaptive response surface is updated with information that comes available from the XBeach evaluations.

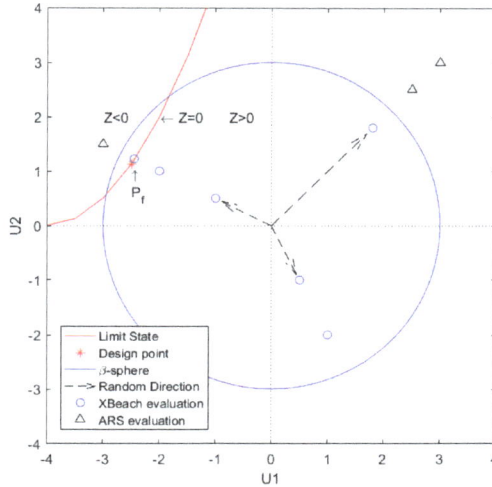

Figure 3. A schematized and arbitrary example of the ADIS method, with two stochastic variables (U1 and U2). The standard normal space is given, together with the limit state (red line), the design point (red asterisk), β-sphere (blue circle), random directions (dashed arrows) and XBeach and ARS evaluations (blue circles and black triangles). The areas where $Z < 0$ and $Z > 0$ and the limit state $Z = 0$ are indicated as well (after Den Bieman et al.; Grooteman [21,36]).

The importance sampling part of ADIS is expressed in the use of the β-sphere, see also Figure 3 (after Den Bieman et al.; Grooteman [21,36]). The reliability index β (= $\Phi^{-1}(1 - P_f)$, with Φ indicating a standard normal distribution) is a commonly used probabilistic measure of safety, in addition to the probability of failure P_f. The value of β indicates the shortest distance from the origin to the failure space. The β-sphere limits the area of the sampling domain for which XBeach calculations are performed, based on a certain β-value. The β-sphere is used when an adaptive response surface with a good fit has been found. In that case, the adaptive response surface is used instead of the limit state function, but still a small number of XBeach (limit state function) evaluations will be made to maintain precision. These XBeach evaluations are performed for the most important points, i.e., the points inside the β-sphere, the points that have a small β-value (and a large contribution to the probability of failure). Hence, the β-sphere defines the area that is excluded from the sampling domain, which reduces the number of exact XBeach limit state simulations needed. Thus, the response surface is applied in the areas outside of the β-sphere, and inside the sphere, the real limit state function using XBeach calculations is used. For a more extensive description of ADIS, refer to Grooteman [36].

The performance of Monte Carlo Simulation (MCS), FORM and ADIS was compared for the case considered here in Oosterlo [32]. The results showed the same patterns as the comparison and validation that was done in Grooteman [36]: MCS requires a very large number of limit state function evaluations, the number of evaluations needed for FORM and ADIS are of the same order of magnitude, with FORM requiring somewhat more evaluations. For a more detailed description of the comparison, refer to Oosterlo [32].

Within ADIS, an accuracy of 95% and a confidence level of 20% are set, which means that the actual probability of failure lies with 95% certainty within 20% difference of the calculated probability of failure. The default ADIS response surface fitting method is used, together with the default margins of the β-sphere (see Grooteman [36]). The minimum number of sampled directions is set to 50, the maximum is set to 1000. In general, the limit state function that is used for wave overtopping is defined as $Z = q_{crit} - q$, with q_{crit} a critical wave overtopping discharge depending on the type of layer and quality of the layer on the landward slope of the dike. For Westkapelle, this limit is $0.1 \, \mathrm{ls^{-1}m^{-1}}$, which was taken as a constant in the calculations. The probabilistic methods showed to have difficulties with fitting to this function. Therefore, the limit state function that is used for this paper is changed into $Z = \log(q_{crit}) - \log(q)$, which has a much smoother behavior. This is consistent with the logarithmic dependence of q on the hydraulic parameters as indicated in the EurOtop [12] equations. Choosing this different limit state function showed to have no influence on the resulting probabilities of failure. The complete set-up of the model framework is schematically presented in Figure 4.

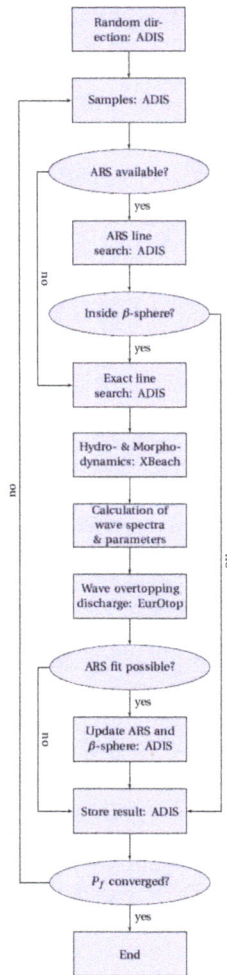

Figure 4. Model framework, using ADIS, XBeach and EurOtop to calculate the probability of dike failure due to wave overtopping.

2.3. Input of Probabilistic Calculations

This subsection describes the input, e.g., water levels and waves, that was used for the probabilistic calculations. The XBeach surfbeat models that were used for the calculations include the long waves, the short waves on the wave group scale, flow, sediment transport and morphological changes. The Kingsday MPI version of XBeach (see www.xbeach.org) was used in combination with cyclic boundary conditions on the lateral boundaries (see e.g., Roelvink et al. [38]). A 1D-model was used, with a rectilinear grid with variable grid cell sizes, ranging from 40 m (offshore) to 1 m (near the coast). The bathymetrical profile with an indication of the stochastic elements is given in Figure 5. The bathymetrical profile runs to approximately −10 m+NAP. The profiles were extended to a depth of −20 m+NAP with a slope of 1:50 to remove boundary effects.

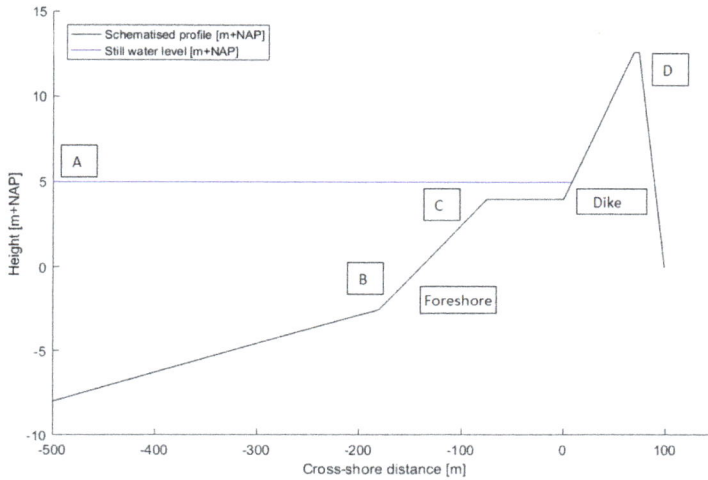

Figure 5. The cross-shore profile, with the locations where the stochastic variables of Table 1 act on, indicated by their tags. Note that only the section between −500 m and 100 m is shown.

Where possible, the default XBeach settings were used, with the exception of the stochastic model parameters as discussed below and as given in Table 1. For a more detailed description of the XBeach models and settings that were used, refer to Oosterlo [32].

For this paper, 15 stochastic variables were chosen, as given in Table 1. The names of the parameters are given in the first column, the second column gives the symbols. A distinction is made between three categories of uncertainties according to Van Gelder [39]. The third column gives the distribution type and parameters of the distribution, where $N(\mu;\sigma)$ indicates normal distributed, $L(\lambda;\zeta)$ lognormal distributed, and $W(\omega;\rho;\alpha;\sigma)$ indicates a Weibull distribution, and $U(a;b)$ a uniform distribution. The references to the parameters and distributions are given in column four, the fifth column gives a tag to each parameter, which indicates at which location the parameter acts (see Figure 5).

Table 1. Stochastic variables as used in the probabilistic calculations and their accompanying probability distributions.

Description	Parameter	Distribution	Reference	Location
Model uncertainty				
XBeach wave breaker index [-]	γ	$N(0.54;0.054)$	[28]	C
XBeach calibration factor wave skewness in sediment transport [-]	$facSk$	$L(-1.024;0.29)$	[40]	C
XBeach calibration factor wave asymmetry in sediment transport [-]	$facAs$	$L(-2.14;0.29)$	[40]	C

Table 1. *Cont.*

Description	Parameter	Distribution	Reference	Location
Model uncertainty				
EurOtop coefficient $\xi_{m-1,0} < 5$ [-]	C1	$N(4.75;0.5)$	[35]	D
EurOtop coefficient $\xi_{m-1,0} < 5$ [-]	C2	$N(2.6;0.35)$	[35]	D
EurOtop coefficient $\xi_{m-1,0} > 7$ [-]	C3	$N(-0.92;0.24)$	[35]	D
EurOtop coefficient zero freeboard [-]	C4	$N(1;0.14)$	[35]	D
Inherent uncertainty: Time				
Storm duration [s]	t_{storm}	$L(3.94;0.34)$	[41]	A
Maximum water level [m]	$z_{s,max}$	$W(\omega,\rho,\alpha,\sigma)$	[42]	A
Wave height [m]	H_{m0}	$N(0;0.6)$, related to z_s (see main text)	[42]	A
Wave period [s]	T_p	$N(0;1)$, related to H_{m0} (see main text)	[42]	A
Wave directional spreading [-]	s	$U(1.5;10)$		A
Location tidal peak relative to surge peak [s]	φ	$U(-t_{tide}/2;t_{tide}/2)$		A
Inherent uncertainty: Space				
Sediment size [m]	D_{50}	$L(-8.12;0.15)$	[43]	B
Manning friction factor [$sm^{-1/3}$]	n	$U(0.01;0.03)$	[44]	B

2.3.1. Boundary Conditions

The probability distributions that are used in the safety assessments for the Dutch dune coast (Dutch Hydraulic Boundary Conditions 2006 for dunes, Rijkswaterstaat; Steetzel et al. [24,42]) are used in this paper for the water level, wave height and wave period (see also Den Heijer et al. [45]). These omnidirectional statistics are valid for the -20 m+NAP depth contour. In these statistics, the water level, significant wave height and wave peak period are correlated, as is further explained below. The parameters of the distributions are given for several measurement stations along the Dutch coast in Steetzel et al. [42]. To determine the parameter values at an arbitrary location, interpolation is used between the two closest measurement stations and the location of interest. For Westkapelle, these stations are Vlissingen and Hoek van Holland. An example of the resulting hydraulic boundary conditions with a return period of 1000 year^{-1} is presented in Figure 6.

For the maximum water level reached during a storm h, the conditional Weibull distribution of Steetzel et al. [42] was used. The wave height H_{m0} is correlated with the water level, as for a given surge level, different wave heights can occur, due to the effects of wind speed, wind direction and wind duration. It was found that this correlation could be approximated by a normal distribution with a mean of 0 and a standard deviation of 0.6 m [46]. The relation between surge level and wave height was taken from Steetzel et al.; Weerts & Diermanse [42,46] and uses several statistical parameters as given in Table 2. The wave period T_p is correlated with the wave height. A normal distribution with a mean of 0 and a standard deviation of 1.0 s was used for the relation between wave peak period and significant wave height, according to Stijnen et al. [47]. The statistical parameters of the conditional Weibull distribution for the water level, and the equations describing the relation between water level and wave height, and wave height and wave period for Westkapelle are given in Table 2. For an overview of the equations and the statistical parameters at all the measurement stations, refer to Steetzel et al. [42].

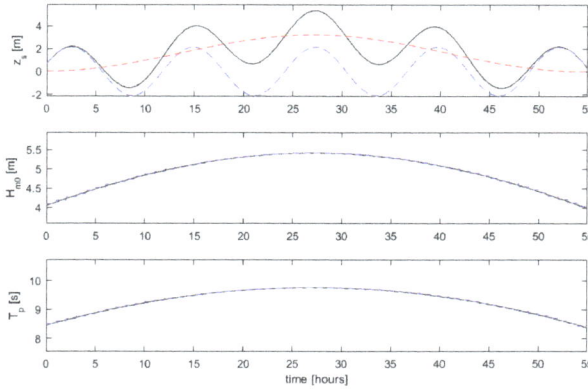

Figure 6. Hydraulic boundary conditions during a storm. The top plot shows the total water level (solid black line), composed of the tide (dashed blue line) and surge (dashed red line). The middle plot shows the significant wave height (solid black line) and this wave height divided into one-hour sections (dashed blue line). The bottom plot shows the same for the wave peak period. Here, the tidal peak coincides with the surge peak. Storm shape based on Steetzel et al. [42].

Table 2. Statistical parameters for water level, wave height and wave period during a storm [42]. α is a shape parameter [-], ω is a threshold above which the conditional Weibull function is valid [m], σ is a scale parameter [m] and ρ is the frequency of exceedance of the threshold level ω [year^{-1}]. a [m], b [-], c [-], d [m] and e [-] are statistical parameters. α [s] and β [sm^{-1}] are statistical parameters as well. All parameters taken from Steetzel et al. [42].

Water Level				
Station	ω [m]	ρ [year^{-1}]	α [-]	σ [m]
Vlissingen	2.97	3.91	1.04	0.280
Hoek van Holland	1.95	7.24	0.570	0.0158

Wave Height					
Station	a [m]	b [-]	c [-]	d [m]	e [-]
Vlissingen	2.40	0.35	0.0008	7	4.67
Hoek van Holland	4.35	0.60	0.0008	7	4.67

Wave Period		
Station	α [s]	β [sm^{-1}]
Vlissingen	3.86	1.09
Hoek van Holland	3.86	1.09

The storm duration t_{storm} is given as a lognormal distribution to prevent negative values from occurring. The surge profile is composed of a tidal effect and a surge effect. For this paper, an adjusted version of the formula from Steetzel [48] was made, with a random phase of the tide at the moment of the maximum surge, see also Figure 6:

$$z_s(t) = h_0 + \hat{h}_a \cos\left(\frac{2\pi\left(t - \frac{t_{storm}}{2}\right)}{t_{tide}} + \phi \right) + \hat{h}_s \cos\left(\frac{\pi\left(t - \frac{t_{storm}}{2}\right)}{t_{storm}} \right)^2 \tag{2}$$

where h_0 is the mean water level [m] (here set to zero), h_a is the tidal amplitude [m], t the time [s], t_{tide} the duration of the tide [s], φ the time shift in the occurrence of the maximum tidal water level opposed to the maximum surge level [s], h_s the surge amplitude [m] and t_{storm} the storm duration [s].

The tidal signal at Westkapelle was schematized as an M2-tide (principal lunar semi-diurnal), with the same amplitude as the in reality occurring maximum spring tide water level, 2.15 m+NAP, and a cycle duration of 12.42 h. No spring-neap tidal cycle was included in the calculations. t_{peak} was set to half the storm duration t_{storm}. Thus, the maximum surge level coincides with a random phase of the tide level, because of the uniform distribution of φ. The variation of the significant wave height and wave peak period during the storm was modelled by the equations of Steetzel [48], see also Figure 6 for the profile.

Data for several measured storms at the measurement station Europlatform (from Rijkswaterstaat, www.waterinfo.rws.nl) were analyzed to determine the probability distribution of the wave directional spreading. A uniform distribution with parameters 1.5 and 10 was chosen for the directional spreading parameter s [-] (for the definition of the parameter, refer to Roelvink et al.; Longuet-Higgins et al. [44,49]), based on the minimum and maximum values during the measured storms.

A normal distribution was chosen for the wave breaker index γ, based on Roelvink [28]. For the XBeach sediment transport calibration parameters $facAs$ and $facSk$ (refer to the XBeach manual [44] for the description), the lognormal distributions according to Van Thiel de Vries [40] were taken. The parameters representing the inherent uncertainty in space were the sediment size D_{50} and Manning bed friction factor n, modelled with a lognormal and uniform distribution, respectively (according to Vrouwenvelder et al.; Roelvink et al. [43,44]).

A sensitivity analysis was performed to determine the importance of the 15 stochastic variables that were chosen. For a detailed description of the sensitivity analysis, refer to Oosterlo [32]. With the results of the sensitivity analysis, the 15 parameters were divided into subsets, where set 1 consists of the parameters with the largest influence on the wave overtopping, and set 4 of the parameters with the smallest influence:

- Set 1, Offshore conditions, 3 stochastic variables: z_s, H_{m0}, T_p.
- Set 2, Model and wave parameters, 7 stochastic variables: $C1$, $C2$, $C3$, $C4$, γ, n, s.
- Set 3, Storm parameters, 2 stochastic variables: t_{storm}, φ.
- Set 4, Morphological parameters, 3 stochastic variables: D_{50}, $FacAs$, $FacSk$.

3. Results

This section gives a description of the probabilistic (ADIS) calculations that were performed with the framework and their results. The influence of the complex hydrodynamics, model and wave parameters, time of the peak of the storm, morphological changes and probabilistic method are described in the sub-sections below.

Table 3 shows an overview of the different probabilistic (ADIS) calculations that were performed with the framework and their results. The table gives the type of hydrodynamic (and morphological) model that was used (column 2), which physical processes were included (columns 4–8), and the resulting number of stochastic variables that were applied (column 3,5–8). Moreover, the resulting probability of failure P_f and the reliability index β are given (columns 9,10), as well as the number of simulations necessary and the calculation duration in days (columns 11,12).

Table 3. Overview of the different probabilistic (ADIS) calculations that were performed with the framework and their results.

Calc. no. [-]	Type of Model	No. of Stoch. Vars.	Complex Hydrodyn.	Set 1, Offsh. Cond.	Set 2, Model Params	Set 3, Tide, Storm	Set 4, Morph. Changes	P_f [year^{-1}]	β [-]	No. of Simulations [-]	Calc. Duration [days]
1	$h/2$	3		X				2.3×10^{-8}	5.47	23	0.004
2	XB 1D	3	X	X				1.6×10^{-5}	4.15	112	1
3	XB 1D	10	X	X	X			2.9×10^{-4}	3.45	341	1
4	XB 1D	12	X	X	X	X		1.2×10^{-4}	3.67	391	5
5	XB 1D	15	X	X	X	X	X	5.3×10^{-4}	3.27	484	7
6	XB 2D	3	X	X				1.6×10^{-6}	4.66	43	1

To determine the influence of using a process-based model with complex hydrodynamics compared to a calculation rule, calculations 1 and 2 were performed and will be compared. To determine the influence of including the model and wave parameters as stochastic variables, calculations 2 and 3 will be compared. To determine the influence of including the storm duration and time of the peak of the storm, calculations 3 and 4 will be compared. The influence of morphological changes of the foreshore on the probability of failure due to wave overtopping is determined by comparing calculation 4 and 5.

To determine the influence of the wave directional spreading (short-crested waves), an extra calculation was performed with XBeach 2D and the three offshore conditions as stochastic variables (calculation 6), which is compared to calculation 2.

To determine the added value of using the advanced and fully probabilistic method ADIS as well, the calculated probabilities of failure by the framework were compared to an approach where the same probability exceedance value for each of the stochastic parameters is used (not in table), see Section 3.6.

3.1. Influence of the Complex Hydrodynamics

The results of calculations 1 and 2 are shown in Figure 7, comparing the calculation rule $h/2$-assumption (solid lines) with the process-based model XBeach (dashed lines). The figure shows the results of calculations at the (closest to the) design point. The top plot shows the development of the significant wave height for both situations (blue lines), as well as the low-frequency part of the wave height (green) and wave set-up (black) for XBeach. The bottom plot shows the water levels (blue) and the cross-shore profile (green). The title of the figure gives the conditions offshore and at the toe of the dike for both situations. Note that not the whole profile is visible, but rather the part from -500 m to 100 m relative to the toe of the dike. Using XBeach leads to a much larger failure probability than when the $h/2$-assumption is used (factor 10^3, see Table 3). The dominant parameter in the calculation with the $h/2$-assumption was the water level, which was very large (7.33 m+NAP), see Figure 7. As can be seen from Figure 7, approximately the same amount of wave overtopping occurs for much less extreme hydraulic conditions (a water level that almost 2 m lower), when including the complex hydrodynamics. The difference in offshore wave height was not as large as for the water level (6.25 m versus 6.09 m). However, the wave set-up in XBeach leads to a somewhat larger wave height at the toe. The main difference lies in the wave period. Offshore, the difference is almost negligible, but due to the inclusion of the wave period transformation and generation of infragravity wave energy, the wave period at the toe is much larger with XBeach. The severe wave breaking on the foreshore leads to a transfer of wave energy to the lower frequencies, which increases the $T_{m-1,0}$ wave period strongly (10.50 s versus 42.82 s, see Figure 7). Hence, compared to a calculation rule, using a process-based model that includes complex hydrodynamics like wave set-up, infragravity waves and wave period transformation leads to larger wave heights and much larger wave periods at the toe of the dike. The ratio between the high frequency and low frequency wave heights at the toe ($H_{m0,HF}/H_{m0,LF}$) as calculated by XBeach 1D is only 0.5 in this case. The infragravity waves have a significant contribution to the total wave energy in this 1D case with foreshore. Thus, it can be concluded that it is of importance to include these hydrodynamic processes for these kinds of situations with severe wave breaking on a shallow foreshore. This means that a complex hydrodynamic model is necessary, which further necessitates an efficient probabilistic model.

3.2. Influence of the Model and Wave Parameters

The inclusion of the model and wave parameters as stochastic (10 stochastic variables, calc. 3) leads to a probability of failure that is approximately 20 times larger than the calculation with only stochastic offshore conditions (calc. 2), see Table 3. Hence, the influence is much smaller than the influence of including the complex hydrodynamics such as infragravity waves. The results of calculations 2 and 3 are shown in Figure 8, again showing the calculations in the design point. As can be seen in the figure, a lower water level with approximately the same offshore wave conditions leads to the same amount of wave overtopping in the calculation with 10 stochastic variables, compared to the situation with 3 stochastic variables. The main difference lies in the breaker index γ which is now a stochastic variable. In the design point, the value for this parameter was higher (0.63 versus 0.54 for the deterministic value). This larger breaker index leads to less wave breaking (larger maximum wave height H_{max} at incipient breaking) and thus larger wave heights at the toe of the dike and more wave overtopping. The difference in probability of failure shows that it can be important to include more stochastic variables than just the offshore conditions. Again, a large wave set-up and strong generation of infragravity wave energy is visible. As a result of this, the $T_{m-1,0}$ wave periods at the toe are again quite large. The ratio between the high frequency and low frequency wave heights at the toe ($H_{m0,HF}/H_{m0,LF}$) as calculated by XBeach 1D is only 0.4 in this case. This again shows that it is of importance to include hydrodynamic processes like infragravity waves and wave set-up in the calculations.

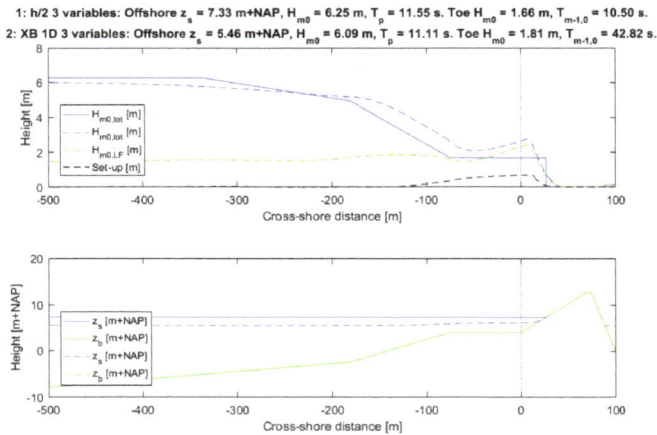

Figure 7. $h/2$-model (solid lines) versus XBeach 1D (dashed lines) with the offshore conditions as stochastic variables and a morphostatic foreshore at the design point. The dike toe is indicated by the dotted line.

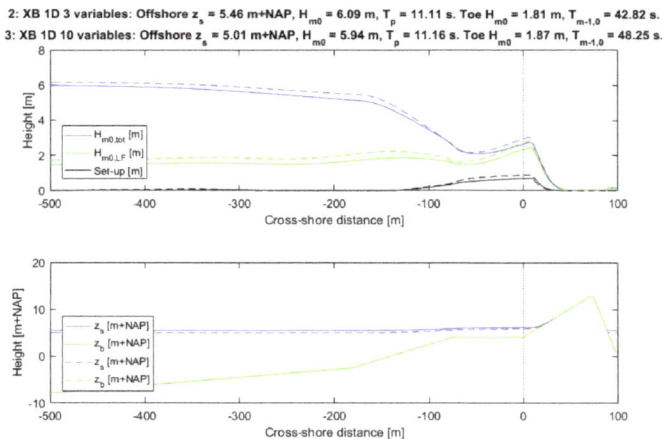

Figure 8. XBeach 1D calculations with 3 (solid lines) and 10 (dashed lines) stochastic variables at the design point. The dike toe is indicated by the dotted line. The parameters offshore and at the toe of the dike are given in the title of the figure.

3.3. Influence of the Storm Duration and Time of the Peak of the Storm

The next step was to include the storm duration t_{storm} and time of the peak of the storm φ as stochastic variables, leading to a calculation with 12 stochastic variables (calc. 4). Including these parameters as stochastic led to a probability of failure that was approximately 2 times smaller than the calculation with 10 stochastic parameters (calc. 3), see Table 3. Hence, the influence of including these parameters is rather small. The reduction in probability of failure can be explained by the time of the peak of the storm relative to the tide. When this parameter is considered as deterministic, it is assumed that the maximum surge occurs at the same moment as tidal high water. When this parameter is considered stochastic, the moment of maximum surge does not necessarily correspond to the moment of the maximum tidal water level, which reduces the probability of failure. The rest of the stochastic parameter values in the design point of the calculations with 10 and 12 stochastic parameters corresponded almost exactly, which explains the small difference in probability of failure. The small difference in probability of failure shows that for this case, the storm duration and time of the peak of the storm can be neglected as stochastic variables.

3.4. Influence of Morphological Changes

The inclusion of morphological changes of the foreshore during a storm and the corresponding three extra stochastic variables led to a failure probability that was somewhat larger (factor four higher with morphological changes included, see Table 3). The extra stochastic variables, compared to the calculations described in the previous subsections, increased the number of calculations necessary to reach a converged probability of failure. This is a direct result of the inclusion of more parameters, which enlarges the search space (more dimensions).

Figure 9 presents the results of the simulations without morphological changes with 12 stochastic variables (calc. 4) and with morphological changes with 15 variables (calc. 5) at the design point. The top plot shows the development of the total significant wave height (blue), the low frequency part of the wave height (green) and set-up (black). The bottom plot shows the water level (blue) and the cross-shore profile at the beginning of the storm (green) and end of the storm (black). The inclusion of morphological changes leads to a somewhat larger probability of failure of about a factor four, as can be seen in Table 3. This change is not very large when compared to the influence of the complex hydrodynamics as infragravity waves and wave set-up for the case considered here. Due to the erosion of the foreshore, the wave dissipation becomes less, which gives larger wave heights at the toe of the dike (2.05 m versus 1.77 m). However, the transfer from high frequency wave energy to the lower frequencies becomes less as well, which gives a smaller $T_{m-1,0}$ wave period (38.85 s versus 51.58 s). In this case, the effects of the difference in wave height and wave period almost cancel one another out. This leads to the small difference in probability of failure, despite the quite large differences in the bathymetry due to erosion of the foreshore. Concluding, for this case, the inclusion of the morphological development of the foreshore and related stochastic variables only led to small differences in the probability of failure.

3.5. Influence of Wave Directional Spreading

To determine the influence of the wave directional spreading (short-crested waves), an extra calculation was performed with XBeach 2D and the three offshore conditions as stochastic variables (calc. 6), which is compared to XBeach 1D (calc. 2).

As shown in the previous subsection, rather large amounts of low-frequency energy and $T_{m-1,0}$ wave periods are found at the toe of the dike when using a 1D simulation. The 1D XBeach calculations can be considered as a wave flume, with long-crested waves without directional spreading. This causes a stronger forcing of the infragravity waves compared to a situation with directional spreading. In reality, wind waves in extreme storms are short-crested. Therefore, an extra calculation was performed with XBeach 2DH with the same three stochastic parameters. With a 2D simulation, the wave directional spreading is included. Thus, the main differences between a 1D and 2D simulation lie within the fact that the 1D simulation does not include the wave directional spreading (hence, long-crested waves), which the 2D simulation does (hence, short-crested waves). For the 2D model, in the alongshore direction, a section of 1 km was modelled with grid cell sizes of 50 m. The wave directional information was solved in 10 degree bins, spanning 110 degrees (55 degrees to both sides of shore normal). Note that the mean wave angle in all simulations was shore-normal and that we use the methodology of Roelvink et al. [34] to resolve the propagation of wave groups in the model domain. The results are shown in Figure 10.

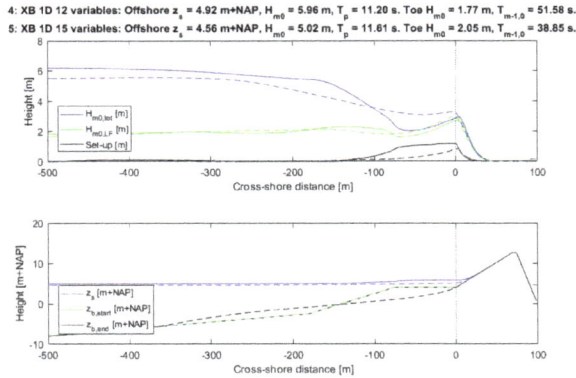

Figure 9. XBeach 1D with 12 stochastic variables and a morphostatic foreshore (solid lines) versus 15 stochastic variables and a morphodynamic foreshore (dashed lines), in the design point. The dike toe is again indicated by the dotted line. The parameters offshore and at the toe of the dike are given in the title of the figure.

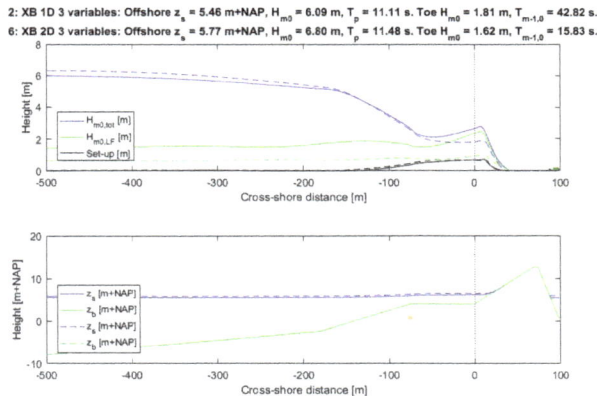

Figure 10. XBeach 1D (long-crested waves) with the three offshore conditions as stochastic variables (calc. 2) versus XBeach 2D (short-crested waves) with the three offshore conditions as stochastic (calc. 6) at the design point. The dike toe is indicated by the dotted line. The parameters offshore and at the toe of the dike are given in the title of the figure.

The calculated probabilities of failure are 1.6×10^{-5} for the 1D case and 1.6×10^{-6} for the 2D case, a difference of a factor 10. The lower probability of failure as found for the 2D case can be explained by the forcing of the infragravity waves, which is less than in the 1D case, because the 2D

XBeach calculations include the wave directional spreading. This leads to less low-frequency energy and therefore a smaller wave period at the toe (15.83 s versus 42.82 s), which results in less wave overtopping. For instance Hofland et al. [11] show as well that the generation of low-frequency wave energy is decreased by the directional spreading. With the inclusion of the wave directional spreading, the results of a 2D simulation are more realistic compared to a 1D simulation, since wind waves in extreme storms are short-crested (see also e.g., Van Dongeren et al. [50]). The ratio between the high frequency and low frequency wave heights at the toe ($H_{m0,HF}/H_{m0,LF}$) is 1.6 for the 2D case (compared to 0.5 for the 1D case). Thus, even in 2D the infragravity waves have a significant contribution to the total wave energy.

3.6. Influence of the Probabilistic Method

The influence of using the level III probabilistic method ADIS on the number of simulations necessary was determined before in Section 2.2.3., as well as in Den Bieman et al.; Grooteman [21,36]. An indication of the added value of using this level III method can be gained by comparing the ADIS results with probabilistic calculations that use the same probability exceedance value for each of the stochastic parameters. This means that the stochastic parameters are considered as fully correlated. This was done for XBeach 1D with the three offshore condition parameters (water level, wave height and wave period) as stochastic variables. The results are given in Table 4. It can be seen that a difference of a factor 12.5 in the probability of failure is found between ADIS and the fully dependent estimate. Hence, when using a level III fully probabilistic approach, a lower probability of failure and a lower required crest height of the dike are found. It can be imagined, that if more stochastic variables would be included in this comparison, the difference between the estimate and the advanced probabilistic approach would be even larger. Thus, using ADIS in the design of dike-foreshore systems can lead to lower required crest levels.

Table 4. Comparison of the probabilities of failure as calculated with ADIS and as calculated with an approach that uses the same probability exceedance value for each of the stochastic parameters (fully correlated). Results for XBeach 1D with the three offshore parameters as stochastic variables.

Calculation	Offshore z_s [m+NAP]	Offshore H_{m0} [m]	Offshore T_p [s]	Toe H_{m0} [m]	Toe $T_{m-1,0}$ [s]	P_f [year^{-1}]	β [-]
XBeach 1D fully correlated	5.16	6.32	11.74	1.84	46.62	2×10^{-4}	3.54
XBeach 1D ADIS	5.46	6.09	11.11	1.81	42.82	1.6×10^{-5}	4.15

4. Discussion

The calculation duration of some of the probabilistic calculations was rather large, in the order of a day without morphological changes (either 1D or 2D), in the order of a week (1D) with morphological changes, on 8 nodes with 16 GB of RAM. For this paper, the calculations with large numbers of stochastic variables and morphological changes were only performed with XBeach 1D. With the current computational capacities, using XBeach 2D and including the morphological changes in the probabilistic framework is not feasible in an engineering context yet, but it is for an academic context. As shown before, for the present case the morphological response has a relatively small influence on the failure probability. As the morphological response is expected to become lower when using a 2D calculation with short-crested waves, this will further decrease the influence of morphology on the failure probability, a consequence of the decreased infragravity wave height. With increasing computational power, the feasibility of XBeach 2D will likely increase in the future. As well, a possible measure could be to use a morphological acceleration factor in XBeach. Also, recently, a prediction formula for the $T_{m-1,0}$ on mildly sloping shallow foreshores was determined [11]. This equation can be used for simple cross-shore profiles and could be implemented in the model framework to save calculation time for such cases. For more complex cross-shore profiles, XBeach will remain necessary.

In the calculations with morphological changes, XBeach created a characteristic storm profile with an offshore bar, the foreshore itself was largely eroded. As mentioned before, XBeach has been extensively calibrated and validated [26,27]. However, even if the calculated offshore bar height would not be completely correct, this would hardly affect the wave conditions at the toe and related overtopping discharge. Van Gent & Giarrusso [15] found that the wave conditions at the toe of the structure are hardly affected by the level of the offshore bar (and/or trough), but are strongly affected by the level of the foreshore.

As mentioned before, the focus of this paper is on the dominant short term storm morphological changes, not on the long term changes by the daily climate. The bathymetrical profile after the storm with conditions corresponding to the parameter values in the design point was compared to the bathymetrical profiles at Westkapelle 1 year after construction of the foreshore and 5 years after construction of the foreshore. The morphological changes of the 'design point storm' were much larger than the changes by the daily climate after five years. This does not mean that the after-storm erosion profile of a pre-storm situation with a foreshore without much sand in it is the same as the erosion profile for a pre-storm situation with much sand in the foreshore. For maintenance, this means that the design foreshore profile should be maintained as much as possible. Furthermore, it is expected that the role of short term (storm) changes of the foreshore become more important when the foreshore is designed smaller (less sand in the profile). Finally, these findings are case-specific. At other locations, there may be more or less dynamic behavior, and also the ratio between the erosion by daily and storm conditions can vary.

Much of the validation for wave overtopping was performed in wave flumes (1D, no directional spreading), which inherently includes the stronger forcing of the infragravity waves, like in XBeach 1D. As was shown, a large part of the difference in the results between 1D and 2D XBeach was caused by differences in the $T_{m-1,0}$ wave period at the toe and uncertainties in the sensitivity of the EurOtop equations with respect to this wave period. Since the $T_{m-1,0}$ wave period is very sensitive to the low frequencies, it is unclear if this wave period is still applicable for situations with large amounts of low frequency energy, such as the ones considered in this paper. Treating the complex combination of infragravity wave energy and wind wave energy with a single parameter as the $T_{m-1,0}$ wave period might be an oversimplification, however the $T_{m-1,0}$ wave period is the wave period that is currently used in all the most commonly used wave run-up and wave overtopping equations. A new equivalent slope concept was determined for wave overtopping at shallow foreshores in Altomare et al. [51] and could be considered a first step. However, still more research is required to be able to validate the influence of the infragravity waves on wave overtopping, for this case, as well as for other cases.

5. Conclusions

This paper presented a probabilistic model framework which is capable of including complex hydrodynamics like infragravity waves and morphodynamics of a sandy foreshore in the calculation of the probability of sea dike failure due to wave overtopping. The method was applied to a test case loosely based on the Westkapelle sea defence, a hybrid defence consisting of a dike with a sandy foreshore. Moreover, the influence of the physical processes like infragravity waves and morphological changes on the probability of dike failure due to wave overtopping were determined. By using the fully probabilistic level III method ADIS, the number of simulations necessary is greatly reduced, which allows for the use of more advanced and detailed hydro- and morphodynamic models. The framework is able to compute the probability of failure with up to 15 stochastic variables and to describe feasible physical processes. Furthermore, the framework is completely modular, which means that any model or equation can be plugged into the framework, whenever updated models with improved representation of the physics or increases in computational power become available.

The model framework as described in this paper, includes more stochastic variables in the determination of the probability of dike failure due to wave overtopping compared to the currently used methods in the Netherlands, where also infragravity waves or morphological changes are not included.

Including a process-based hydrodynamic model led to much larger failure probabilities for the hybrid defence considered in this paper (factor 10^3). This difference was mainly caused by the difference in wave period at the toe of the dike, which is not accounted for by the simple calculation. Hence, this indicates that it can be important to include the complex hydrodynamics such as infragravity waves and wave set-up for these kinds of situations.

The inclusion of 7 extra stochastic model and wave parameters led to a probability of failure that was approximately 20 times larger than the calculation with 3 offshore condition stochastic parameters. The influence is much smaller than the influence of including the complex hydrodynamics such as infragravity waves, but the difference in probability of failure shows that it can still be important to include more stochastic variables than just the offshore conditions in the calculations. Furthermore, for this case the storm duration and time of the peak of the storm have a negligible influence as stochastic variables.

The inclusion of the morphological changes led to a failure probability that was only somewhat larger (factor 4) for the case considered here (relative to e.g., the much larger influence of the complex hydrodynamics that was found). Hence, for this case, the influence of the morphological changes of the foreshore on the failure probability was not that large. In this case, the effects of the difference in wave height and wave period at the toe almost cancelled one another out.

More research is required to be able to validate the influence of the infragravity waves on wave overtopping, for this case, as well as for other cases, especially in connection to short-crested waves. The results of a 2D simulation are more realistic compared to a 1D simulation, see also e.g., Van Dongeren et al. [50], as wind waves in extreme storms are short-crested. The long-crested waves in XBeach 1D create a stronger forcing of the infragravity waves. With the current computational capacities, a 2D calculation and including the morphological changes in the level III probabilistic framework is not feasible in an engineering context yet, but it is for an academic context. With increasing computational power, the feasibility will likely increase in the future.

Using the framework offers more insight into the physics, and can be used to determine the contribution of foreshores to flood protection. It is recommended to apply the framework to other cases as well, to determine if the effects of complex hydrodynamics as infragravity waves and morphological changes on the probability of sea dike failure due to wave overtopping as found in this paper hold for other cases as well. Furthermore, it is recommended to investigate broader use of the method, e.g., for safety assessment, reliability analysis and design.

Author Contributions: R.T.M., V.V. and P.O. conceived and designed the research; Based on this design, the research was mainly conducted by P.O.; R.T.M., V.V., B.H., J.W.M. and S.N.J. provided many insightful discussions and contributed to analyzing and interpreting the data; P.O. wrote most of the paper; R.T.M., V.V. and B.H. helped with the conceptual design of the article; R.T.M., V.V., B.H., J.W.M. and S.N.J. provided critical reviews and revisions of the contents of the paper.

Acknowledgments: The authors would like to thank Joost den Bieman of Deltares for providing the ADIS Matlab scripts and for his guidance with the probabilistic aspects of this paper.

Conflicts of Interest: The authors declare no conflict of interest.

Appendix A. EurOtop (2007) Equations

The EurOtop [35] equations use the (fictitious deep water) surf similarity parameter, defined as:

$$\xi_{m-1,0} = \frac{\tan \alpha}{\sqrt{H_{m0}/L_{m-1,0}}} \tag{A1}$$

where α is the average structure slope [degrees], H_{m0} is the significant wave height [m] and $L_{m-1,0}$ is the deep-water wave length [m]. The equations that are used for this paper are given by:

$$\frac{q}{\sqrt{gH_{m0}^3}} = \frac{0.067}{\sqrt{\tan\alpha}}\gamma_b\xi_{m-1,0}\,exp\left[-C1\frac{R_c}{\xi_{m-1,0}H_{m0}\gamma_b\gamma_f\gamma_\beta\gamma_v}\right] \tag{A2}$$

with a maximum of

$$\frac{q}{\sqrt{gH_{m0}^3}} = 0.2\,exp\left[-C2\frac{R_c}{H_{m0}\gamma_f\gamma_\beta}\right] \tag{A3}$$

to be used for gentle dike slopes and $\xi_{m-1,0} < 5$, with q the average overtopping discharge [m^3 s^{-1} m^{-1}], g the gravitational acceleration [ms^{-2}], R_c the crest freeboard [m], γ_f a correction factor for roughness and permeability [-], γ_b a correction factor for a berm [-], γ_β a correction factor for oblique wave attack [-], γ_v a correction factor for a vertical wall on top of the crest [-]. The reliability is described by assuming the coefficients $C1 = 4.75$ and $C2 = 2.6$ to be normal distributed with a standard deviation of 0.5 and 0.35 respectively. The different correction factors are described in more detail in [12]. Furthermore, for shallow and very shallow foreshores ($\xi_{m-1,0} > 7$):

$$\frac{q}{\sqrt{gH_{m0}^3}} = 10^{C3}\,exp\left[-\frac{R_c}{H_{m0}\gamma_f\gamma_\beta(0.33 + 0.022\xi_{m-1,0})}\right] \tag{A4}$$

in which $C3$ is normal distributed with a mean of -0.92 and a standard deviation of 0.24.

References

1. Van Slobbe, E.; de Vriend, H.J.; Aarninkhof, S.; Lulofs, K.; de Vries, M.; Dircke, P. Building with Nature: In search of resilient storm surge protection strategies. *Nat. Hazards* **2013**, *66*, 1461–1480. [CrossRef]
2. Fiselier, J.; Jaarsma, N.; Van der Wijngaart, T.; De Vries, M.; Wal, M.; Stapel, J.; Baptist, M.J. *Perspectief Natuurlijke Keringen: Een Eerste Verkenning Ten Behoeve Van Het Deltaprogramma*; EcoShape: Dordrecht, The Netherlands, 2011. (In Dutch)
3. Wainwright, D.J.; Ranasinghe, R.; Callaghan, D.P.; Woodroffe, C.D.; Jongejan, R.; Dougherty, A.J.; Rogers, K.; Cowell, P.J. Moving from deterministic towards probabilistic coastal hazard and risk assessment: Development of a modelling framework and application to Narrabeen Beach, New South Wales, Australia. *Coast. Eng.* **2015**, *96*, 92–99. [CrossRef]
4. Symonds, G.; Huntley, D.A.; Bowen, A.J. Two dimensional surf beat: Long wave generation by a time varying breakpoint. *J. Geophys. Res. Ocean.* **1982**, *87*, 492–498. [CrossRef]
5. Longuet-Higgins, M.S.; Stewart, R.W. Radiation stress and mass transport in gravity waves, with application to "surf beats". *J. Fluid Mech.* **1962**, *13*, 481–504. [CrossRef]
6. Munk, W.H. Origin and generation of waves. *Coast. Eng. Proc.* **1950**, *1*, 1. [CrossRef]
7. Suzuki, T.; Verwaest, T.; Veale, W.; Trouw, K.; Zijlema, M. A Numerical Study on the Effect of Beach Nourishment on Wave Overtopping in Shallow Foreshores. In Proceedings of the 33rd International Conference on Coastal Engineering, Santander, Spain, 1–6 July 2012.
8. Holterman, S.R. Golfoploop op Dijken Met Ondiep Voorland. Master's Thesis, Delft University of Technology, Delft, The Netherlands, 1998. (In Dutch)
9. Van Gent, M.R.A. *Physical Model Investigations on Coastal Structures with Shallow Foreshores: 2D Model Tests with Single and Double-Peaked Wave Energy Spectra*; Technical Report; Deltares (WL): Delft, The Netherlands, 1999.
10. Van Gent, M.R.A. Wave Runup on Dikes with Shallow Foreshores. *J. Waterw. Port. Coast. Ocean Eng.* **2001**, *127*, 254–262. [CrossRef]
11. Hofland, B.; Chen, X.; Altomare, C.; Oosterlo, P. Prediction formula for the spectral wave period Tm-1,0 on mildly sloping shallow foreshores. *Coast. Eng.* **2017**, *123*, 21–28. [CrossRef]
12. EurOtop. EurOtop: Manual on Wave Overtopping of Sea Defences and Related Structures. An Overtopping Manual Largely Based on European Research, but for Worldwide Application, Pre-release 2nd ed. van der Meer, J.W., Allsop, N.W.H., Bruce, T., De Rouck, J., Kortenhaus, A., Pullen, T., Schüttrumpf, H., Troch, P., Zannutigh, B., Eds.; 2016. Available online: www.overtopping-manual.com (accessed on 31 January 2018).

13. TAW. *Technical Report Wave Run-Up and Wave Overtopping at Dikes*; Technical Report; TAW: Delft, The Netherlands, 2002.

14. Roeber, V.; Bricker, J.D. Destructive tsunami-like wave generated by surf beat over a coral reef during Typhoon Haiyan. *Nat. Commun.* **2015**, *6*. [CrossRef] [PubMed]

15. Van Gent, M.; Giarrusso, C. Influence of foreshore mobility on wave boundary conditions. In Proceedings of the International Conference on Ocean Waves Measurements and Analysis, Madrid, Spain, 3–7 July 2005; pp. 1–10.

16. Bakker, W.T.; Vrijling, J.K. Probabilistic design of sea defences. *Coast. Eng.* **1980**, *17*, 2040–2059.

17. Madsen, H.O.; Krenk, S.; Lind, N.C. *Methods of Structural Safety*; Prentice-Hall, Inc.: Englewood Cliffs, NJ, USA, 1986; ISBN 0-486-44597-6.

18. Joint Committee on Structural Safety: RILEM—CEB—CECM—CIB—FIP—IABSE—IASS. *General Principles on Reliability for Structural Design*; IABSE: Lisbon, Portugal, 1981; Volume 35.

19. Van de Graaff, J. Probabilistic design of dunes; an example from The Netherlands. *Coast. Eng.* **1986**, *9*, 479–500. [CrossRef]

20. Callaghan, D.P.; Ranasinghe, R.; Roelvink, D. Probabilistic estimation of storm erosion using analytical, semi-empirical, and process based storm erosion models. *Coast. Eng.* **2013**, *82*, 64–75. [CrossRef]

21. Den Bieman, J.P.; Stuparu, D.E.; Hoonhout, B.M.; Diermanse, F.L.M.; Boers, M.; Van Geer, P.F.C. Fully probabilistic dune safety assessment using an advanced probabilstic method. *Coast. Eng. Proc.* **2014**, *1*, 9. [CrossRef]

22. Stripling, S.; Panzeri, M.; Blanco, B.; Rossington, K.; Sayers, P.; Borthwick, A. Regional-scale probabilistic shoreline evolution modelling for flood-risk assessment. *Coast. Eng.* **2017**, *121*, 129–144. [CrossRef]

23. Vuik, V. *Effect of Nourishment on Failure Probability of Westkapelle Sea Defence*; Technical Report; HKV: Lelystad, The Netherlands, 2014. (In Dutch)

24. Rijkswaterstaat. *Hydraulische Randvoorwaarden Primaire Waterkeringen, Voor de Derde Toetsronde 2006–2011 (HR 2006)*; Technical Report; RWS DWW: Delft, The Netherlands, 2007. (In Dutch)

25. Wolters, A.F. *Inventarisatie Effecten Wijzigingen Golfrandvoorwaarden*; Technical Report; RWS DWW: Delft, The Netherlands, 2002. (In Dutch)

26. Roelvink, D.; Reniers, A.; van Dongeren, A.; van Thiel de Vries, J.; McCall, R.; Lescinski, J. Modelling storm impacts on beaches, dunes and barrier islands. *Coast. Eng.* **2009**, *56*, 1133–1152. [CrossRef]

27. Smit, P.; Stelling, G.; Roelvink, J.A.; Van Thiel de Vries, J.; McCall, R.; Van Dongeren, A.; Zwinkels, C.; Jacobs, R. *XBeach: Non-Hydrostatic Model: Validation, Verification and Model Description*; Delft University of Technology and Deltares: Delft, The Netherlands, 2010.

28. Roelvink, D. Dissipation in random wave groups incident on a beach. *Coast. Eng.* **1993**, *19*, 127–150. [CrossRef]

29. Reniers, A.J.H.M.; Roelvink, J.A.; Thornton, E.B. Morphodynamic modeling of an embayed beach under wave group forcing. *J. Geophys. Res. Ocean.* **2004**, *109*. [CrossRef]

30. Galappatti, G.; Vreugdenhil, C.B. A depth-integrated model for suspended sediment transport. *J. Hydraul. Res.* **1985**, *23*, 359–377. [CrossRef]

31. Soulsby, R. *Dynamics of Marine Sands: A Manual for Practical Applications*; Thomas Telford: London, UK, 1997; ISBN 072772584X.

32. Oosterlo, P. A Method to Calculate the Probability of Dike Failure Due to Wave Overtopping, Including the Infragravity Waves and Morphological Changes. Master's Thesis, Delft University of Technology, Delft, The Netherlands, 2015.

33. Guza, R.T.; Thornton, E.B.; Holman, R.A. Swash on steep and shallow beaches. In *Coastal Engineering Proceedings*; ASCE: Houston, TX, USA, 1984; Volume 19.

34. Roelvink, D.; McCall, R.; Mehvar, S.; Nederhoff, K.; Dastgheib, A. Improving predictions of swash dynamics in XBeach: The role of groupiness and incident-band runup. *Coast. Eng.* **2017**, *134*, 103–123. [CrossRef]

35. EurOtop. *EurOtop. Wave Overtopping of Sea Defences and Related Structures: Assessment Manual*, 1st ed.; Pullen, T., Allsop, N.W.H., Bruce, T., Kortenhaus, A., Schüttrumpf, H., van der Meer, J.W., Eds.; Die Küste: Hamburg, Germany, 2007.

36. Grooteman, F. An adaptive directional importance sampling method for structural reliability. *Probab. Eng. Mech.* **2011**, *26*, 134–141. [CrossRef]

37. OpenEarth OpenEarth: Open Source Initiative. Available online: www.openearth.eu (accessed on 31 January 2018).

38. Roelvink, D.; Den Heijer, C.; Van Thiel De Vries, J.S.M. Morphological modelling of strongly curved islands. In Proceedings of the 7th International Conference on Coastal Dynamics, Arcachon, France, 24–28 June 2013; Bordeaux University: Bordeaux, France, 2013.

39. Van Gelder, P.H.A.J.M. Statistical Methods for the Risk-Based Design of Civil Structures. Ph.D. Thesis, Delft University of Technology, Delft, The Netherlands, 2000.

40. Van Thiel de Vries, J.S.M. Dune Erosion during Storm Surges. Ph.D. Thesis, Delft University of Technology, Delft, The Netherlands, 2009.

41. Den Heijer, F.; Vos, R.J.; Diermanse, F.L.M.; Groeneweg, J.; Tonis, R. *De Veiligheid van de Primaire Waterkeringen in Nederland: Achtergrondrapport HR2006 voor de Zee en Estuaria*; Technical Report; Rijkswaterstaat, RIKZ: Lemmer, The Netherlands, 2006. (In Dutch)

42. Steetzel, H.J.; Diermanse, F.L.M.; van Gent, M.R.A.; Coeveld, E.M.; Van de Graaff, J. *Dune Erosion-Product 3: Probabilistic Dune Erosion Prediction Method*; Technical Report; Deltares(WL): Delft, The Netherlands, 2007.

43. Vrouwenvelder, A.C.W.M.; Steenbergen, H.M.G.M.; Slijkhuis, K.A.H. *Theoriehandleiding PC-RING. Deel B: Statistische Modellen*; Technical Report; TNO Bouw: Rijswijk, The Netherlands, 1999. (In Dutch)

44. Roelvink, D.; Van Dongeren, A.; McCall, R.; Hoonhout, B.M.; Van Rooijen, A.; Van Geer, P.; De Vet, L.; Nederhoff, K.; Quataert, E. *Xbeach Manual*; Deltares and Delft University of Tecnhology: Delft, The Netherlands, 2015.

45. Den Heijer, C.K.; Baart, F.; Van Koningsveld, M. Assessment of dune failure along the Dutch coast using a fully probabilistic approach. *Geomorphology* **2012**, *143*, 95–103. [CrossRef]

46. Weerts, A.H.; Diermanse, F.L.M. *Golfstatistiek op Relatief Diep Water 1979–2002*; Technical Report; Deltares (WL): Delft, The Netherlands, 2004. (In Dutch)

47. Stijnen, J.W.; Duits, M.T.; Thonus, B.I. *Deep Water Boundary Conditions (ELD, EUR, YM6, SCW and SON). (Diepwater Randvoorwaarden (ELD, EUR, YM6, SCW en SON))*; Technical Report Pr841. 40; HKV Consultants: Lelystad, The Netherlands, 2005. (In Dutch)

48. Steetzel, H. Cross-Shore Sediment Transport during Storm Surges. Ph.D. Thesis, Delft University of Technology, Delft, The Netherlands, 1993.

49. Longuet-Higgins, M.; Cartwright, D.; Smith, N. Observations of the directional spectrum of the sea waves using the motions of a floating buoy. In *Proceedings Conference Ocean Wave Spectra*; Prentice Hall: Upper Saddle River, NJ, USA, 1963; pp. 111–132.

50. Van Dongeren, A.; Lowe, R.; Pomeroy, A.; Trang, D.M.; Roelvink, D.; Symonds, G.; Ranasinghe, R. Numerical modeling of low-frequency wave dynamics over a fringing coral reef. *Coast. Eng.* **2013**, *73*, 178–190. [CrossRef]

51. Altomare, C.; Suzuki, T.; Chen, X.; Verwaest, T.; Kortenhaus, A. Wave overtopping of sea dikes with very shallow foreshores. *Coast. Eng.* **2016**, *116*, 236–257. [CrossRef]

Journal of
*Marine Science
and Engineering*

MDPI

Article

Failure of Grass Covered Flood Defences with Roads on Top Due to Wave Overtopping: A Probabilistic Assessment Method

Juan P. Aguilar-López [1,2,*], Jord J. Warmink [2], Anouk Bomers [2], Ralph M. J. Schielen [2,3] and Suzanne J. M. H. Hulscher [2]

[1] Water Management Department, Delft University of Technology, 2628 CN Delft, The Netherlands
[2] Marine and Fluvial Systems Group, University of Twente, 7522 NB Enschede, The Netherlands; j.j.warmink@utwente.nl (J.J.W.); a.bomers@utwente.nl (A.B.); ralph.schielen@rws.nl (R.M.J.S.); s.j.m.h.hulscher@utwente.nl (S.J.M.H.H.)
[3] Ministry of Infrastructure and Water Management (Rijkswaterstaat), 8224 AD Lelystad, The Netherlands
* Correspondence: j.p.aguilarlopez@tudelft.nl; Tel.: +31-(0)15-278-5559

Received: 22 April 2018; Accepted: 19 June 2018; Published: 22 June 2018

Abstract: Hard structures, i.e., roads, are commonly found over flood defences, such as dikes, in order to ensure access and connectivity between flood protected areas. Several climate change future scenario studies have concluded that flood defences will be required to withstand more severe storms than the ones used for their original design. Therefore, this paper presents a probabilistic methodology to assess the effect of a road on top of a dike: it gives the failure probability of the grass cover due to wave overtopping over a wide range of design storms. The methodology was developed by building two different dike configurations in computational fluid dynamics Navier–Stokes solution software; one with a road on top and one without a road. Both models were validated with experimental data collected from field-scale experiments. Later, both models were used to produce data sets for training simpler and faster emulators. These emulators were coupled to a simplified erosion model which allowed testing storm scenarios which resulted in local scouring conditioned statistical failure probabilities. From these results it was estimated that the dike with a road has higher probabilities ($5 \times 10^{-5} > P_f > 1 \times 10^{-4}$) of failure than a dike without a road ($P_f < 1 \times 10^{-6}$) if realistic grass quality spatial distributions were assumed. The coupled emulator-erosion model was able to yield realistic probabilities, given all the uncertainties in the modelling process and it seems to be a promising tool for quantifying grass cover erosion failure.

Keywords: wave overtopping; dike; levee; failure; emulation; roads; flood risk; erosion

---

1. Introduction

Structural embedment of roads on top of dikes will generate stability effects in the flood defence during normal operation and during a flood event [1]. Yet, these effects are not explicitly considered in the current probabilistic assessment methods. It is expected that climate change increases sea level rise and that it will affect the wave climate around the world [2,3]. For rivers, extreme water levels occur more frequently as shown for the River Rhine basin [4], for example. Such climate change effects increase the risk of flooding of low lying, highly populated areas around world [5,6]. In the last decades, flood defence design has moved towards a risk-based approach [7,8] in which it is possible to express the design parameters uncertainty as probabilistic distributions [8]. Such methods allow inclusion of the expected future effects due to climate change as modifications in the parameter statistical distributions. Nevertheless, there is a lack of methods for assessing non-convectional flood defences so that their interaction effects are also captured and reflected in the estimated failure probabilities [1].

In case of grass covered dikes the failure mechanism known as wave overtopping is influenced by the presence of non-water retaining functions [1]. This failure mechanism consists in the intermittent surpass of water excess volumes over the dike derived from the 'attacking' waves during a storm. Only the waves that surpass the crest height will overtop and 'attack' the landward slope of the dike, potentially eroding the cover until an eventual dike breach.

For dike cover erosion assessment, reliability methods based on partial safety factors validated by a fully probabilistic Monte Carlo method are available [9]. These methods are based on the classical overtopping approach of defining dike safety based on an allowable overtopping discharge q_{max} [m^3/s/m]. Other studies concluded that overtopping erosion is better estimated from individual wave overtopping volumes V [m^3/m] [10–13]. In Ref. [14], three different erosion predictors (velocity excess, shear stress excess, and work excess) were tested for calculating the scour depths per wave volume and they found that work excess was the best predictor. Simultaneously, the cumulative overload method [15] was developed which relies on the shear stress principle, using the total overtopping duration per wave as input while acknowledging that the "real" erosion time per wave may be shorter. Additional research [16,17] allows inclusion of the effects of obstacles and transitions in both the hydraulic load and resistance terms in the form of calibration coefficients for the cumulative overload method from Ref. [15].

With respect to the grass cover resistance, a conceptual geo-mechanical approach [18] describes the physical process of grass cover erosion based on the vertical and horizontal stresses that act upon a turf element model. In addition, two main studies [19,20] recommended grass and soil quality values expressed in terms of their erodibility rates. Based on these studies and concepts, Ref. [21] aimed to determine the grass cover failure of conventional dikes in terms of probabilities, while considering the spatial distribution of grass quality. The method used in this last study is fully probabilistic and could be adapted to more complex dike profiles. Nevertheless, it relies on empirical formulas for estimating the water depths and flow velocities on top of the crest and along the landside slope. These empirical formulas [22] simplified the estimation of water depth and flow velocity for the design of sea dikes by proving a set of equations. More holistic approaches [23,24] are available for including the effects of bottom irregularities in the flow. Both studies used one dimensional (1D) models, which allowed to simulate the time-dependent erosion process of a grass covered coastal dike. Yet, both models estimate the exerted bottom shear stresses as a function of uniform flow models, such as Chezy or Manning, which are only applicable for smooth bed slope transitions. If abrupt bottom slope variations occur, the vertical velocity distribution tends to have a larger variability [25], thereby violating the uniform flow assumption. This variability might result in more turbulent flows, higher energy dissipation and hence less accurate bed shear stress estimations. The work presented in Ref. [26] used a more detailed computational fluid dynamics (CFD) model and showed that the presence of a road leads to higher scour depths in places with little to non-existent grass cover. In this paper, we extend the deterministic CFD approach of Ref. [26] towards a fully probabilistic assessment method for failure of the grass cover due to wave overtopping erosion.

Emulation of detailed models is as an efficient way to reduce computational times, for implementation in probabilistic failure studies. Emulation consists of building a computationally inexpensive model and training it with data sets generated from the input-output data sets obtained from more complex models [27,28]. Hydrodynamic model emulation has proven to be a powerful tool for water level predictions while improving the calculation speed significantly [29,30]. Also, for the reliability assessment of flood defences, emulation has already been implemented for other failure mechanisms, such as backward piping erosion [31–33]. For wave overtopping, [34] used the results of 10,000 physical model tests of different types of coastal defences to train a neural network.

The aim of this study was to develop an emulator-based dike failure assessment methodology for estimating the effects of a road over the crest including the spatial variation of the grass quality in the failure probability of the grass cover. This failure is estimated as a function of a storm surge wave overtopping event composed of a set of single overtopped water volumes due to wind generated gravity waves, which flow over the dike profile. The methodology was based on two different CFD transient models; one with a road on top [35] and one fully covered with grass. The research question we are answering in this paper is: What is the effect of a road on the grass cover failure probability of a dike due to overtopping waves?

The article is composed of 6 main sections. Section 2 explains the theoretical background of the shear stress excess (*T*). Section 3 explains the experiment in Millingen from which the collected data was used for construction and calibration of the models. Section 4 presents a detailed explanation of the steps of the developed methodology and its correspondent assumptions. Sections 5 and 6 present the results and discussion and Section 7 gives the conclusions to highlight the major findings.

2. Theoretical Background

Wave overtopping erosion is a transient process in which tensile and shear stresses are exerted along the dike grass cover over time because of the overtopped water volume's momentum. While the erosion is generated by one single wave, it has been observed that failure of the dike cover is mostly achieved by the overtopping of numerous waves during a storm event [12]. The time span in which erosion caused by a single wave takes place at a specific location is in the order of seconds [26] and it is termed total wave overtopping time (t_{total}). This duration differs from wave to wave depending on its volume, flow depth, and flow velocity and it is spatially influenced by the surface irregularities and bottom slope. If the generated stresses exceed the mechanical resistance of the grass cover elements (e.g., grass leaves, roots, substrate), erosion occurs by pulling, scouring, and flushing the eroded material downstream.

2.1. Shear Stress Excess (T)

Most of the overtopping scour erosion methods are based on definitions of critical threshold values for average overtopping discharges in steady flow conditions [14]. For combined grass and soil covers, the scour threshold can also be defined by a critical shear stress τ_c, which depends on the grass cover quality [19]. This value only represents the threshold which defines if scouring occurs or not. To include two other important scour characteristics (erosion rate and real erosion duration), the time dependent shear stress excess [N·s/m^2] concept from [36] was adopted in the present study. It represents the surplus of shear stress during the period of time in which erosion occurs. Only during this period does erosion take place. This implies that for larger values of τ_c, the difference between the total overtopping duration (t_{total}) and the real erosion time span (e.g., $f(\tau_{c2}) = t_4 - t_2$ in Figure 1) is significant. The value of *T* is calculated for a given location along the profile as the integral of the bottom shear stress function $\tau(t)$ minus the critical erosion stress threshold τ_c, over the erosion time span (e.g., area *A* for time span $t_4 - t_2$ and τ_{c2} in Figure 1). A lower value of τ_c represents lower resistance to erosion and implies a longer time span of erosion and a greater.

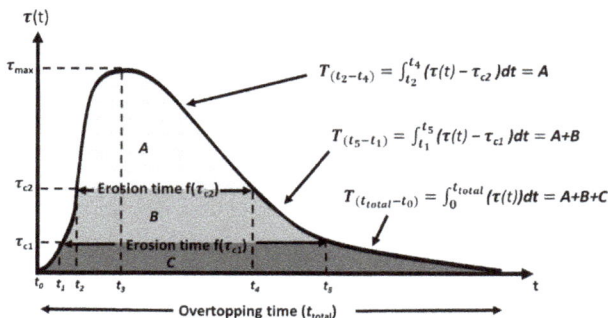

Figure 1. Excess shear stress integral over time for different critical thresholds.

The fact that the erosion time span $(t_{i+1} - t_i)$ and the total wave overtopping duration time (t_{total}) differs may have a significant influence on the estimated erosion was acknowledged in Ref. [37], who included the "real" erosion excess time by assuming a triangular shape of the wave overtopping discharge hydrograph. However, this assumption may be too general for representing the shear stress excess over a highly irregular bottom profile. The overtopping integration bounds $(t_{i+1}$ and $t_i)$ vary between locations due to the presence of bottom irregularities which may either accelerate or decelerate the overtopping flows. In addition, their values are also determined by the value of τ_c (See Figure 1).

2.2. Erosion Model as a Function of T

The erosion model used in the present study is derived from the time dependent mass erosion rate equation for cohesive soils presented in Ref. [38] as:

$$\frac{dm}{dt} + M_p \frac{(\tau - \tau_c)}{\tau_c} = 0 \tag{1}$$

In which M_p [kg/(m²s)] corresponds to the characteristic soil mass transport coefficient, $\frac{dm}{dt}$ corresponds to the mass transport rate in time, τ [N/m²] corresponds to the exerted bottom shear stress, and τ_c [N/m²] corresponds to the critical shear stress threshold.

According to [36], an overall erosional strength parameter C_E [m/s⁻¹] for soil and grass together was derived [19] as:

$$C_E \equiv \frac{M_p}{\tau_c} \tag{2}$$

Note that the C_E parameter is directly proportional to M_p and inversely proportional to τ_c. Based on this equivalence and by dividing Equation (1) by the dike cover relative density per unit of width $(\rho'_{cover}$ [kg/m²]), the erosion rate at any specific location for a unitary width of dike can be expressed as:

$$\frac{d\varepsilon}{dt} + \frac{C_E}{\rho'_{cover}}(\tau(t) - \tau_c) = 0 \tag{3}$$

Since C_E or ρ'_{cover} are independent of time, the scour depth (ε) in a particular location can be calculated by integrating Equation (3) as:

$$-\int_{x_i}^{x_{i+1}} d\varepsilon = \frac{C_E}{\rho'_{cover}} \int_{t_i}^{t_{i+1}} (\tau(t) - \tau_c)dt \tag{4}$$

The integral on the right side of Equation (4) represents the T value for a given τ_c for a single overtopped wave. The integral bounds of the right-side integral are defined by the specific moments in time where $\tau(t) \geq \tau_c$, as erosion will occur during this condition (Figure 1). The integrand bound x_i represents the initial bottom elevation before the scouring process and x_{i+1} the resultant bottom elevation after the scouring process. This model includes the time-dependent variability of the shear stress, but it only estimates the erosion process as a function of the bottom shear stresses. The vertical tensile resistance of grass and roots of the grass cover [36] are accounted for in the parameterization by the C_E coefficient. These tensile stresses are described by the turf element method [36] but were too complex to be included in the model. Therefore, the shear stress excess erosion concept presented in Section 2.1 is selected as it requires less input data and allows to use the erodibility value C_E [19]. This simplification makes the implementation more feasible for a probabilistic model as it represents the grass quality and resistance uncertainty in one single stochastic variable.

3. Millingen aan de Rijn Wave Overtopping Experiment with a Road on Top

Experiments for wave overtopping were conducted on top of a riverine dike along the river Waal nearby Millingen aan de Rijn (The Netherlands) in February and March of 2013 [39]. Random wave volumes V [m^3/m] were released by the wave overtopping simulator (WOS) [15] during a period of time equivalent to the duration of a storm. The WOS was located on top of the crest of the dike close to the riverside vertex (see Figure 2).

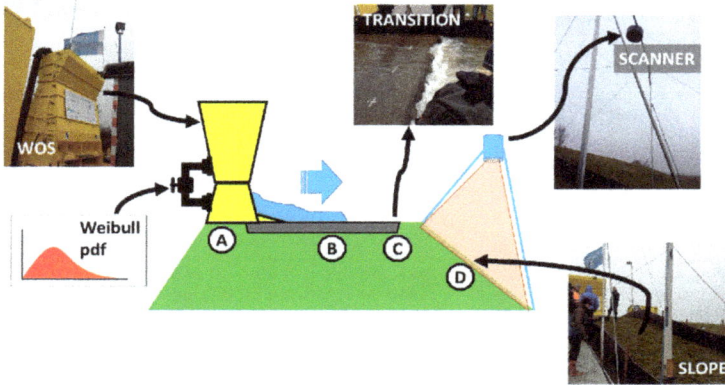

Figure 2. Millingen wave overtopping simulator (WOS) experiment with a road on top of the crest. Volumes released by the WOS are obtained from random sampling list from a fitted Weibull distribution.

For the present study, a storm condition is defined by a crest height (R_c [m]), a water level and the total number of waves N_w [-] generated in the foreshore, with significant wave height, H_s [m], during the storm duration period D [s]. In addition, a storm condition is often characterized by its mean overtopping discharge per unit of width q_m [m^3/s/m]. This discharge is calculated as the total overtopped volume of water per storm divided by storm duration as shown in Equation (5):

$$q_m = \frac{\sum_{j=1}^{j=N_{ow}} V_j}{D} \tag{5}$$

Note that the same storm event may generate a different q_m depending on the crest level of the dike as only a limited number of wave volumes will overtop the dike N_{ow} [-]. Hence, for this study, a combination of factors (e.g., wave conditions, dike geometry, and roughness) that originate a certain q_m is referred to a storm condition. In the WOS experiment, 10 storm conditions were simulated consecutively. After each simulated storm condition, the dike was scanned using a 3D laser scanner to estimate the scour depths and spatial scouring patterns. This experiment included a road on top of the dike to analyze the effect of transitions on the scouring process due to wave overtopping. The experiment was divided into two parts performed in two adjacent locations. Each test location was 4 m wide with lateral walls that prevented leakage during the tests.

3.1. Millingen Experiment Part I: Scour Measurements

During the first part of the experiment, the WOS was located near the riverside edge of the dike (zone A, Figure 2). From this location, a series of volumes were released during a period of six hours, which is equivalent to the duration of the design storm. The released volumes were randomly sampled following a Weibull distribution which represents the stochastic nature of the waves that overtop the dike during a storm. The former Dutch safety standards, for example [11], allow storm conditions of maximum overtopping discharge q_{max} between 0.001 m^3/s/m and 0.01 m^3/s/m depending on the dike location, cross sectional design, and materials [12]. The first part of the experiment [39], was conducted for different consecutive storm conditions (q_m) as presented in Table 1.

Table 1. Experimental storm conditions for part I of Millingen experiment.

Experiment Part I			
Overtopping Discharge	Interval [1] Interval	Scanning Instant	Profile Scan Label
q_m [m^3/s/m]	[min]	[min]	[min]
Initial state	0	0	q0_t0
0.001	72	72	q1_sc1
0.010	180	252	q10_sc1
0.010	180	432	q10_sc2 [2]
0.050	60	492	q50_sc1
0.050	60	552	q50_sc2
0.050	60	612	q50_sc3
0.050	60	672	q50_sc4
0.050	180	852	q50_sc5 [2]
0.100	100	952	q100_sc1
0.100	120	1072	q100_sc2

[1] The intervals shows the actual test durations and scanning instant shows the exact moment in which the laser scanning takes place. The [2] profiles are used for model the CFD model construction and validation.

While the experiment's durations range from 0 to 3 h for each mean discharge, dikes are often designed for storms that range between 2 and 6 h depending on their boundary conditions. High resolution 3D laser scan images were taken (Figure 2) every 2 h with an accuracy of 2 mm. For some storm conditions, it was necessary to accelerate or decelerate some of the tests due to the WOS pumping constraints and release capacity restrictions. The different storm average discharges are conditioned by the relative position of the crest of the dike with respect to the still water level, the dike geometry, dike cover, and revetment type.

In the initial state profile, already significant damage due to traffic was observed in the transition gap (zone C, Figure 2). This gap increased and propagated landward with each test, as bare soil erodes faster than grass covered soil. Besides this zone, no significant erosion was observed in the grass cover for the 0.001 and 0.010 m^3/s/m tests. The scan labels presented in Table 1 include the average overtopping discharge and the scan number of that experiment.

3.2. Millingen Experiment Part II: Flow Depths and Velocity Measurements

The second part of the experiment consisted of measuring flow depth time series and velocity time series of individual overtopping wave volumes along the dike profile. Paddle wheel devices (PW) were used for measuring flow velocities and surfboard meters (SB) for measuring flow depths in 8 locations along the crest and landward slope (see [39] for more details). To exclude the effects on the measurements of the landside road transition and roughness change, a geotextile cover was placed over the transition between the asphalt road and the remaining crest part (Figure 2, zone C). Additionally, the WOS was located directly over the road (Figure 2, zone B), to avoid the induced effects on the flow measurements from the riverside transition. Velocities and depths time series were measured at least two times for wave volumes (V) of 0.4, 0.6, 0.8, 1.0, 1.5, 2.0, 2.5, 3.0, 4.0, 5.0, and 5.5 m^3/m.

4. Methodology

The complete methodology is divided in 4 steps as shown in the flow chart presented in Figure 3. The first step consists of constructing and validating two CFD models built for producing accurate and detailed time series of bottom shear stresses (τ) along the dike profile as a function of different wave volumes (V). The first model includes an asphalt road at the crest of the dike and corresponds to the profile of the Millingen dike initial state scan. This CFD model is referred to as "Road on crest dike" model (RCD). The second model represents the same dike but the road on top was replaced by a smooth grass cover. This CFD model is referred to as "Grass crest dike" model (GCD).

In the second step (Section 4.2) the bottom stress time series (τ) are integrated at each location for different ranges of critical shear stress values (τ_c). The obtained set of values represent the shear stress excess integral T, which are used to build two computationally inexpensive models (emulators) which imitate both CFD models. These emulators are capable of estimating T values as a function of combinations of V and τ_c values.

In the third step (Section 4.3), the "grass cover quality" is determined for each location and represented by a C_E function. These functions are calibrated based on the previously built emulators and the scanned profiles of the Millingen experiment part I. One of the main assumptions of the methodology is that the estimated curves are representative for both dike conditions.

Finally, in the fourth step (Section 4.4) the emulators of the RCD and GCD are used to predict the erosion depth at each location along the dike profile, using stochastic input for V and τ_c and the calibrated C_E functions from step 3. This yields a random set of erosion profiles from which the failure probabilities in each location for both the road and grass covered dike are derived for each storm condition (q_m). Note that the use of emulators allowed us to reduce the computational times so that crude Monte Carlo simulations could be used that captured the high level of detail of the hydrodynamics instead of using approximate reliability methods such as first order reliability method (FORM) or second order reliability method (SORM) [31].

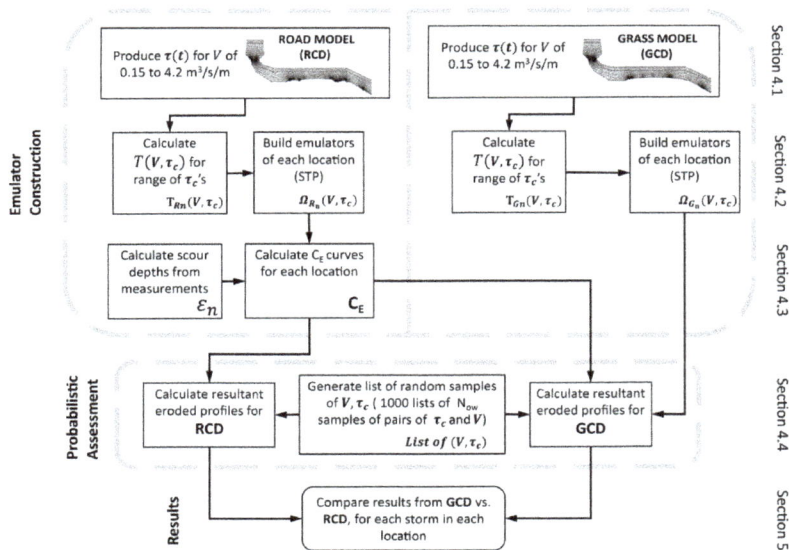

Figure 3. Flow chart of the steps of the proposed methodology.

4.1. CFD Models

Detailed two dimensional (2D) CFD Reynolds averaged Navier-Stokes equations (RANS) with k-ε turbulence closure solution models are considered better approximations of flow models where transient bed shear stresses are required with respect to uniform flow-based methods. They simulate turbulent flows for which rapid dissipation of kinematic energy is converted into internal energy in the form of turbulence. The inclusion of turbulence derived effects in the bed shear stresses becomes more important when abrupt geometric bottom changes and local surface roughness variations are present (e.g., RCD bottom profile, see Figure 4).

The RCD and GCD models were built using the COMSOL Multiphysics® software (COMSOL Inc., Stockholm, Sweden) for solving the 2D RANS k-ε equations for a two-phase field domain (air/water) based on a two equation (k-ε) turbulence closure model. This model accounts for turbulence effects by including two additional unknowns; the turbulent kinetic energy (k) and the rate of dissipation of kinetic energy (ε). Both are used for representing the transfer of momentum inside the flow due to turbulent eddies represented by the eddie viscosity. In addition, a phase field two equation transport approach is used for the interface tracking. For further information of the numerical solution see Ref. [40,41]. Details of the construction of the grid are given in Ref. [35].

The first CFD model (RCD), originally built in Ref. [26,35], was based on the scanned profile obtained as a final condition after the 0.01 m³/s/m experiment (q10_sc2: where q implies the mean overtopping discharge in l/s/m and sc2 refers to scan 2, the first profile scan after the q10 event, see Table 1). This profile includes an asphalt road of 3.1 m width with adjacent eroded transition gaps on each side of approximately 0.5 m each (Figure 4). Additionally, 2.5 m of the landside slope (approx. 1:3) is included in the model domain. For a detailed description of the RCD model setup, the accuracy of the mesh refinement and selected turbulence model we refer to Ref. [26].

A second model (GCD) was built to recreate a dike which could be tested with the same hydrodynamic conditions, but without including the asphalt road and transition gaps present in the original model. This model was built based on the original Millingen experiment dike profile (q10_sc2, Section 3.1), but the transition gaps were removed with a line smoothing procedure. The asphalt cover roughness was replaced by the roughness of grass (Figure 4). The rest of the profile remained unmodified in terms of the original elevation and slopes with respect to the RCD model.

The simulation principle consists of defining a water domain inside the WOS equivalent to the wave volume to be routed and run the model by letting the water to flow free from the domain during 15 s to ensure the whole volume has left the system (besides water retained in concave zones).

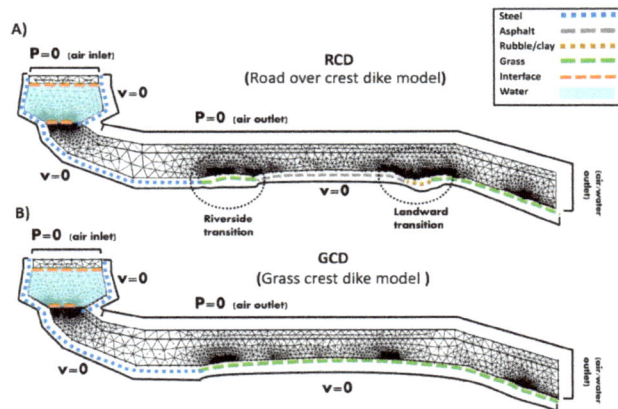

Figure 4. CFD models boundary conditions of (**A**) Road over crest dike model and (**B**) Grass over crest dike model (schematic meshing not in actual size on both models).

Both GCD and RCD models are used for producing the bottom shear stress time series ($\tau(t)$) along the bottom profile by releasing a certain volume contained in the WOS (see Figure 4). The bottom shear stress is calculated as the product of the flow density times the boundary shear velocity (u^*) squared. This shear velocity is calculated parallel to the bottom surface which allows to account for the fact that the erosion model was developed assuming a flat bottom. This model also allows one to estimate the tensile stresses in the normal direction of the bottom, but for the present method it is assumed that the erosion is a product of only the tangential shear stresses. The Manning's roughness coefficients used for both RCD and GCD models were obtained from the measured values reported in Ref. [39]. The steel roughness was used as the calibration value for achieving the measured velocities in the outlet of the WOS. Further information about the calibration may be found in Ref. [1]. The transformation from

Manning values (n) to equivalent sand roughness (K_s) values required as input by the software was done with Ref. [42]:

$$n = 0.04 K_s^{\frac{1}{6}} \tag{6}$$

The values used for the different boundary material roughness are presented in Table 2.

Table 2. Roughness coefficients used for both road and crest models from [39].

	n	K_s [m]	Source
Surface	**[s/m$^{1/3}$]**	**[m]**	**-**
Asphalt	0.016	0.0047	[39]
Grass	0.025	0.0680	[39]
Steel	0.017	0.0068	[43]
Rubble/Clay	0.026	0.0670	[43]
Geotextile [1]	0.024	0.0660	[44]

[1] This value was not used as input in any of the two CFD models, because its value is similar to grass.

Simulations for 0.15, 0.4, 0.7, 1.0, 1.5, 2.5, 3.2, and 4.5 m^3/m wave volumes (V) were performed in both RCD and GCD models. These values were chosen such that different orders of magnitude could be covered in the emulator training set while also including wave volumes which were measured in the Millingen experiment (0.15, 0.4, 1.0, and 2.5 m^3/m) for the CFD models validation (presented in appendices A, B, and C). From each simulation, bottom shear stress time series were generated in 33 different locations along the profiles of both models (Figure 5). From now on these locations are referred to as STP1 until STP33.

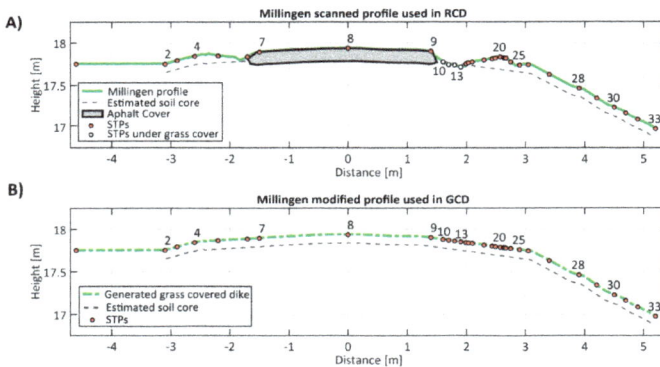

Figure 5. Cross section of (**A**) Road Covered Dike RCD; (**B**) Grass Covered Dike GCD model domains with study points (STPs) and estimated grass cover layer thickness (0.1 m).

These 33 study points (numbered dots in Figure 5) were selected for estimating the erosion depths at the locations where the bottom slope line segment changed significantly with respect to the previous line segment. All 33 STPs are located in the same horizontal position in both RCD and GCD models. The manual smoothing procedure of the RCD profile for generating the GCD profile consisted of placing straight lines between STPs 5 and 7, 9 to 20, and 23 to 26 (Figure 5). According to the field measurements [39,45], the average thickness of the grass cover was around 10 cm (dashed lower line in Figure 5). Points that are below this line in the RCD are located inside the clay zone (e.g., STPS 10 to 13 in the RCD, Figure 5).

The RCD model and its validation are presented in Ref. [35] and Appendixs A–C of the present study. As the Millingen experiment included a road at the crest, no experimental measurements were available for validating the GCD model. Yet, maximum value data from two other WOS experiments performed in the Vechtdijk in Zwolle [46] and the Tholen dike near Nieuw-Strijen [39] were used for validating the GCD model for which good agreement was obtained (See Appendixs A and B).

4.2. Emulator Surfaces Construction

Bed shear stress time series produced for each STP location from each CFD model and each wave volume (0.15, 0.4, 0.7, 1.0, 1.5, 2.5, 3.2, and 4.5 m^3/m) are used to generate the training data of the emulators. To do that, each of these time series was integrated in time for one given value of τ_c in steps of 1 N/m^2 until 300 N/m^2. The τ_c range was defined based on the recommended values for grass and soil presented in Ref. [36]. As a result, lists of training data sets of three columns (V, τ_c, and T) for each STP location were obtained and used to train the emulators.

A data-based emulation approach was selected [27], as we are only interested in the final T values for estimating the erosion depth in each location. A data-based emulator reproduces input-output relations from the input-output datasets produced by the CFD models. A 3D linear interpolation surface was used to build the emulators [47]. Each emulator on each location is defined as Ω_{R_n} and Ω_{G_n} according to the referred dike model. The Greek letter Ω denotes emulator, letters R or G denote either "RCD" or "GCD" and the sub index n denotes the STP location. These emulators were built based on the Matlab® (MathWorks Inc., Natick, MA, U.S.) interpolation scientific package and were designed as 3D surfaces for estimating the $T_{G_n}(V, \tau_c)$ and $T_{R_n}(V, \tau_c)$ functions in each STP. The emulators on STP8 and STP28 ($\Omega_{R_8}, \Omega_{G_8}, \Omega_{R_{28}}, \Omega_{G_{28}}$) are plotted as an example in Figure 6 to show what the surfaces look like. In total, 66 emulators were built to predict T as a function of V, τ_c; 33 for the RCD and 33 for the GCD. These emulators can calculate the set of T values of 1000 storms in less than 10 s on a standard personal computer. Note that the surface that corresponds to the emulator of a dike with a road on top ($\Omega_{R_{28}}$) presents a less smooth behaviour along the slope with respect to the grass emulation surface in the same location ($\Omega_{G_{28}}$), which demonstrates that there is indeed a significant effect derived from the presence of the road upstream. This also shows that the overtopping bed shear stress process is non-linear and may be under- or over-estimated when irregularities are not explicitly considered.

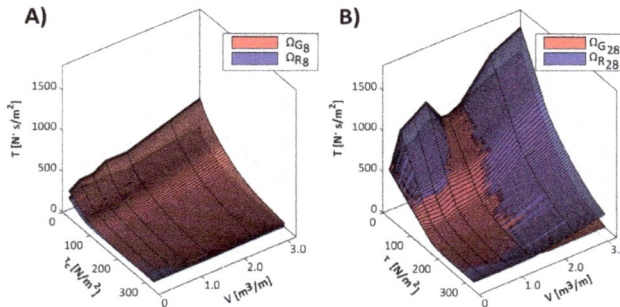

Figure 6. Example of the 3D linear interpolation surfaces (emulators) of STP8, Ω_{R_8} and Ω_{G_8} (**A**) and STP28, $\Omega_{R_{28}}$ and $\Omega_{G_{28}}$ the landward transition (**B**).

For estimating the scour depth per wave, the emulators become very powerful as they are equivalent to the solution of the excess shear stress integrals present in the right side of Equation (4) (Figure 1), as presented in Equations (7) and (8) for both GCD and RCD cases, respectively.

$$\Omega_{G_n}(V, \tau_c) \approx \int_{t_i}^{t_{i+1}} (\tau_{G_n}(V, t) - \tau_c)dt \qquad (7)$$

$$\Omega_{R_n}(V, \tau_c) \approx \int_{t_i}^{t_{i+1}} (\tau_{R_n}(V, t) - \tau_c) dt \tag{8}$$

These emulators are valid for estimating the T values of any given wave volume (V) value between 0.150 and 3.2 m^3/m and for any given τ_c value between 1 and 300 N/m^2. The τ_c value reflects the grass quality and hence when erosion occurs. At the same time, the resistance is also affected by the coefficient of erodibility C_E (see Equation (4)) which represents the resistance of the cover to be eroded which is referred in this manuscript as the grass quality.

4.3. C_E Functions

The C_E coefficient determines the resistance of the grass to be eroded as a function of the τ_c in the selected erosion model as shown in Equation (2) [19]. In the present methodology, it is assumed that C_E is a function of τ_c instead of a constant value which ensures that the same grass quality is reflected in both the estimation of T and the erosion depth estimation so that Equation (2) still remains valid for any τ_c. These C_E functions were built using the pre-trained emulators, the WOS released volume list from the experiment Part I, and the q10_sc2 and q50_sc5 scanned profiles [39]. These two scans represent the initial and final state of the 0.05 $m^3/s/m$ storm condition experiment. The functions were calculated for each STP as:

$$C_{E_n}(\tau_c) = \rho'_{cover} \frac{-\int_{x_i}^{x_{i+1}} d\varepsilon}{\Omega_{R_n}(V, \tau_c)} \tag{9}$$

This equation was obtained by substituting Equation (8) into Equation (4). This equation allowed us to generate C_E curves for each SPT for an arbitrary τ_c range between 1 and 300 N/m^2 in steps of 1 N/m^2. In each step, the actual release list of the volumes corresponding with the 0.050 $m^3/s/m$ storm condition was used as input for the RCD emulators to obtain a cumulative value of T for evaluating Equation (9). The initial and final profiles of the 0.050 $m^3/s/m$ storm condition were chosen for building the C_E functions, as for storm conditions with a lower q_m no significant scouring was obtained during the experiment. The scour depths (ε) per location due to the $q_m = 0.050$ $m^3/s/m$ experiment were obtained in each STP by subtracting both profile scans (Figure 7 lower plot).

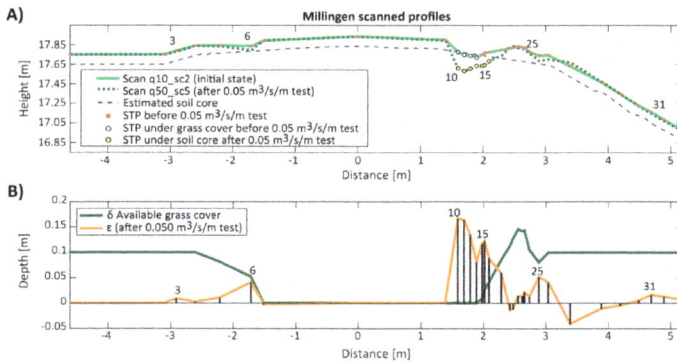

Figure 7. (**A**) Cross section of the GCD showing two scanned profiles of the Millingen dike before (q10_sc2) and after (q50_sc5) the 0.05 $m^3/s/m$ experiment; (**B**) Available grass cover thickness (δ) calculated as the difference initial state q10_sc2 and estimated soil core, scour depths (ε) calculated as the difference between the q10_sc2 and q50_sc5 scanned profiles.

They are equivalent to the solution of the negative integral present in Equation (9) for each location. For STPs 18, 19, 27, 28, and 29 where accretion was observed, the results were discarded and replaced by the closest one where erosion was found as the erosion model assumes that all material is fully washed to the downstream part.

Different grass cover relative densities (ρ'_{cover}) per unit of width are required for each C_E curve depending on whether the final profile is located in the grass layer or inside the deeper clayey soil layer. As no field data was collected regarding this parameter, density values for each type of cover (grass or soil), reported in Ref. [48] were used as reference. They correspond to a saturated sample extracted from a Danish dike which was composed of 0.17 m of soil and 0.03 m of grass with a total average density of 1870 kg/m^3. For this study, we assumed a reference density per unit of width value of soil ($\rho_{cover_{soil}}$) of 2000 kg/m^3. Based on these values, it is estimated that the saturated density of grass solely ($\rho_{cover_{grass}}$) is 1100 kg/m^3. The resultant C_E curves obtained by the use of Equation (9) are presented in the results Section 5.2. It is acknowledged by the authors that these values may differ from the actual ones of Millingen. However, they are of good order of magnitude and even result in very similar C_E values with respect to the ones reported in Ref. [19].

4.4. Probabilistic Safety Assessment

In general, failure mechanisms can be expressed as a limit state equation:

$$Z = R - S \tag{10}$$

where the variable S represents the load (solicitation) and R represents the resistance of the structure. The Z term represents the marginal resistance which defines the state of the system as safe when positive and in failure when equal to zero or negative. Wave overtopping failure is a major threat to flood defence safety, because when the grass cover is completely lost, a rapid scouring process of the soil core may initiate a dike breach. Hence for the present study, failure is defined as the complete loss of the available grass cover in any location (STP). The grass cover layer thickness (δ) is defined as the top 10 cm (Figure 5), based on Ref. [32,38].

The dike grass cover limit state function is obtained by rearranging Equation (4) and making it equal to the marginal strength term for each overtopped wave and then summing over all overtopping waves (Equation (11)). Here, δ represents the resistance term (R) and the second term describes the erosion that represents the solicitation (S).

$$Z = \delta - \sum_{i=0}^{N_{ow}} \frac{C_E(\tau_c)}{\rho'_{cover}} \int_{t_i}^{t_{i+1}} (\tau(V,t) - \tau_c)dt \tag{11}$$

The failure state is reached if the grass cover is completely removed ($Z \leq 0$). The emulators were used to compute the excess shear stress per overtopping wave. All overtopping waves during a storm condition together yield the accumulated scour depth, which is evaluated for every STP along the dike profiles for the cases with (Z_R) and without (Z_G) a road on top. The δ values in each location and their corresponding emulators were replaced in Equation (11) such that the limit state equation of the grass cover for both cases with and without a road in each location could be evaluated. The resultant limit state equations per STP are defined as the accumulated scour depth per storm condition (Z_R and Z_G) as shown in Equations (12) and (13):

$$Z_R = \delta_{R_n} - \sum_{i=0}^{N_{ow}} \frac{C_E(\tau_c)}{\rho'_{cover}} \Omega G_n(V, \tau_c) \tag{12}$$

$$Z_G = \delta_{G_n} - \sum_{i=0}^{N_{ow}} \frac{C_E(\tau_c)}{\rho'_{cover}} \Omega R_n(V, \tau_c) \tag{13}$$

The available grass cover per location (δ_{R_n} and δ_{G_n}) is defined as a deterministic variable and was calculated as the difference in elevation between the surface profile and the soil core line (dashed lower lines in Figure 5). The cover relative densities of grass (ρ'_{cover}) were assumed to be constant all along the dike profile.

For every storm condition, possible storm realizations that consist of N_{wo} sets of values of V, τ_c were randomly sampled via crude Monte Carlo. These values are used to evaluate Equations (12) and (13) which result in one single value of Z_R and Z_G per realization for each STP. Statistical distributions of Z_R and Z_G per location were built from 1000 realizations of overtopping waves. For each tested storm condition (see Table 3, q_m), the random sampling was done following a pre-fitted Weibull distribution. The failure probabilities per storm condition are then calculated from the distributions as P_{f_R} ($Z_R \leq 0$) and P_{f_G} ($Z_G \leq 0$) in each location. For each realization, a final eroded profile was also estimated by subtracting the accumulated scour depth (second term of Equations (12) and (13)) from the initial elevation profile (green lines of Figure 5). Note that each estimated value is conditioned to the event of the simultaneous occurrence of the boundary conditions of wind speed and water level. For the present study, the same boundary conditions were used for each storm condition but the crest level was modified so that different q_m values could be obtained.

Individual wave volumes, V, were sampled from the two parameter (a and b) Weibull distribution [49]. For the present study, it is intended to represent storm/wind surge overtopping events of duration D [h] composed of individual single wind generated waves V_i that result in an overtopped volume distribution [50]. The cumulative exceedance function of this distribution is expressed as:

$$P_V = P(V_i \leq V) = 1 - exp\left[-\left(\frac{V}{a}\right)^b\right] \tag{14}$$

For the scale (a) and shape (b), the method presented in [50] improved the estimation of b for wave overtopping by fitting an empirical equation to experimental results (Equation (16)).

$$a = \left(\frac{1}{\Gamma\left(1+\frac{1}{b}\right)}\right) \cdot \left(\frac{q_m \, T_m}{\%_{ov}}\right) \tag{15}$$

$$b = \left[exp\left(-0.6\frac{R_c}{H_{m0}}\right)\right]^{1.8} + 0.64 \tag{16}$$

where q_m represents the average overtopping discharge [m³/s/m] of the storm, $\%_{ov}$ represents the overtopping percentage [-], R_c represents the dike free crest height [m], T_m represents the mean wave period [s], and H_{m0} represents the incident energy based significant wave height [m]. Former Dutch guidelines [11] define the allowable q_m between 0.0001 and 0.01 m³/s/m depending on the type of cover, the outer slope revetment and the importance of the structure in the system [11]. This range plus four larger storm conditions were tested (Table 3).

Table 3. Characteristic values of dike configurations and 7 tested storm boundary conditions for riverine dikes for the tested range of overtopping discharges.

	Mean Overtopping Discharge q_m [m³/s/m]						
	0.0001	0.001	0.010	0.020	0.050	0.075	0.100
D [h]	6	6	6	6	6	6	6
R_c [m]	2.99	2.20	1.40	1.17	0.85	0.71	0.61
Slope [-]	1:3	1:3	1:3	1:3	1:3	1:3	1:3
N_w [-]	6545	6545	6545	6545	6545	6545	6545
N_{ow} [-]	65	458	2291	3142	4451	4974	5367
$\%_{ov}$ [-]	0.01	0.07	0.35	0.48	0.68	0.76	0.82
V_{max}^{1} [m³/m]	0.21	0.46	0.94	1.27	1.91	2.35	2.72

[1] Maximum expected volume from Weibull distribution [12].

All the storm conditions tested in this study (Table 3) corresponded to a wave ($H_s = 1$ m and T_p = 4 s), wind speed, and water level event with exceedance probability $P(H_s, T_p, q_m) = 1 \times 10^{-4}$, which is the design event (in the former legislation) for this particular dike location. All estimated failure probabilities obtained from the emulators are conditioned to this event ($P_{f_R}(Z_R \leq 0 | H_s, T_p, q_m)$ and ($P_{f_G}(Z_G \leq 0 | H_s, T_p, q_m)$). For calculating the failure probability per location on each model we rely on the use of the conditional probability so that $P_f(Z \leq 0 \cap H_s, T_p, q_m) = P_f(Z \leq 0 | H_s, T_p, q_m) \times P(H_s, T_p, q_m)$.

For the τ_c random variable, there is no available literature to our knowledge that studied or suggested the stochastic nature (distribution or moments) of this variable for grass. However, from the values reported in Ref. [39], an equivalence curve between critical erosion velocities U_c and τ_c was built based on the table presented in Ref. [36], as shown in Figure 8. This equivalence curve allowed the transformation to define the C_E as a stochastic random variable.

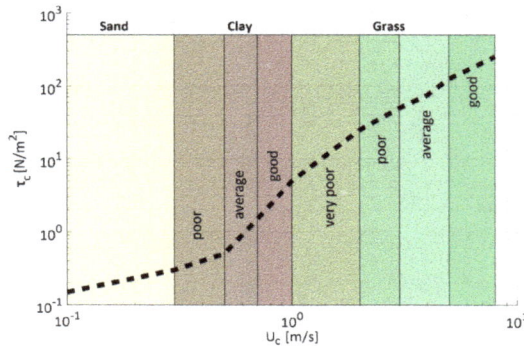

Figure 8. Equivalence curve between critical velocity, U_c, and critical shear stress, τ_c, built from values reported in [14]. This curve is used to include C_E as a stochastic random variable by transforming the stochastic U_c from Table 4 to τ_c values, which is required as input for C_E using Equation (9).

The stochastic distributions for the U_c of grass and soil, were taken from Ref. [21]. This work concluded that U_c can be represented by a log-normal distribution for the grass cover and by a generalized extreme value (GEV) distribution for bare soil locations. The parameters for the GEV distribution could not be estimated from the experimental data collected during the Millingen experiment. Hence, a log-normal distribution was also assumed for the bare soil spots, as presented in Table 4.

Table 4. Stochastic random variables of cover quality U_c used as input for C_E calculation. CoV is the coefficient of variation and QCF is grass quality correction factor for poor, average, and good.

		Clay U_c	Grass U_c		
		Good	Poor	Average	Good
Distribution	[-]	Log-norm	Log-norm	Log-norm	Log-norm
mean	[m/s]	0.85	3	4	6.5
CoV	[-]	0.1	0.3	0.3	0.3
QCF	[-]	-	1.5	1	0.1

Based on the indicative average values of τ_c presented by Ref. [36], it was possible to define mean values of U_c per grass and soil qualities. The coefficients of variation (CoV) were calculated from the U_c statistical distributions presented by Trung [21]. The grass quality correction factor (QCF) is obtained by estimating the value which shifts the "average" C_E grass quality function towards the recommended values in Ref. [19].

5. Results

5.1. CFD Calibration and Validation

Calibration and validation of the RCD model were carried out as in Ref. [26], using the observations of the Millingen experiment part II. In Ref. [26], it is concluded that although the depth simulations show some deviations, the simulated velocities along the crest and slope are within 20% of the observations. A detailed analysis of the effects and estimated errors derived from the mesh refinement, the CFD RANS parametrization, and the temporal effects in the shear stress time series error can be found in Ref. [26]. The flow depths and velocities for the GCD case are in good agreement with the ones fitted for the Vechtdijk experiment (Appendixs A and B). Also, in this case, the simulated velocities are more accurate than the depth simulations. In the modelling approach, erosion is computed based on excess shear stress. Appendix C shows that for most wave volumes, both the GCD and RCD models simulate the duration of the overtopping waves quite well. Especially, the magnitude and timing of the velocity and depth at the wave fronts are simulated well.

5.2. Effects of Turbulence on the Excess Shear Stress $\overline{T}$

Both the RCD and GCD model results show highly turbulent flow over the dike profile with Reynolds values between 10,000 and 200,000 for the 0.15 and 3.2 m³/m volumes, respectively, with supercritical flow (Froude between 10 and 2 for 0.15 and 3.2 m³/m volumes, respectively). Low Froude numbers are observed for the RCD model in STP 14 Fr $\approx$ 1 at the adverse bottom slope. For abrupt changes in the bottom geometry, the use of a single depth-averaged turbulence factor along the entire profile may either underestimate or overestimate the locally exerted bottom shear stress as the turbulence intensity changes significantly along the dike profile. This can be observed in Figure 9A, where the average kinematic energy in time ($\overline{k(t)}$) for waves of 0.4 m³/m and 3.2 m³/m in both GCD and RCD models is shown. These plots show that the highest values of turbulence occur at the locations where abrupt bottom changes (irregularities) are observed.

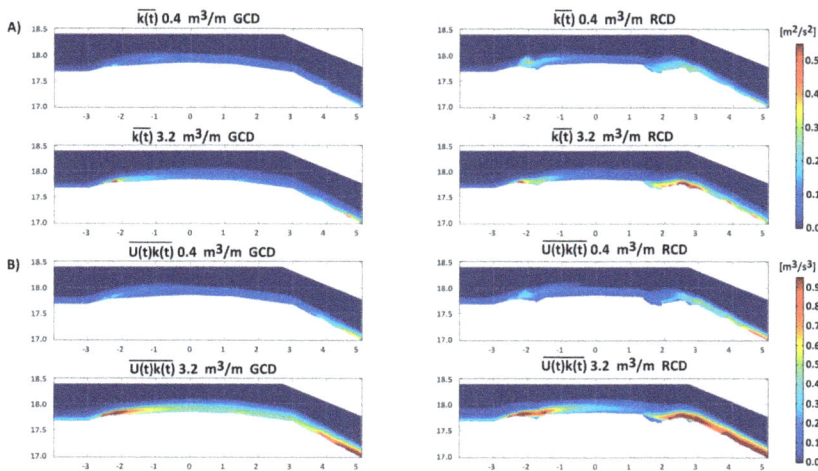

Figure 9. (**A**) Average kinetic energy ($\overline{k(t)}$) of single wave volumes of 0.4 m³/m and 3.2 m³/m and (**B**) average turbulent erosive potential ($\overline{U(t)k(t)}$) for single wave volumes of 0.4 m³/m and 3.2 m³/m.

Nonetheless, the amount of erosion does not solely depend on the amount of turbulence but also on the momentum of the fluid. This implies that high erosion potential may also occur at low turbulence locations due to the acceleration of the flow. Hence, the average of the product between mean velocity ($U(t)$) and kinematic energy ($k(t)$) averaged in time allows one to determine the highest potential erosion locations, as shown in Figure 9B. Note that these plots do not directly imply which locations are more prone to erosion as the cover resistance is also a determining factor.

The bottom shear stress $\tau(t)$ intrinsically contains the effects of turbulence. Hence, they are influencing the probabilistic estimation as they also influence T. To observe this effect, the release volume list used as input for the 0.050 m³/s/m test is used to calculate a T value per volume while assuming $\tau_c = 0$. From the produced set of T values, the average shear stress excess ($\overline{T}$) value and 95% confidence bounds per location for each of the 33 locations in both cases were calculated and plotted in Figure 10.

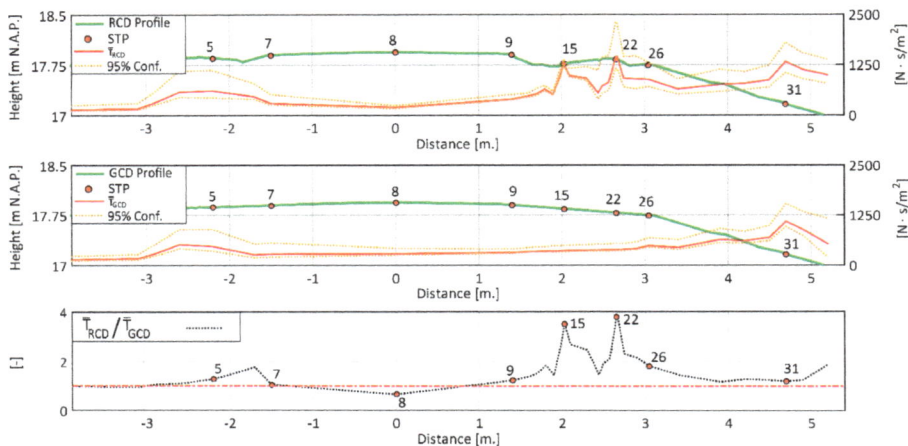

Figure 10. Mean values of T (red line, right axis) and 95% confidence intervals (dashed lines) for RCD (upper plot) and GCD (middle plot). Lower plot shows ratios of RCD and GCD mean T values at each location.

The ratio between the $\overline{T}$ values of the RCD and GCD cases (Figure 10, bottom) shows that $\overline{T}$ is around 2 to 4 times larger in the road transition zones, which correspond to the high turbulent kinetic energy regions on Figure 9. Additionally, the $\overline{T}$ values in the upper part of the landward slope (between STPs 26 and 33) were larger in the RCD with respect to the GCD. This behavior was surprising as the landward transition gap was expected to dissipate energy. However, the turbulence generated in the transition propagates downstream (see $\overline{U(t)k(t)}$ in Figure 9) and probably induces an increase in the potential erosion of the landward upper slope.

From these results, it is also possible to identify the effects of surface roughness and bottom surface irregularities separately. On one hand, due to form roughness, the $\overline{T}$ values increase significantly after STP 9 where evident abrupt bottom changes are present in the RCD case. On the other hand, the $\overline{T}$ in the centerline location (STP8) of Figure 10 is lower in the RCD case, which can be attributed to the smoother surface. This can be corroborated from Figure 9 at the same location where the bottom slope is almost the same in both models. At this location, the turbulent kinetic energy is also very similar and yet the erosive potential is much greater in the crest for the GCD.

Except for STP8, all locations present equal or larger $\overline{T}$ values in the RCD with respect to the GCD. From a probabilistic point of view this means that on average, the probability of scouring at these locations will be higher if grass quality and cover thickness remain constant along both profiles.

Interestingly, the 95% confidence bounds show a wider spreading with respect to the mean values in all locations with high erosive potential. This is concluded by correlating the spreading of the confidence bounds per location of Figure 10 with respect to the locations of highest values of

$\overline{U(t)k(t)}$ in Figure 9B. Similarly, it can be observed that at locations where bottom slope changes are minimal (e.g., STP8), less spreading of the confidence bounds is observed which indicates that there is less turbulence.

5.3. CE Curves Calculated from Millingen Measurements

The C_E curves obtained from the experiment data are presented in Figure 11. The results are presented in Figure 5 and correspond to an estimation of the corresponding curve of grass quality of each of the places where erosion was found along the dike cross sectional profile during the Millingen experiment. This figure only includes the resulting CE curves for STPs located after the asphalt cover (STPs > 9), which were influenced by the presence of the road. The right plot of Figure 11 shows the CE curves obtained for SPTs inside the gap where no grass cover was present and consequently correspond to soil erodability values.

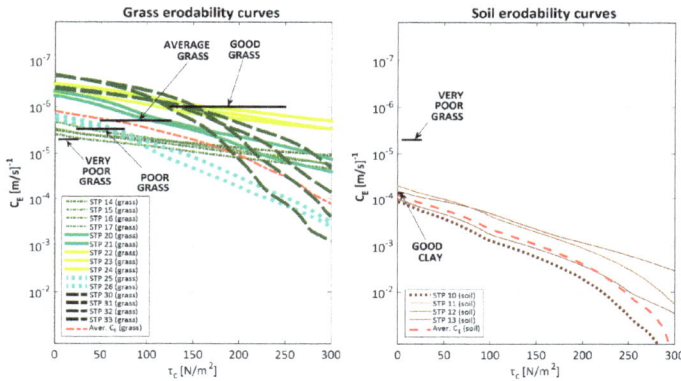

Figure 11. C_E curves for each STP for grass (left) and bare soil (right). Horizontal lines represent the grass quality reference intervals for very poor, poor, average, and good grass cover quality reported in [19].

The C_E curves from the STPs located closer to the road (STPs 14 to 17 plotted as dense dashed lines in Figure 11) cross the grass quality reference intervals suggested by [19]. Thus, we classified them between the very poor and poor grass quality. For STPs located over the vertex plotted as continuous thicker lines (STPs 20 and 26) coincide with average and good grass quality reference intervals. For STPs 25 and 26 which are located in the dent after the vertex, poor grass quality C_E curves were found. The presence of such a dent may be a good indication of deterioration due to traffic which explains such low quality just after spots of good quality. The rest of the profile (mostly along the inner slope in STPs 30 to 33) present good grass quality. These results are in good agreement with the field findings for grass quality reported by [39]. For the STPs located initially in the soil zone, good clay quality values were obtained (right plot of Figure 11). For the GCD case, no information about grass quality was available. Therefore, we used the average C_E curve, corresponding to average grass quality. It is acknowledged that a dike without a road may have better quality in reality than what we assumed as we used the transition zone data in the averaging procedure for obtaining the C_E "average curve".

5.4. Scour Depth Profiles

The results of the simulations for the storm conditions of q_m 0.0001, 0.001, and 0.010 m³/s/m showed almost no scouring with the 1000 samples. Therefore, only the scour profiles for the largest q_m (0.02–0.1 m³/s/m) were analyzed. Each profile was produced by plotting the mean value of the final scour depths for each STP after the 1000 realizations generated for each q_m (Table 3). These values are equivalent to the mean value of the probability density function obtained for the second term of Equations (12) and (13) for the 1000 sampled realizations in each STP. The profiles were only analyzed for the STP locations after the asphalt cover.

For the first part of the analysis, the simulations were done assuming a uniform average grass quality along the whole profile. The obtained scoured profiles (see Figure 12) show that the large failure spot in the lower part of the RCD may be explained from the additional energy available due to the presence of a smoother upstream surface (asphalt cover) and the momentum gain when flowing downslope. Moreover, an eventual increase of the turbulent flow due to the slope bottom irregularity around $x = 3.8$ m near STP 28 may also contribute. Another significant scour zone is observed on the transition between bare soil (STP12) and the dike crest vertex (STP23). These deep erosion spots coincide with the locations where the initial profile has an adverse slope. These zones are identified as the initiators of turbulence in the flow as it can be observed in Figure 9.

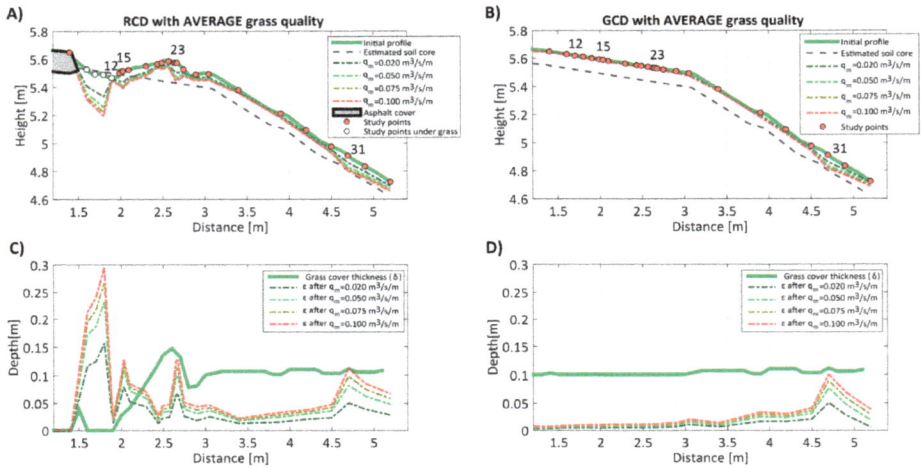

Figure 12. Results of scoured profiles for different q_m conditions on average grass quality for (**A**) RCD and (**B**) GCD. Lower plots show their initial grass cover thickness (δ) and the eroded depths (ε) after each storm condition for (**C**) RCD and (**D**) GCD.

From these results, the most interesting finding is that STP23 represents a weak spot in the profile despite the fact that the available grass cover is even thicker than most of the other locations. This can be explained based on Figures 9 and 10, where STP23 corresponds to one of the most turbulent locations. Again, this supports the importance of including a highly detailed geometry and its derived turbulence effects in modelling of dike cover erosion.

For a more realistic representation of the grass cover erosion, the grass qualities of each point were replaced by one of the three estimated qualities presented in Table 4, that seemed closer to the local condition per STP found in Section 5.2. The resultant scoured profiles (Figure 13) become safer in the lower part of the slope (STP 31) for both RCD and GCD as better grass quality than average was obtained in that zone.

However, two new close to failure locations were identified in STPs 25 and 26 which were not observed when assuming all grass quality as average. For the 0.1 m^3/s/m RCD storm profile, it can be observed how STP 25 is very close to failure as the scoured depth is very close to the available grass cover thickness (lower left plot in Figure 13). This location corresponds to the most prone to fail spot in the GCD given the poor grass quality present in that particular location.

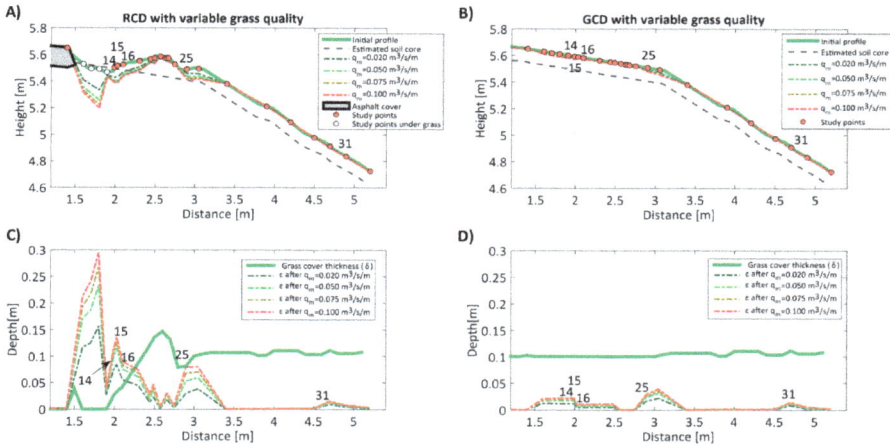

Figure 13. Results of scoured profiles for different q_m conditions on with the estimated grass quality for Millingen at each STP for (**A**) RCD and (**B**) GCD. Lower plots show their initial grass cover thickness (δ) and the eroded depths (ε) after each storm condition for (**C**) RCD and (**D**) GCD.

5.5. Probability of Failure

The failure probabilities for both dike cases are presented in Figure 14. For this study, any failure probability larger than 9.9×10^{-5} is assumed as an already failed spot. This value was assumed since STPs 10, 11, 12, and 13 (located in bare soil) presented values above this threshold and were already in failure state (no grass cover) before routing each different storm condition and where not taken into account for the dike assessment.

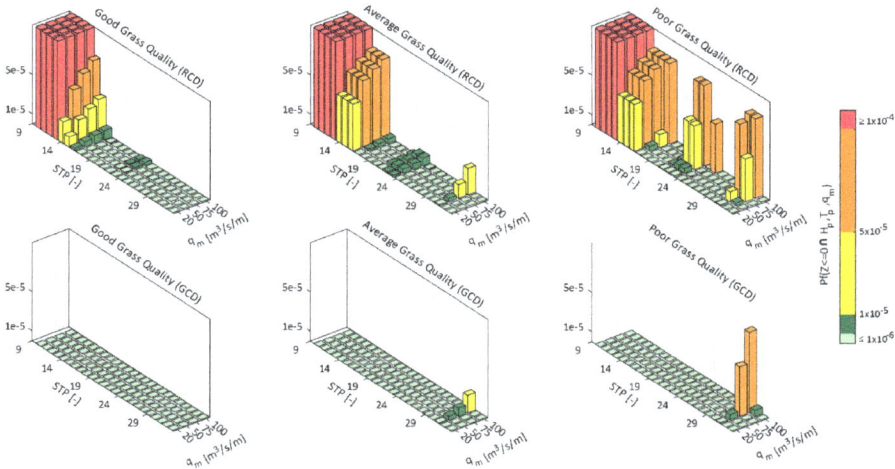

Figure 14. Failure probabilities per storm at each STP for different grass qualities for the RCD (upper three bar plots) and for the GCD (lower three bar plots). Bars in read coincide with locations in failure state before the storm condition was routed.

For the present study, it is assumed that the failure probability of the configuration to be studied is equal to the largest value found along the profile. This allows us not only to define the failure probability but also the spatial location of the failure which can later be related to the analysis of the origin of the failure.

Figure 14 shows that the dike without a road presents significant failure probabilities in the slope on STP31 with Pfs of 5×10^{-5} and 8.2×10^{-5} for the q_m 0.075 and 0.1 m^3/s/m. The results for this last configuration show that the probabilities of failure along the landward slope are almost identical for both RCD and GCD cases for the largest tested q_m. This means that when using the failure criteria of assigning the maximum failure probability along the profile for a single storm it could be concluded that both dikes with and without road have almost the same failure probability. Yet, for lower q_m (<0.075 m^3/s/m), the probabilities of failure of the landward slope are always higher in the dike with a road. It is also observed on STP25, that a sudden potential failure spot is created for a q_m larger than 0.050 m^3/s/m. It is described as sudden because the failure probability was less than 1×10^{-4} for lower storms with poor grass quality. This can be explained due to the highly turbulent flows expected in this location (see Figure 9).

Also, the assessment was performed with the spatially varying grass quality based on observed grass quality in the Millingen experiment to represent a more realistic estimation. Here, a sudden failure spot is observed in STP25 for a 0.1 m^3/s/m storm (see Figure 15).

These results show that for the dike with a road, the maximum probability of failure is equal to 5×10^{-5} for the largest storm in STP25. Based on all previous analyses, STP25 presented a high $\overline{T}$ value combined with poor grass quality and medium to high grass cover thickness ($\delta = 7.8$ cms). This means that for this particular place, turbulence plays a more important role than cover thickness as STP25 represents the highest failure potential spot on the dike while also presenting the highest value of average kinetic energy (see Figure 9). From this last figure, it is expected that this dike will present its first failures occurring nearby the already scoured zones (STP14 to STP18) with Pf range between (1×10^{-5} and 1×10^{-4}) for the case where the road is present. For the no road configuration, it is expected to have a Pf lower than 1×10^{-6}.

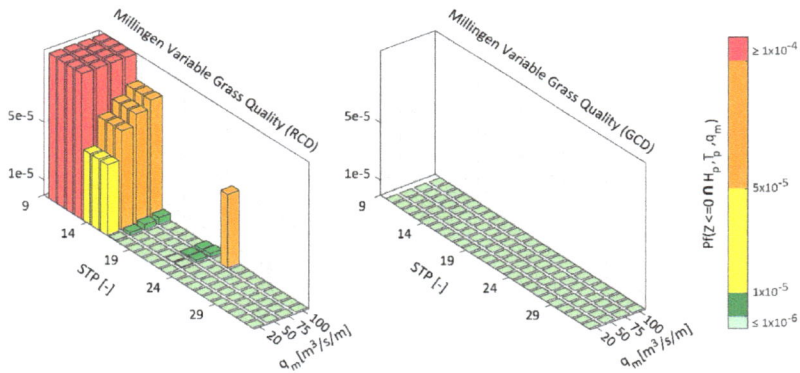

Figure 15. Failure probabilities per storm at each STP for the Millingen estimated grass quality per location for the RCD (Left side bar plot) and for the GCD (Left side bar plot). Bars in read coincide with locations in failure state before the storm condition was routed.

6. Discussion

6.1. Uncertainties in the Modelling Process

In this paper, a methodology is presented to predict the spatially varying probability of failure along the grass cover using emulation of a highly detailed CFD model. The P_f values obtained for each location are conditional to the upstream and the downstream locations. This means that the turbulence effects are captured in the emulators as they are trained based on the time series from the CFD model which includes them by solving for the kinematic energy variable in time and space. One of the main assumptions in this study is that the CFD models can be validated by checking the order of magnitude and trends of water depths and maximum velocities in two points along the profile (See Appendixs A–C). However, the main output of the CFD models used in this study is the produced bed shear stress which cannot be validated at this stage, because no data was collected regarding this quantity during the experiments. Yet, we are confident that the model is giving consistent results in terms of order of magnitude. The uncertainties of the CFD model and assumptions in the erosion model determine the accuracy of the emulators and therefore the probabilities. Therefore, in order to increase the accuracy of the final failure probabilities, the first step is to improve and better validate the CFD models. In general, the presented methodology can be applied to other CFD models as well.

Additionally, the scouring process is highly time-dependent as the dike profile changes in time (not only due to erosion, but also due to accretion). Consequently, hydrodynamics change in time as well. In that sense, the emulators built for this study become more uncertain for larger storms as the chances of significant change in the dike profile become larger. This means that the updating of the bottom profile should be done after each routed wave and not after the complete storm. However, for the present methodology, the resultant profiles are a function of the emulators, which use a set of 1000 realizations for each storm, generated independently from the set of volumes used in the Millingen experiment and yet they result in very similar scour patterns compared to the one observed during the Millingen experiment. This indicates that even though the CFD model and its emulators were not perfect, and the profiles were not updated after each wave overtopping, the bed shear stress stochastic nature is fairly well represented by the emulators, plus it allows one to make scenario assessments without rerunning the original models.

Regarding the erosion model adopted for this study, it is also acknowledged that it fully attributes the complete erosion process to the bottom shear stress, whereas, in reality, normal stresses also contribute. This choice was made to actually reduce the uncertainty as using more complex erosion models requires a larger amount of data with their associated uncertainties potentially making the model even more uncertain. Still, the results obtained for the erodibility functions (C_E) also show good agreement with the suggested reference values deducted from field measurements and literature. This may be a good indicator that the simpler adopted erosion model is sufficiently robust for probabilistic modelling.

6.2. Sensitivity to the Grass Quality Estimation

Grass quality is highly variable in space and difficult to determine [36] in field experiments. Therefore, we inferred the grass quality for the test location from the observed erosion depths and emulators and we subsequently compared different configurations of grass quality. The inferred grass quality was better along the slope than on the crest where thickness was smaller, but the results showed that grass quality is not necessarily correlated to the available grass cover thickness which is another factor that determines the resistance to erosion. An example of this can be found for STP21 and STP26 where ε is larger than 0.10 m for both locations and still the estimated grass qualities were good and poor. These examples show that grass quality also has a significant effect on the occurrence of reduced failure probabilities. For example, the failure probability in the RCD with more realistic grass quality distribution was 5×10^{-5} while presenting a poor local grass quality in STP 25. The failure probability in this same location for the same dike case is close to zero by just changing the grass quality to average (see Figure 14). This shows that the safety assessment cannot be done solely from the hydrodynamic point of view as the erosion pattern differs from the expected one using highest $\overline{T}$ values only. Poor grass quality spots have a high probability of failure despite the magnitude of q_m. Additionally, the RCD more realistic model also presents a failure probability of 6×10^{-5} in the landward vertex (STP 25, Figure 15) during the most severe storm, whereas the realistic GCD has almost a negligible probability of failure in this same location. The variability on the grass cover may be attributed to an external factor such as traffic which, for the case of a dike with a road on top, is inevitable.

Accurate observations of grass quality were therefore essential in our study case. However, we used a relatively simple approach to infer the grass quality from the erosion scanned profiles. For further applications of the methodology, detailed grass quality data is required, which can be derived from dike inspections or, possibly, remote sensing methods.

It is acknowledged that grass quality will also vary in the longitudinal direction. The present study assumes that the dominant direction of the generated shear stresses occurs in the cross-sectional direction (from crest to landside) and, therefore, the most probable way to determine failure is in that direction. However, with the constant modification of the cross section during the erosion process it can happen that grass quality and shear stresses importance vary their dominant direction. These effects cannot be considered with the present method and may be worth further studying.

6.3. Applications of the Probabilistic Method

The results show that the probability of failure if a road is present is higher along the simulated dike section, with respect to the case without a road, for all storms and grass qualities tested. However, for the GCD case with variable grass quality, all estimated failure probabilities (Figure 15) were less than 1×10^{-6}. This result does not imply that the dike is 100% safe at these spots, because the small failure probabilities are predicted less accurately. The reason for the latter is that the 1000 samples of the Monte Carlo assessment were probably not sufficient to explore the tail of the distribution in depth as their associated P_f can be smaller than the ones obtained for a Monte Carlo standard error >3%. The methodology presented in this paper enables the use of such a high number of computations within a feasible time period. However, a more extensive probabilistic analysis, including the uncertainties in the modelling process and increasing the number of samples (lower standard error in the Monte Carlo) and using smart sampling techniques to cover the tail will greatly increase the accuracy of the predicted probability of failure.

The method can be extended by including the wave parameters and water levels during the random sampling by making them stochastic variables fitted to data series of storm wave conditions and water levels [51]. This allows one to extend the method to a fully probabilistic analysis by including the wind speeds water levels uncertainties explicitly. The resulting overtopping volumes are used as input for the emulators presented in this study, allowing one to derive fully probabilistic failure values along the dike profiles. Nevertheless, it remains a challenge to fully analyze the implications of such an extensive analysis and this is recommended for further study.

One of the main assumptions in the methodology was the definition of failure of the dike cover if the erosion depth exceeded the grass cover thickness of 10 cm. This assumption is highly arbitrary and debatable. However, no definition of failure for wave overtopping is available, other than a dike breach, which is a highly complex process caused by many interacting failure mechanisms and therefore too difficult to model. The adopted 10 cm threshold has obviously large consequences for the resulting probabilities. In order to achieve realistic values for the failure probability due to wave overtopping, we recommend more research to achieve a more advanced definition of the threshold of failure.

To summarize, the precise quantification of the low failure probabilities is more uncertain, but it is expected that they will still be in the obtained order of magnitude, which is sufficient to draw a good conclusion from the study in terms of trends and locations. The spatially distributed failure probabilities become useful, not only to determine the most prone to failure location, but also to stress their analysis during the yearly visual inspections for Dutch dikes. The quantification of the local probabilities of failure is input for any probabilistic dike safety assessment or probabilistic design. These results become even more relevant in cases of dike concepts such as the unbreachable dike and the multi-functional flood defence [51] which are designed to cope with climate change [5] and to withstand large amounts of overflow and storms similar or larger than the largest ones tested in this study.

7. Conclusions and Recommendations

This study shows a methodology to assess the spatially distributed probability of failure of a dike, with and without a road on the crest, due to wave overtopping erosion. In our case, the presence of a road on the dike increased the erosion of the grass cover and increased the probability of failure locally and for the entire dike profile. Probabilities were derived using a novel emulation method. Based on the obtained results of the study, specific conclusions for the two cases can be summarized as:

1. The dike with a road showed higher probabilities ($5 \times 10^{-5} > P_f > 1 \times 10^{-4}$) of failure with respect to a dike without a road ($P_f < 1 \times 10^{-6}$) if realistic grass quality distribution was assumed. The probability of failure in the zones where higher initial deterioration was observed (e.g., locations immediately next to the road and over the vertex) is significantly higher with respect to the remaining part of the dike. Yet, these locations also correspond to spots where higher turbulence was observed with respect to the remaining part of the locations along the dike profile. However, if good quality grass was present all along the dike, both dike cases will present very low failure probabilities along the slope and the vertex ($P_f < 5 \times 10^{-5}$).
2. Local erosion depth is highly influenced by the momentum of the water volume, which is reduced (energy dissipation) by a rough surface such as grass. If a road is present, both the smoother asphalt surface and the resultant bottom irregularities (possibly derived by traffic and material change) increase the scour potential for failure along the crest and the slope with respect to the dike without a road as they influence the generated and dissipated turbulence (kinetic energy) and its variability (turbulence intensities).

Finally, the coupled emulator-erosion model was able to yield realistic probabilities and trends, given all uncertainties in the modelling process. Yet, further improvements are required, especially for the CFD model validation, grass quality estimation, and erosion modelling. The present method can be extended by including model uncertainty coefficients as stochastic variables, to account for the errors originated from the imperfection of both the erosion and the hydrodynamic models. This is especially important to improve the erosion estimation of wave volumes located in the tail of the distribution.

Author Contributions: Conceptualization, J.P.A.-L., J.J.W., R.M.J.S. and S.J.M.H.H.; Formal analysis, J.P.A.-L.; Investigation, J.P.A.-L. and A.B.; Methodology, J.P.A.-L; Software, J.P.A.-L and A.B.; Supervision, J.J.W., R.M.J.S. and S.J.M.H.H.; Writing—original draft, J.P.A.-L and J.J.W.; Writing—review & editing, A.B., R.M.J.S. and S.J.M.H.H.

Funding: This research was funded by the Dutch Technology Foundation STW, part of the Netherlands Organization for Science NWO with project number [P10-28/12178].

Acknowledgments: This research is supported by the Dutch Technology Foundation STW, which is part of the Netherlands Organization for Scientific Research (NWO), and which is partly funded by the Ministry of Economic Affairs. The authors would like to thank J.R. and A.F.M. from Twente University, G.-J. S. and R. M. from INFRAM, Andre van Hoven from Deltares and J.V.d.M. from V.d.M. Consulting B.V. for providing the data, and the Integral and sustainable design of Multi-functional flood defences program for the help and support. Finally, we thank the reviewers and editor for their detailed comments on drafts of the manuscript that greatly improved the quality of the final paper.

Conflicts of Interest: The authors declare no conflict of interest.

Appendix A. CFD Validation on the Crest

For the validation of the RCD model presented in Ref. [35] and the GCD model, we use a set of measurements of velocity (PW) and depth (SB) that was taken at the measuring devices SB1, PW1 and PW2 (see [39]) as reference. The location of these measurements (at the midpoint of the road), correspond to the exact STP8 location (see Figure 5). In addition, the fitted curves for the results of other WOS experiments over the Tholen dike and the Vechtdijk are also included in the validation of the GCD model.

The maximum values obtained for different wave volumes are plotted as squares for the GCD case and diamonds for the RCD case. The RCD model results differ significantly with respect to the measurements performed in Millingen (dots) and the fitted curve of the Tholen dike. The latter results were expected to correspond with the Millingen experiment (see Ref. [39]), as a geotextile cover was placed between the midpoint of the road and the foreland slope to cover the damaged transition. Additionally, it was also reported that the water depth measurements at the SB1 were larger than expected when compared to the obtained values from other experiments. However, the results obtained for the GCD (see Figure A1) case in terms of flow depths and velocities are in good agreement with the ones fitted for the Vechtdijk experiment (upper dashed line). Furthermore, they also show lower velocities and larger flow depths with respect to the RCD model as the grass is significantly rougher than asphalt.

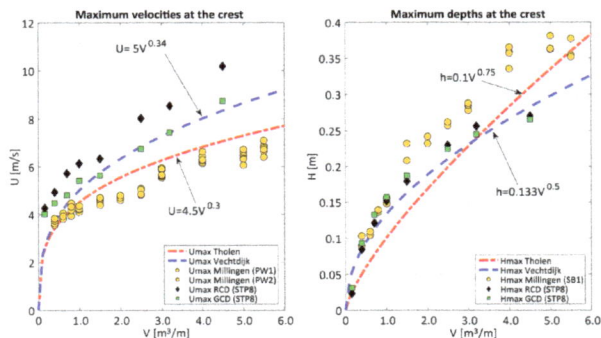

Figure A1. Validation of crest maximum velocity (**Left**) and maximum flow depths (**Right**) at PW1, SB1, and STP8. Both figures show the results from simulations of RCD and GCD, the measurements from the Millingen experiments, the fitted curve obtained from the measurements of the Vechtdijk experiment and the fitted curve obtained from the measurements of the Tholen experiment.

Appendix B. CFD Validation for the Slope

The second set of measuring devices (SB4 and PW4) during the Millingen experiment were located on the slope at almost one meter distance from the vertex of the crest and landward slope (STP28). The validation results for both models for this location are presented Figure A2.

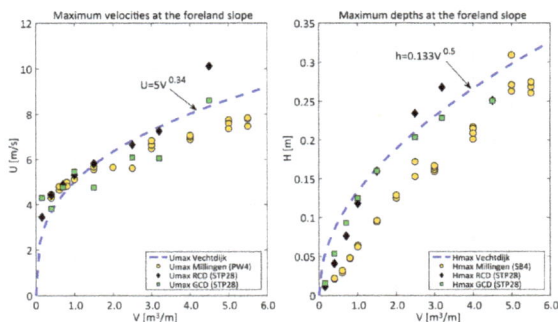

Figure A2. Validation of slope maximum velocity (**Left**) and maximum flow depths (**Right**) at PW4, SB4, and STP28. Both figures show the results from simulations of RCD and GCD, the measurements from the Millingen experiments and the fitted curve obtained from the Vechtdijk experiment.

The velocity measurements from the Millingen experiment and the results from the RCD and GCD models are in fairly good agreement with each other for waves with volumes which are less than 3.5 m^3/m. The flow depths for both GCD and RCD models differ significantly from the recorded values during the experiment. Again, these differences could be attributed to the presence of the geotextile. The accuracy of the measuring devices should also be included in this analysis but unfortunately, no data was available in the literature. Yet, it is concluded that maximum velocities and maximum flow depths of the GCD model are in good agreement with the measurements taken at the Vechtdijk presented in [15]. Additionally, the expected behavior presented by the GCD model (i.e., lower velocities and larger flow depths) with respect to the RCD model gives sufficient confidence for using the model in the present study.

Appendix C. CFD Validation for Overtopping Duration

The simulated overtopping durations presented in Figure A3 are in good agreement with the ones measured by the surf boards on the crest and slope but not with the ones computed from the paddle wheel velocity meters. Apparently, there is a discrepancy between both measuring devices. No thorough analysis of the sensitivity of the CFD model was carried out in this study, but based on validation results, a 20% error is estimated. In a further study, a thorough sensitivity analysis and extended validation of the CFD model is required. Yet, as we concluded in the analysis of the validation of depth and velocity on the crest and slope, the order of magnitude and trends are reasonably well predicted given all uncertainties in the model and measurement data. This suggests justified validation of the model for both crest and slope in terms of overtopping durations.

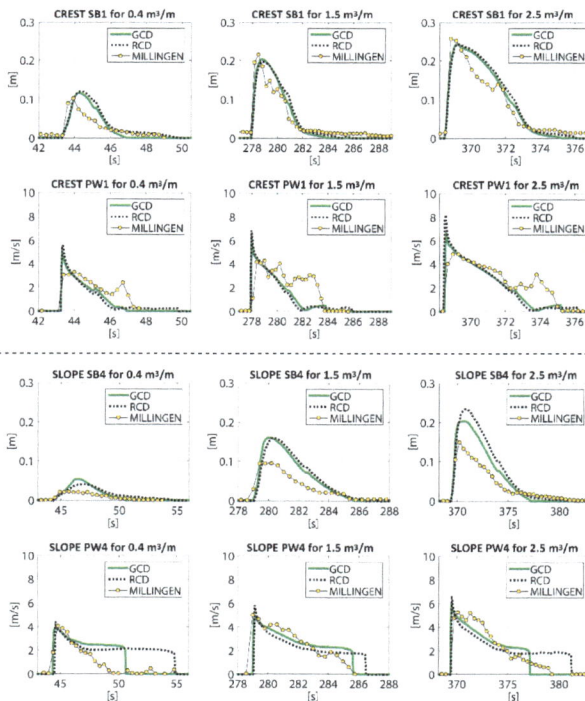

Figure A3. Depth and velocity time series for the location on the crest (**upper** 6 plots) and the side slope (**lower** 6 plots) of different wave volumes from the RCD and GCD model and the measured values for the same waves during the Millingen experiment.

References

1. Aguilar-López, J.P. *Probabilistic Safety Assessment of Multi-Functional Flood Defences*; Twente University: Enschede, The Netherlands, 2016.
2. Hemer, M.A.; Fan, Y.; Mori, N.; Semedo, A.; Wang, X.L. Projected changes in wave climate from a multi-model ensemble. *Nat. Clim. Chang.* **2013**, *3*, 471–476. [CrossRef]
3. Stocker, T.F.; Qin, D.; Plattner, G.-K.; Alexander, L.V.; Allen, S.K.; Bindoff, N.L.; Bréon, F.-M.; Church, J.A.; Cubasch, U.; Emori, S. Technical summary. In *Climate Change 2013: The Physical Science Basis. Contribution of Working Group I to the Fifth Assessment Report of the Intergovernmental Panel on Climate Change*; Cambridge University Press: Cambridge, UK, 2013; pp. 33–115.
4. Middelkoop, H.; Daamen, K.; Gellens, D.; Grabs, W.; Kwadijk, J.C.; Lang, H.; Parmet, B.W.; Schädler, B.; Schulla, J.; Wilke, K. Impact of climate change on hydrological regimes and water resources management in the rhine basin. *Clim. Chang.* **2001**, *49*, 105–128. [CrossRef]
5. Ranasinghe, R.; Callaghan, D.; Stive, M.J. Estimating coastal recession due to sea level rise: Beyond the bruun rule. *Clim. Chang.* **2012**, *110*, 561–574. [CrossRef]
6. Jongejan, R.; Ranasinghe, R.; Wainwright, D.; Callaghan, D.P.; Reyns, J. Drawing the line on coastline recession risk. *Ocean Coast. Manag.* **2016**, *122*, 87–94. [CrossRef]
7. Naulin, M.; Kortenhaus, A.; Oumeraci, H. Reliability-based flood defense analysis in an integrated risk assessment. *Coast. Eng. J.* **2015**, *57*. [CrossRef]
8. Vrijling, J.K. Probabilistic design of water defense systems in the netherlands. *Reliabil. Eng. Syst. Saf.* **2001**, *74*, 337–344. [CrossRef]
9. Lee, C.-E.; Kwon, H.J. Reliability analysis and evaluation of partial safety factors for random wave overtopping. *KSCE J. Civ. Eng.* **2009**, *13*. [CrossRef]
10. Victor, L.; Van der Meer, J.W.; Troch, P. Probability distribution of individual wave overtopping volumes for smooth impermeable steep slopes with low crest freeboards. *Coast. Eng.* **2012**, *64*, 87–101. [CrossRef]
11. Van der Meer, J.W. *Technical Report Wave Run-Up and Wave Overtopping at Dikes*; Rijkswaterstaat, DWW: Utrecht, The Netherlands, 2002.
12. Pullen, T.; Allsop, N.; Bruce, T.; Kortenhaus, A.; Sch, H.; Van der Meer, J. *Eurotop: Wave Overtopping of Sea Defences and Related Structures: Assessment Manual*; Environment Agency: Bristol, UK, 2007.
13. Franco, L.; De Gerloni, M.; Van der Meer, J. Wave overtopping on vertical and composite breakwaters. *Coast. Eng. Proc.* **1994**. [CrossRef]
14. Dean, R.G.; Rosati, J.D.; Walton, T.L.; Edge, B.L. Erosional equivalences of levees: Steady and intermittent wave overtopping. *Ocean Eng.* **2010**, *37*, 104–113. [CrossRef]
15. Van der Meer, J.W.; Hardeman, B.; Steendam, G.J.; Schuttrumpf, H.; Verheij, H.J. Flow depths and velocities at crest and landward slope of a dike, in theory and with the wave overtopping simulator. *Coast. Eng. Proc.* **2010**, *1*. [CrossRef]
16. Steendam, G.J.; Van Hoven, A.; Van der Meer, J.; Hoffmans, G. Wave overtopping simulator tests on transitions and obstacles at grass covered slopes of dikes. *Coast. Eng. Proc.* **2014**, *1*. [CrossRef]
17. Hoffmans, G.J.C.M.; van Hoven, A.; Harderman, B.; Verheij, H. Erosion of grass covers at transitions and objects on dikes. In Proceedings of the 7th International Conference on Scour and Erosion, Perth, Australia, 2–4 December 2015; Cheng, L., Drapper, S., An, H., Eds.; Taylor and Francis Group: Perth, Australia, 2015.
18. Hoffmans, G.J.C.M.; Akkerman, G.J.; Verheij, H.; Van Hoven, A.; Van der Meer, J. The erodibility of grassed inner dike slopes against wave overtopping. *ASCE* **2008**, 3224–3236. [CrossRef]
19. Verheij, H.J.; Meijer, D.G.; Kruse, G.A.M.; Smith, G.M.; Vesseur, M. *Onderzoek Naar de Sterkte van Graszoden van Rivierdijken*; Deltares (WL): Delft, The Netherlands, 1995.
20. Hewlett, H.; Boorman, L.A.; Bramley, L. *Design of Reinforced Grass Waterways*; Construction Industry Research and Information Association: London, UK, 1987.
21. Trung, L. *Overtopping on Grass Covered Dikes: Resistance and Failure of the Inner Slopes*; TU Delft, Delft University of Technology: Delft, The Netherlands, 2014.
22. Schüttrumpf, H.; Oumeraci, H. Layer thicknesses and velocities of wave overtopping flow at seadikes. *Coast. Eng.* **2005**, *52*, 473–495. [CrossRef]
23. Kobayashi, N.; Weitzner, H. Erosion of a seaward dike slope by wave action. *J. Waterw. Port Coast. Ocean Eng.* **2014**, *141*. [CrossRef]

24. Quang, T.T.; Oumeraci, H. Numerical modelling of wave overtopping-induced erosion of grassed inner sea-dike slopes. *Nat. Hazards* **2012**, *63*, 417–447. [CrossRef]

25. Ribberink, J.S. Bed-load transport for steady flows and unsteady oscillatory flows. *Coast. Eng.* **1998**, *34*, 59–82. [CrossRef]

26. Bomers, A.; Lopez, J.A.; Warmink, J.J.; Hulscher, S.J. Modelling effects of an asphalt road at a dike crest on dike cover erosion onset during wave overtopping. *Nat. Hazards* **2018**, 1–30. [CrossRef]

27. Castelletti, A.; Galelli, S.; Ratto, M.; Soncini-Sessa, R.; Young, P.C. A general framework for dynamic emulation modelling in environmental problems. *Environ. Model. Softw.* **2012**, *34*, 5–18. [CrossRef]

28. Razavi, S.T.; Bryan, A.; Burn Donald, H. Review of surrogate modeling in water resources. *Water Resour. Res.* **2012**, *48*. [CrossRef]

29. Duncan, A.; Chen, A.S.; Keedwell, E.; Djordjevic, S.; Savic, D. *Urban Flood Prediction in Real-Time from Weather Radar and Rainfall Data Using Artificial Neural Networks*; International Association of Hydrological Sciences: Exeter, UK, 2011.

30. Aguilar-Lopez, J.P.; Van Andel, S.J.; Werner, M.; Solomatine, D.P. Hydrodynamic and water quality surrogate modelling for reservoir operation. In Proceedings of the 11th IWA/IAHR International Conference on Hydroinformatics, New York, NY, USA, 17–21 August 2014.

31. Kingston, G.B.; Rajabalinejad, M.; Ben, P.G.; Van Gelder, P.H.A.J.M. Computational intelligence methods for the efficient reliability analysis of complex flood defence structures. *Struct. Saf.* **2011**, *33*, 64–73. [CrossRef]

32. Aguilar-López, J.P.; Warmink, J.J.; Schielen, R.M.J.; Hulscher, S.J.M.H. Data-driven surrogate models for flood defence failure estiamtion: "Jarillon de calí". In Proceedings of the 11th IWA/IAHR International Conference on Hydroinformatics, New York, NY, USA, 17–21 August 2014.

33. Aguilar-López, J.P.; Warmink, J.J.; Schielen, R.M.J.; Hulscher, S.J.M.H. Piping erosion safety assessment of flood defences founded over sewer pipes. *Eur. J. Environ. Civ. Eng.* **2016**, 1–29. [CrossRef]

34. Van Gent, M.R.A.; van den Boogaard, H.F.P.; Pozueta, B.; Medina, J.R. Neural network modelling of wave overtopping at coastal structures. *Coast. Eng.* **2007**, *54*, 586–593. [CrossRef]

35. Bomers, A.; Aguilar-López, J.P.; Warmink, J.J.; Hulscher, S.J.M.H. Modelling erosion development during wave overtopping of an asphalt road covered dike. In Proceedings of the FLOODrisk 2016 3rd European Conference on Flood Risk Management, Innovation, Implementation and Integration, Lyon, France, 16–21 October 2016.

36. Hoffmans, G.J.C.M. *The Influence of Turbulence on Soil Erosion*; Eburon Uitgeverij BV: Delft, The Netherlands, 2012.

37. Hughes, S.A. *Adaptation of the Levee Erosional Equivalence Method for the Hurricane Storm Damage Risk Reduction System (Hsdrrs)*; DTIC Document: Fort Belvoir, VA, USA, 2011.

38. Partheniades, E. Erosion and deposition of cohesive soils. *J. Hydraul. Div.* **1965**, *91*, 105–139.

39. Verheij, H.J.; Hoffmans, G.J.C.M.; van der meer, J. *Evaluation and Model Development, Grass Erosion Test at the Rhine Dike*; Deltares: Delft, The Netherlands, 2015.

40. Launder, B.E.; Spalding, D.B. The numerical computation of turbulent flows. *Comput. Methods Appl. Mech. Eng.* **1974**, *3*, 269–289. [CrossRef]

41. COMSOL. *The Comsol CFD Module User's Guide (Version 5.2)*; COMSOL: Stockholm, Sweden, 2015.

42. Marriott, M.J.; Jayaratne, R. *Hydraulic Roughness–Links Between Manning's Coefficient, Nikuradse's Equivalent Sand Roughness and Bed Grain Size*; UEL: London, UK, 2010.

43. Te Chow, V. *Open Channel Hydraulics*; McGraw-Hill: London, UK, 1959.

44. Jansen, R.B. *Advanced dam Engineering for Design, Construction, and Rehabilitation*; Springer Science & Business Media: Berlin/Heidelberg, Germany, 2012.

45. Bakker, J.; Melis, R.; Mom, R. *Factual Report: Overslagproeven Rivierenland*; INFRAM: Dutch, The Netherlands, 2013.

46. Van der Meer, J.W.; Van Hovern, A.; Paulissen, M.; Steendam, G.J.; Verheij, H.J.; Hoffmans, G.J.C.M.; Kruse, G.A.M. *Handreiking Toetsen Grasbekledingen op Dijken TBV Het Opstellen van het Beheerdersoordeel (bo) in de Verlengde Derde Toetsronde*; Rijkswaterstaat, Waterdienst: Utrecht, The Netherlands, 2012.

47. Forrester, A.; Sobester, A.; Keane, A. *Engineering Design via Surrogate Modelling: A Practical Guide*; John Wiley & Sons: Hoboken, NJ, USA, 2008.

48. Piontkowitz, T. *Erograss: Failure of Grass Cover Layers at Seaward and Shoreward Dike Slopes. Design, Construction and Performance*; Danish Coastal Inspectorate: Copenhagen, Denmark, 2009.

49. Van der Meer, J.W.; Janssen, J.; Hydraulics, D. *Wave Run-Up and Wave Overtopping at Dikes and Revetments*; Delft Hydraulics: Delft, The Netherlands, 1994.

50. Hughes, S.A.; Thornton, C.I.; Van der Meer, J.W.; Scholl, B.N. Improvements in describing wave overtopping processes. *Coast. Eng. Proc.* **2012**, *1*. [CrossRef]

51. Van Loon-Steensma, J.M.; Vellinga, P. Robust, multifunctional flood defenses in the dutch rural riverine area. *Nat. Hazards Earth Syst. Sci.* **2014**, *14*, 1085–1098. [CrossRef]

Journal of
Marine Science and Engineering

MDPI

Article

An Effective Modelling Approach to Support Probabilistic Flood Forecasting in Coastal Cities—Case Study: Can Tho, Mekong Delta, Vietnam

Hieu Ngo [1,*], Assela Pathirana [1], Chris Zevenbergen [1,2] and Roshanka Ranasinghe [1,3,4]

[1] Department of Water Science and Engineering, IHE Delft Institute for Water Education, P.O. Box 3015, 2601 DA Delft, The Netherlands; a.pathirana@un-ihe.org (A.P.); c.zevenbergen@un-ihe.org (C.Z.); r.ranasinghe@un-ihe.org (R.R.)
[2] Department of Hydraulic Engineering, Faculty of Civil Engineering and Geosciences, Delft University of Technology, P.O. Box 5048, 2628 CN Delft, The Netherlands
[3] Department of Water Engineering and Management, University of Twente, P.O. Box 217, 7500 AE Enschede, The Netherlands
[4] Harbour, Coastal and Offshore Engineering, Deltares, P.O. Box 177, 2600 MH Delft, The Netherlands
* Correspondence: h.ngo@un-ihe.org; Tel.: +31-068-769-9329

Received: 12 March 2018; Accepted: 4 May 2018; Published: 11 May 2018

Abstract: Probabilistic flood forecasting requires flood models that are simple and fast. Many of the modelling applications in the literature tend to be complex and slow, making them unsuitable for probabilistic applications which require thousands of individual simulations. This article focusses on the development of such a modelling approach to support probabilistic assessment of flood hazards, while accounting for forcing and system uncertainty. Here, we demonstrate the feasibility of using the open-source SWMM (Storm Water Management Model), focussing on Can Tho city, Mekong Delta, Vietnam. SWMM is a dynamic rainfall-runoff simulation model which is generally used for single event or long-term (continuous) simulation of runoff quantity and quality and its application for probabilistic riverflow modelling is atypical. In this study, a detailed SWMM model of the entire Mekong Delta was built based on an existing ISIS model containing 575 nodes and 592 links of the same study area. The detailed SWMM model was then systematically reduced by strategically removing nodes and links to eventually arrive at a level of detail that provides sufficiently accurate predictions of water levels for Can Tho for the purpose of simulating urban flooding, which is the target diagnostic of this study. After a comprehensive assessment (based on trials with the varying levels of complexity), a much reduced SWMM model comprising 37 nodes and 40 links was determined to be able to provide a sufficiently accurate result while being fast enough to support probabilistic future flood forecasting and, further, to support flood risk reduction management.

Keywords: coastal cities; SWMM; simplified model; Mekong Delta; Can Tho city

---

1. Introduction

Coastal cities are among the most urbanised and populated areas of the world [1–5]. Small and Nicholls [1] estimated that, in 1990, 23% of the global population lived within 100 km of the coast and less than 100 m above sea level. Thus, flooding can cause serious effects on human activities and properties in coastal cities which is amply reflected by Hallegatte et al [3], who predicted that the economic losses due to flooding alone in coastal cities are expected to be around US $1 Trillion by 2050.

This escalation of damage will be caused by a number of reasons. On the one hand, cities, particularly in the global south, are undergoing rapid landuse change due to population growth and migration and increasing industrialization [6–8]. On the other, the coastal and estuarine cities are threatened by increasing water levels due to both sea-level rise and changes in the upstream flow patterns [9]. On top of these there is the possibility that flooding might increase due to the local rainfall regime connected to both global climate change and the local land-use driven microclimate changes [10].

The ability to predict the changes in the flood hazard and risk under variety of internal (e.g., landuse) and external (e.g., climate) forcing scenarios plays an important role in flood risk management under a rapidly changing environment. Traditional modelling applications in the domain of flood risk management typically involved developing detailed flood models and conducting a handful of flood simulation exercises. Such an approach works adequately under the assumption that the forcing variables and the model conditions can be represented by such a limited repertoire of scenarios—in other words ignoring the uncertainty of those parameters. Due to rapid and uncertain changes in the forcing parameters, flood risk management in the modern context, is demanded to embrace much more statistically robust approaches. In this context, probabilistic forecasts are a fundamental requirement for quantitative flood risk assessments, which aid urban planners and decision makers to develop informed risk reduction strategies that minimize the damage caused by floods. This is especially important for coastal cities which are not only facing the impact of upstream flow changes due to human interventions (upstream dam construction) and the effects of climate change, but also climate change driven sea level rise, storm surges as well as other natural processes such as land subsidence in deltas.

Hydraulic models that simulate river flow and flooding are needed to estimate water level changes under various future scenarios. With the rapid development of computing facilities over the last few decades, many different hydrodynamic models (e.g., MIKE11, MIKEFLOOD, ISIS, HEC-RAS, SOBEK, LISFLOOD-FP, Delft 3D) have been developed to address complex real-world hydraulic problems including flood forecasting [11,12]. Most of these developed in a direction of more and more complex and sophisticated modelling approaches in order to achieve highly precise and (oftentimes) spatially-explicit, results. In addition, many of these models are complex models that are proprietary and commercial. These factors limit their appeal for wide-spread use in flood forecasting, especially in developing countries with limited financial and technological resources. Moreover, the way in which simulation models are applied has undergone a paradigm shift in recent years: traditionally these were strictly limited to the domain of modelling specialists. However, with the emergence of wider-stakeholder engagement in co-learning, co-designing and co-solving of problems and resulting social-learning have become commonplace [13]. There is a need for simulation models which are more accessible and versatile. There is thus a need for non-restrictively licensed (e.g., open source), simple models for the wide-spread deployment in contemporary co-learning environments.

Probabilistic flood forecasting requires flood models that are simple and fast. Taking into account the large uncertainties in the future forcings and different model parameters, high degree of accuracy of models, which are sought after in deterministic applications, often become secondary to their simplicity in use and rapidity in execution in probabilistic applications. This demands models that are fit-for-purpose by being fast in execution and simple to be deployed in iterative contexts to realize thousands of models runs.

This study focuses on the development of such a 'fit-for-purpose' modelling approach suitable for probabilistic flood forecasting which also accounts for forcing and system uncertainty. Herein, we describe the application of this modelling approach to forecast flood at a selected developing country case study: Can Tho city in Mekong Delta, Vietnam.

In the following sections, we describe the study area, followed by a review of previous modelling studies in the Mekong Delta including an overview of the hydrodynamic models that have been used in those studies. This is followed by the research methodology used in this study. Then we present the results obtained and discussion. The final section of this article is main conclusions arising from the study.

2. Study Area

2.1. Mekong Delta

The Mekong River originates from Tibet, and flows through China, Laos, Myanmar, Thailand, Cambodia and into the East Sea of Vietnam (Figure 1a). With a length of 4800 km and a mean annual flow of 475 km^3, the Mekong River ranks twelfth and tenth in the world in term of length and flow, respectively [14,15].

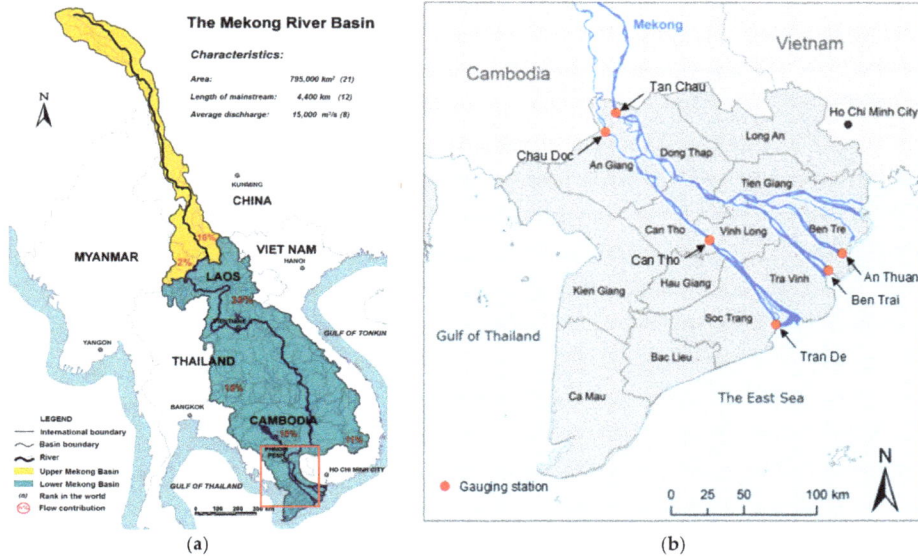

Figure 1. (a) Location of the Mekong Delta (Source: Mekong River Commission, Phnom Penh, Cambodia); (b) Detailed descriptions of the Mekong Delta include Can Tho city and 12 provinces as well as the location of hydrological gauging stations (revised from [16]).

The Mekong Delta is the largest delta of Vietnam, and is located in the lower Mekong River Basin (Figure 1), spanning latitudes 8°33′ N and 11°01′ N and longitudes 104°26′ E and 106°48′ E. It includes Can Tho city and the 12 provinces: Long An, Dong Thap, Vinh Long, Tien Giang, Tra Vinh, Ben Tre, Hau Giang, Soc Trang, Bac Lieu, Kien Giang, An Giang, and Ca Mau, with a total land area of about 4 million ha. The Mekong Delta has a population of approximately 17.5 million, accounting for 19% of the country's population, while this region accounts for only 13% of the country's area. The livelihoods of a majority of the population (85%) in the region depend on agricultural activities [17]. The Mekong Delta is known as the granary of the nation and is also a key area for the production of fishery, fruit and agricultural products. Annually, it contributes about 90% of rice, 70% of fruit, and 60% of fishery products in the national export turnover for each class. The economic growth of the region reached 7.39% in 2017, up 0.49% compared to 2016 (6.9%). The per capita income in the Mekong Delta is about 40.2 million VND (around 1770 USD). Furthermore, the Mekong Delta is also known as one of the most biologically diverse in the world with abundant fauna and flora (e.g., fish, lizards, mammals, etc.), including rare species such as Laotian rock rat, is thought to be extinct.

The Mekong Delta has low and flat terrain, with an average elevation of between 0.7 and 1.2 m above mean sea level [18]. Mekong Delta is also thought to be one of areas that are globally most sensitive to the impacts of climate change [19,20]. The most prominent reason for the climate sensitivity of the Mekong Delta is the strongly felt influence of the sea-level rise. For example the Can Tho city (explained below) that is some 80 km upstream from the ocean, is still impacted by tidal variation (and hence any changes in the sea level in the future). For example, the cities drainage system cannot discharge water to the river during the high tide periods [9].

Besides sea level rise, climate change is likely to also affect the riverflow in the Mekong Delta, due to projected changes in temperature and rainfall [21,22]. Future projections of the upstream flow in the Mekong Delta presented by, among others, [23–25] indicate increases in riverflow for all IPCC RCPs. In addition to climate change, riverflow is also affected by human activities such as land-use change, upstream dam construction. The foreshadowed construction of new upstream hydropower dams is expected to vary the riverflow regime [24,26], with the flow varying from year to year depending on hydropower operations [27].

While flooding (which usually occurs between July and November) in the Mekong Delta has many negative impacts (e.g., damage to property, infrastructure, and crops, loss of life) on inhabitants and economic development of the area it also has some positive impacts such as washing away salt and alum from soil, providing supplementary fertiliser for rice fields, and increasing fish resources [17,28]. The threat of flooding in future will be undoubtedly be affected by both climate change and human interventions as described above [18,29,30].

2.2. Can Tho city

Can Tho is the largest city within the Mekong Delta, with a population of around 1.6 million as of 2016, is the city located on the south bank of the Hau River, one of two major branches of the Mekong River in Vietnam, at a distance of about 80 km upstream of the sea (Figure 2). Can Tho has a dense system of rivers and canals, and therefore, it is also known as "the municipality of river water region".

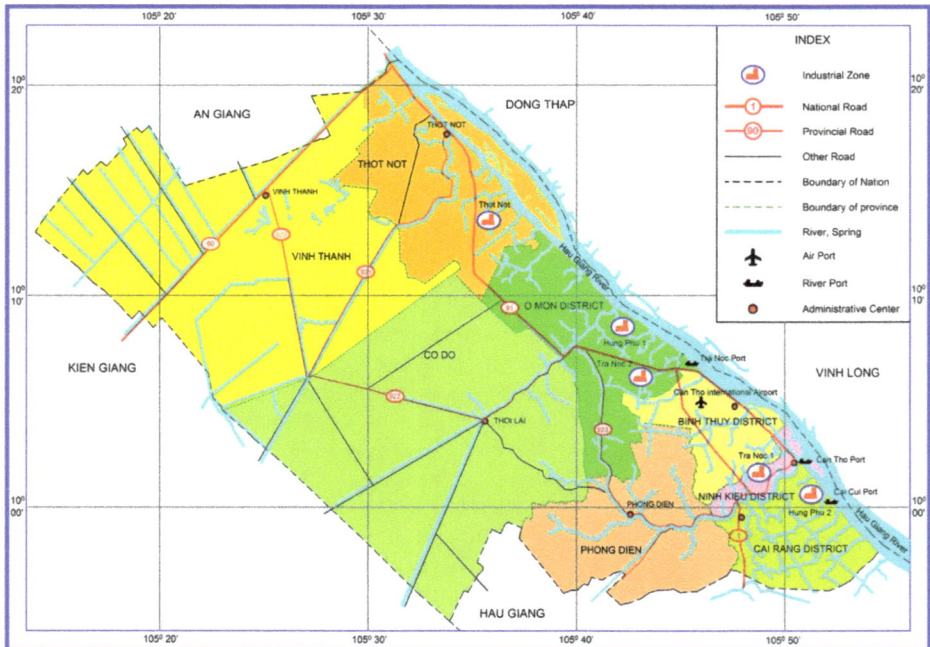

Figure 2. A map of Can Tho city. The area shown here extends much beyond the urban center. The urban area is largely in lower right corner of the diagram, north of the Can Tho river.

Can Tho is dynamic city that is emerging as an economic centre and is expected to play an important role in the Mekong Delta and the adjacent international regions in future [9,31]. With its strategic location, Can Tho city is expected to witness growth exponentially in the next several decades [9,32]. The rapid urbanization and growth of population in Can Tho will likely lead to a significant change in land use within and surrounding area of the city [9]. This may result in substantial changes in the urban water cycle, increasing flood frequency and water pollution, all of which will further increase the city's already high risk of flooding in the coming decades.

Serious flooding is frequent in Can Tho City, which leads to significant impacts on environment, economy and society. Flooding occurs at least 2–3 times a year during the monsoon season, with flood water depths ranging from a few centimetres up to 20–30 cm (even up to 50 cm in some places such as Ninh Kieu district). According to Southern Hydrometeorology Station, in 2011, the maximum water level reached 2.15 m relative to mean sea level. Table 1 shows the water level in Can Tho city in flooding years.

Table 1. Water level and discharge in flooding years in Can Tho (relative to MSL) since 2000 (Source: Adapted from [33]).

Year	Date	Highest Water Level (cm)	Average Discharge in Can Tho (m³/s)	
			On Day of Flooding	Highest Flow in Month of Flooding
2000	30/09	179	13,000	(23/09) 17,700
2011	27/10	215	16,100	(05/10) 19,600
2013	20/10	213	12,290	(30/10) 18,180
2014	10/10	208	-	-

Flooding in the city is significantly impacted by three factors: (1) sea level (tide, storm surge, sea-level rise) has a direct impact on the river level near Can Tho and hence the dynamics of flooding; (2) Changes in the upstream river flow, which is impacted by landuse change, construction of hydraulic structures such as dams [27,34–36]; and (3) Urban hydrology (changes in local rainfall, landuse, etc.). In order to represent the first two factors above, a physically-based simulation model of the river system of the Mekong Delta is essential. With such a model it is possible to use sea-level changes and upstream flow changes as boundary conditions to ascertain their impact on the river water level near Can Tho city and therefore the impact on floods in the city.

In the next section, therefore, we summarize previous attempts to develop such river system models for the Mekong delta.

3. Previous Flood Modelling Studies in the Mekong Delta

Wassmann et al. [30] used the "Vietnam River System and Plains" (VRSAP) model to assess the water level changes in the Vietnamese Mekong Delta (VMD) due to the impacts of sea level rise (SLR). The VRSAP model was developed by the Sub-Institute for Water Resources Planning, Ministry of Agriculture and Rural Development, Vietnam (Khue, 1986). The VRSAP model for whole VMD included 1505 nodes, 2111 segments, and 555 storage plains. It was calibrated with hydrological data obtained in 1996, and then used to predict water levels for years 2030 and 2070. This model projected water levels for VMD when flooding is presently high (from August to November) for SLR projections of 20 cm (in 2030) and 45 cm (in 2070), respectively. The results of this study indicated that the average increase in August water levels corresponding to the two considered scenario were 14.1 cm and 32.2 cm, respectively, relative to 1996. In October, the increase in water levels were found to be weaker due to high discharge from upstream, but the average increases for the two considered scenarios were still high at 11.9 cm and 27.4 cm, respectively.

Le et al. [37] used the HydroGis model, which combines a hydrodynamic model and GIS tools [38], was developed by the Ministry of Natural Resource and Environment of Vietnam (MONRE) to evaluate changes in flooding in the VMD due to the combined effect of upstream river flows, storm surge, SLR, estuarine siltation, and hydraulic structures. The HydroGis model for the entire VMD comprised

13,262 cross sections, 2535 flood cells, and 467 sewers, sluices, and bridges. This study indicated that the flood levels depend on the combined impacts of Mekong river flows, storm surges and SLR as well as the construction of upstream dams. Water levels in the delta were predicted to increase from 5 cm to 200 cm corresponding each scenario. Additionally, this study predicted that the construction of upstream dams would cause siltation the Hau river estuary and followed by the increase by up 2 m in peak flood levels in the VMD.

HR Wallingford and Halcrow (UK) developed a 1-D ISIS model for the VMD to assist the Mekong River Commission (MRC) in managing and using water resources in the Mekong River Basin (MRB). Currently, this ISIS model is maintained by the MRC. MRC's detailed modelling studies for Cambodia and the VMD have used the results of this model as boundary conditions. Van et al. [39] used this model to study changes of flood characteristics under the impacts of upstream development and SLR. The 1-D ISIS model for the VMD comprised 3036 cross-sections representing 8619 km system of river and canals, 538 junctions, 749 floodplain units, 193 spillways, 409 reservoirs, and 29 sluices. This study used the flood event data of the year 2000 to validate this model, and subsequently used it to predict potential changes in flooding for the year 2050 under scenario 3a and 3b. The results showed that the flood hazard in 2050 may become more severe along the coastal area as a result of the tidal regime. Additionally, future (2050) inundation levels in upstream VMD were projected to be lower and shorter than in 2000, while along the coastal areas would be higher and longer.

All the models applied in the aforementioned studies are advanced process based hydrodynamic models with superior features, and as such the model outcomes are expected to be highly accurate and reliable. However, these models do take a long time for extended period simulations, owing to the large and complex application domain. For example, the time for the aforementioned ISIS model of VMD to complete a one-year simulation is about 90 min (on a single core on an Intel(R) Core(TM) i5-4210M CPU @ 2.6 GHz 2.6 GHz processor in a computer with 8.0 GB of memory (RAM)). With such a computational time, it is impractical to execute the thousands of individual simulations required for probabilistic flood forecasting with this model.

In the next section we describe the methodology that was used to develop a fit-for-purpose model that is fast and simple enough to be deployed in probabilistic water level calculations near Can Tho city, which in-turn can be used to force urban flood models.

4. Methodology

4.1. Data

The data for this study were collected from two sources: (i) upstream flow (discharge) from 2000 to 2006, measured water level in 2000 at Chau Doc, Tan Chau, Can Tho, Tran De, Ben Trai and An Thuan stations (Figure 1b), cross-section data and the Manning's roughness coefficient of links were taken from the aforementioned Mekong Delta 1-D ISIS model of the MRC. Additionally, discharge in 2011 was collected from MRC as well; (ii) measured water level of the years 2001, 2002 and 2011 at six above gauging stations were collected from the National Hydro-meteorological Service of Viet Nam (NHMS).

4.2. Model Selection

To circumnavigate the complexities arising due to simulation time and proprietary nature of sophisticated process based models such as those used in the studies summarized in Section 3 above, here, to achieve the objective of this study, we used the open-source SWMM model. SWMM is a dynamic rainfall-runoff simulation model used for single event or long-term (continuous) simulation of runoff quantity and quality [40].

The application of the SWMM model for this task was unconventional. This decision was made based on a number of considerations. Firstly, the SWMM implementing full St. Venant's equation in one dimensional conduits [40], is technically capable of simulating the required flow conditions in this complex river system. It can manage complex flow conditions such as downstream forcing by tide. Secondly, it provides a simple, uncluttered user interface that is simple enough to be deployed in multi-stakeholder co-design sessions. Finally, being a public-domain, open source model, and it is not bound by restrictions associated with commercial licenses. Although commercial licensing is not always a significant barrier for a model to be used by experts in, for example, a consulting environment, delivering a model for wider use by non-expert stakeholders is sometimes severely hampered by such restrictions.

4.3. Model Application

4.3.1. SWMM Model Development and Simplification

In the first step of this study, a very detailed SWMM model of the entire Mekong Delta, comprising 575 nodes and 592 conduits, was developed (Figure 3). This model covered only the important tributaries of the Mekong, ignoring small canals and other waterways.

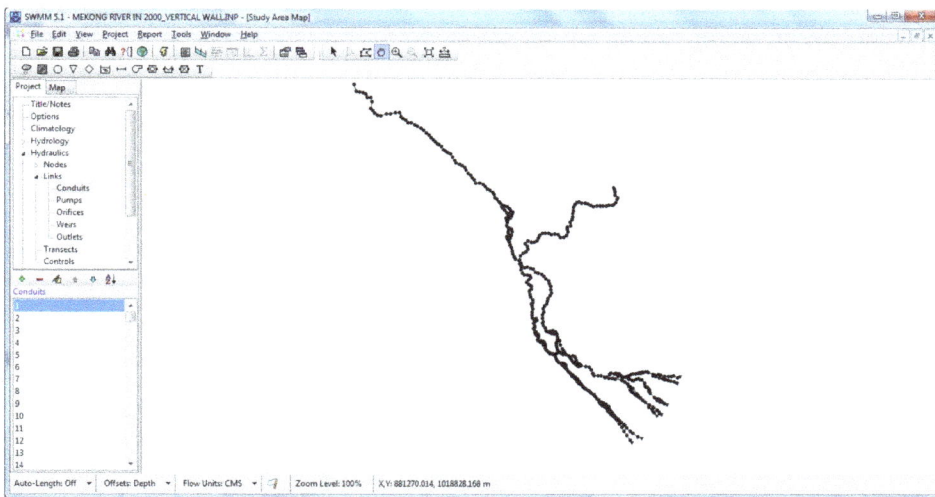

Figure 3. The detailed SWMM model for Mekong Delta with 575 nodes and 592 conduits.

The input data used were the daily upstream flow and hourly downstream sea level in 2000. This model is already simplified compared to many of the models described in the section above and took about 10 min for a one year simulation (for details see Table 11). The next step is to reduce the model complexity such that very fast individual simulations are possible. To this end, a series of simulations (hourly temporal resolution) of the detailed SWMM model were undertaken where the level of detail in the model was gradually decreased (i.e., systematic removal of small to medium size tributaries, based on their size and/or distance from target area). In doing this here nodes that appeared to be at non-critical locations were sequentially removed from the system while retaining nodes at critical locations where there were river branch divisions, changes in the direction of flow, significant changes in cross-sectional area, hydrological stations, etc. The model obtained after each stage in the reductions process was run for the year 2000, and results compared with measured water levels at Chau Doc, Tan Chau and Can Tho stations in 2000. If the post-reduction model results were still good enough, the process of reduction was continued until the simplest level of detail that provides sufficiently accurate predictions of water levels in the local study area (Can Tho, Vietnam) was obtained. The final model thus obtained is referred to herein as the simplified SWMM model. The Simplified SWMM model took only a minute to complete one year period of simulation, representing a 10 and

90 fold reduction in run time relative to the detailed SWMM model and the detailed ISIS model described above (see also Table 11).

4.3.2. Model Calibration and Validation

Model Calibration

An automatic calibration of the simplified SWMM model was undertaken for the year 2000 using observed hourly water levels at Chau Doc, Tan Chau and Can Tho stations. The SWMM5-EA software [41], which uses evolutionary algorithms to optimize drainage networks, was used to optimise the Manning's roughness coefficient of conduits of the simplified SWMM model.

Water levels at the gauging stations in the Mekong Delta sharply change between flood season and dry season. Therefore, monthly NSE (Nash-Sutcliffe efficiency) and RMAE (Relative Mean Absolute Error) values of the 3 gauging stations Can Tho, Chau Doc and Tan Chau were used for model calibration.

- NSE indicator

The NSE is a normalized statistic that determines the relative magnitude of the residual variance ("noise") compared to the observed data variance ("information") [42]. Model performance is commonly classified for various ranges of the NSE as shown below [42].

-    0.75 < NSE ≤ 1.00: Very Good;
-    0.65 < NSE ≤ 0.75: Good;
-    0.50 < NSE ≤ 0.65: Satisfactory;
-    NSE ≤ 0.50: Unsatisfactory.

- RMAE indicator

The RMAE is a commonly used error statistic [43].
Model performance is commonly classified for various ranges of the RMAE as shown below (Sutherland et al., 2004).

-    RMAE < 0.20: Excellent;
-    0.20 ≤ RMAE < 0.40: Good;
-    0.40 ≤ RMAE < 0.70: Reasonable;
-    0.70 ≤ RMAE ≤ 1.00: Poor;
-    RMAE > 1.00: Bad.

Model Validation

Model validation was performed manually for the years 2001, 2002 and 2011, using the model calibrated with the year 2000 data, to gain confidence in model predictions. Results of the validation simulations were compared with hourly water level data acquired at Chau Doc, Tan Chau and Can Tho stations.

5. Results and Discussion

5.1. Calibration of the Simplified SWMM Model

After a number of trials with varying levels of complexity, a simplified model where the number of nodes and conduits were 37 and 40 respectively (Figure 4), was determined to have a positive result (see Tables 4 and 5) while being sufficiently fast (1 min run time for a 1 year simulation period at hourly temporal resolution). Figures 5 and 6 show the comparison between simulated water levels with observed water levels at Can Tho station in 2000.

The results of the simplified SWMM calibration for year 2000 is shown in Table 2. Scatter plots of simulated and observed hourly water level of each month in 2000 at Chau Doc, Tan Chau and Can Tho stations are shown in Figures 7–9, respectively.

Table 2. Data sources.

Data Type	Source	Data Description
Discharge	MRC	Daily discharge data at upstream (2000–2006) and 2011
Cross-section	MRC	3036 cross-sections representing 8619 km system of river and canals in the Mekong Delta
Manning's roughness	MRC	10 different values of Manning's roughness coefficient
Measured water level	MRC	Hourly water level data at Chau Doc, Tan Chau, Can Tho, Tran De, Ben Trai and An Thuan stations in 2000
	NHMS	Hourly water level data at Chau Doc, Tan Chau, Can Tho, Tran De, Ben Trai and An Thuan stations in 2001, 2002, and 2011

Based on the results in Table 3 and the classifications for NSE as well as RMAE, error classification for NSE and RMAE indicators for each station in 2000 are shown in Tables 4 and 5 respectively.

Table 3. General performance rating for the model for the calibration period (year 2000) using the NSE and RMAE indicators (hourly resolution).

Station / Month	Chau Doc NSE	Chau Doc RMAE	Tan Chau NSE	Tan Chau RMAE	Can Tho NSE	Can Tho RMAE
January	−0.89	0.05	-0.89	0.12	0.67	0.04
February	−0.14	0.01	0.16	0.09	0.71	0.12
March	0.02	0.07	0.37	0.13	0.72	0.10
April	0.04	0.16	0.51	0.21	0.75	0.80
May	0.27	0.19	0.34	0.28	0.74	0.17
June	0.54	0.07	0.41	0.16	0.68	0.40
July	0.03	0.16	0.76	0.05	0.71	0.09
August	−16.78	0.01	−8.54	0.02	0.72	0.10
September	−2.41	0.09	−3.15	0.07	0.81	0.08
October	−1.05	0.04	−0.92	0.03	0.76	0.08
November	0.77	0.04	0.59	0.05	0.71	0.04
December	−0.53	0.13	−1.17	0.14	0.68	0.05

Table 4. Error classification (%) for NSE indicator.

Station	Classification for NSE Very Good	Good	Satisfactory	Unsatisfactory
Chau Doc	8.33	0.00	8.33	83.33
Tan Chau	8.33	0.00	16.67	75.00
Can Tho	16.67	83.33	0.00	0.00

Table 5. Error classification (%) for RMAE indicator.

Station	Classification for RMAE Excellent	Good	Reasonable	Poor	Bad
Chau Doc	100.00	0.00	0.00	0.00	0.00
Tan Chau	83.33	16.67	0.00	0.00	0.00
Can Tho	83.33	0.00	8.33	8.33	0.00

Based on the results in Tables 4 and 5, almost RMAE values at Chau Doc, Tan Chau and Can Tho stations are in excellent and good classifications, especially Chau Doc station with 100% values is in excellent classification, while Can Tho is 83.33%. Nonetheless, Can Tho also has 8.33% values in total is poor classification. For NSE indicator, only Can Tho station has classifications are in very good and good with 100% in total, while Chau Doc and Tan Chau are 8.33%. The main classifications of Chau Doc and Tan Chau stations are unsatisfactory with 83.33% and 75%, respectively.

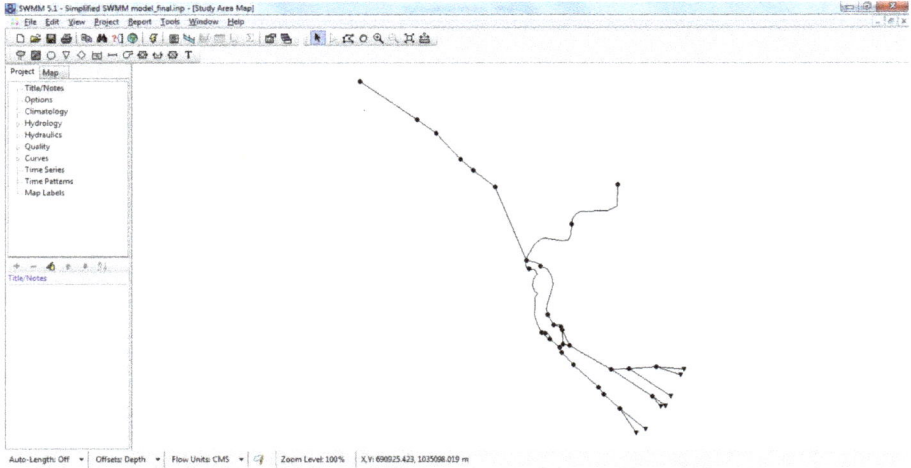

Figure 4. The simplified model for Mekong delta with 37 nodes and 40 conduits.

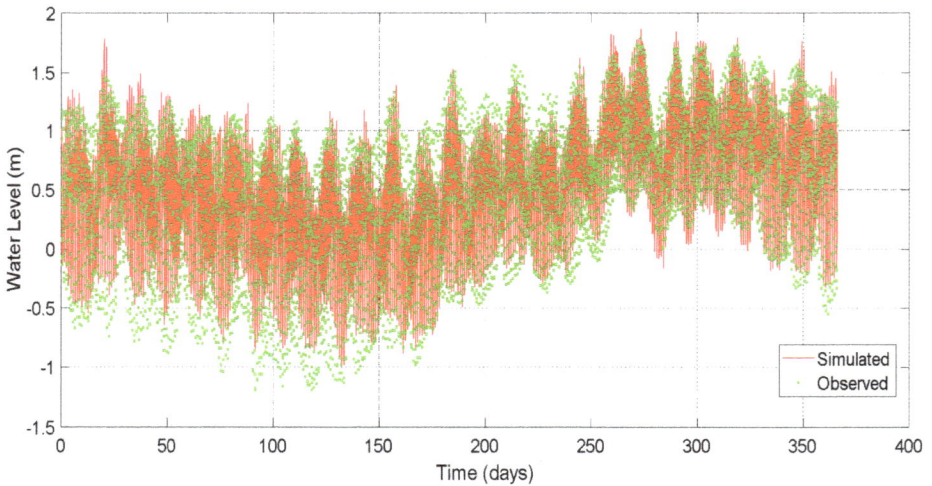

Figure 5. Comparison between simulated and observed water level (relative to MSL) time series at Can Tho station in 2000.

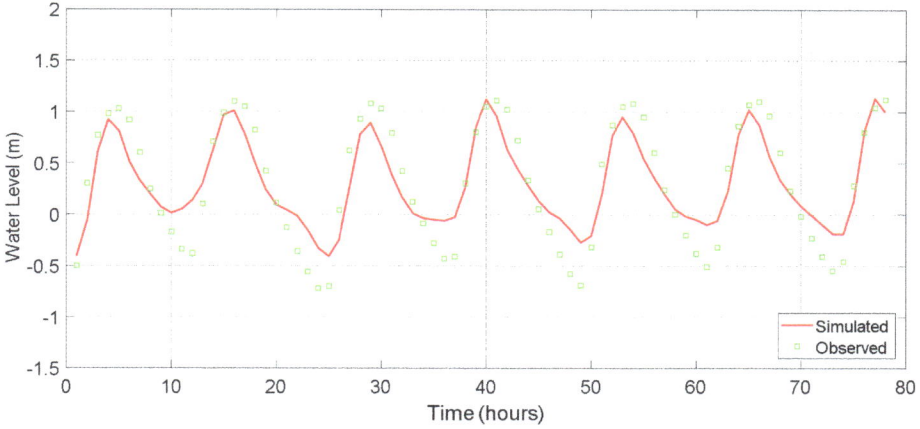

Figure 6. Zoom in of the comparison between simulated and observed water levels (relative to MSL) at Can Tho station from 07/03/2000 to 10/03/2000.

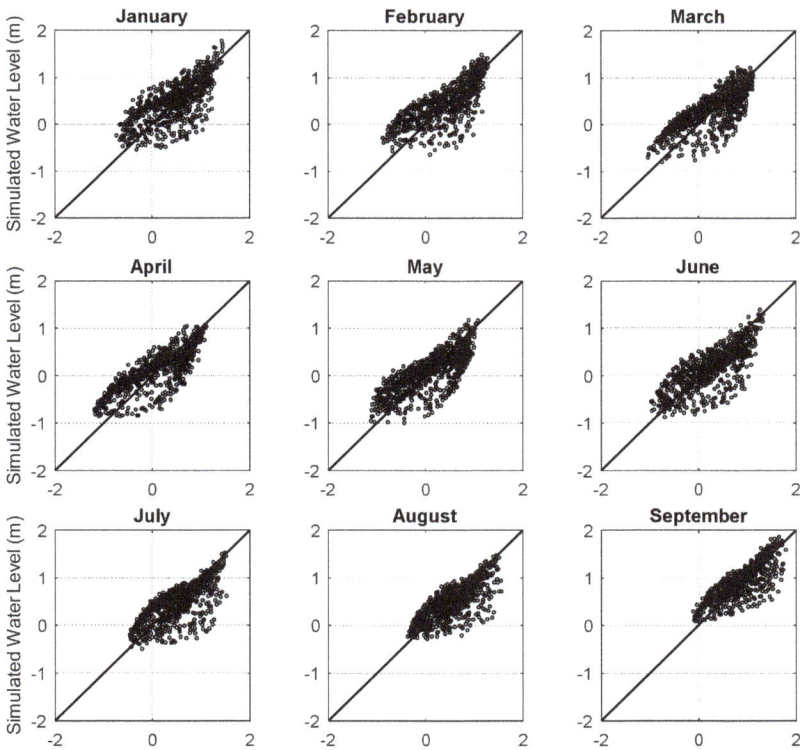

Figure 7. *Cont.*

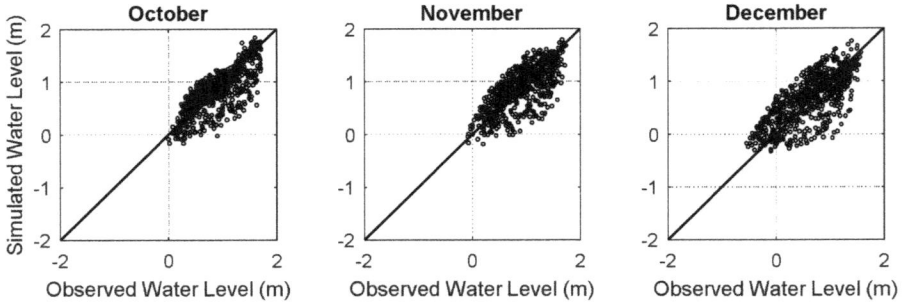

Figure 7. Scatter plots of simulated and observed hourly water level (relative to MSL) of each month in 2000 at Can Tho station.

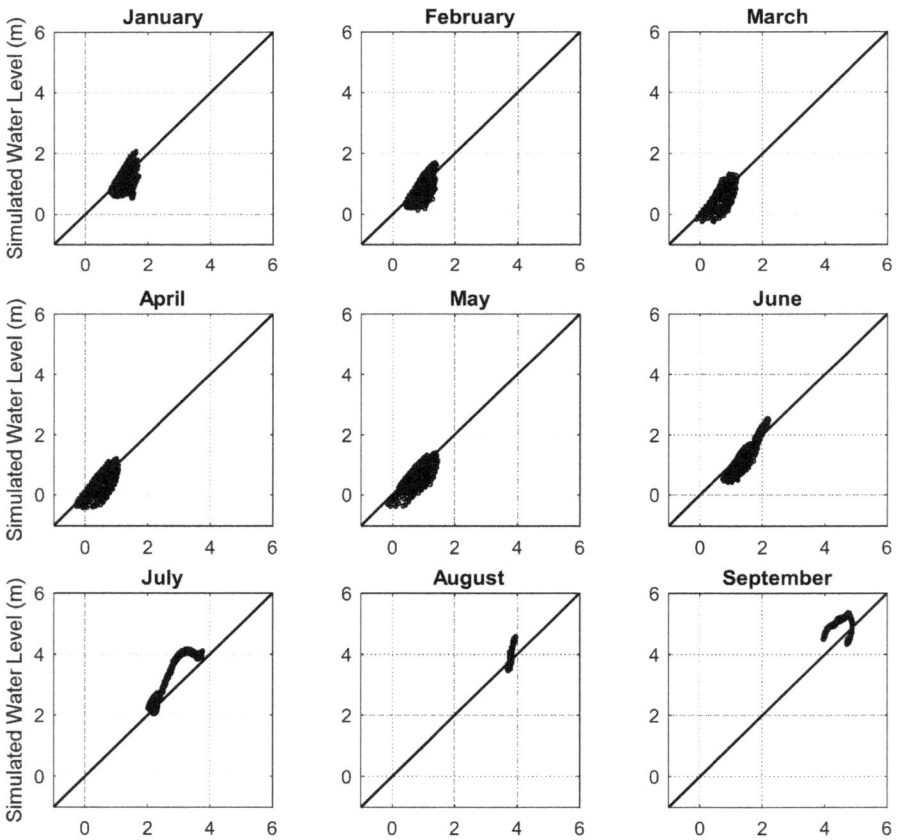

Figure 8. *Cont.*

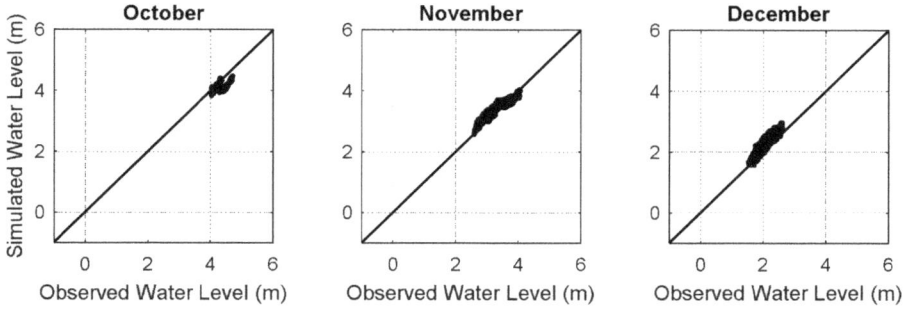

Figure 8. Scatter plots of simulated and observed hourly water level (relative to MSL) of each month in 2000 at Chau Doc station.

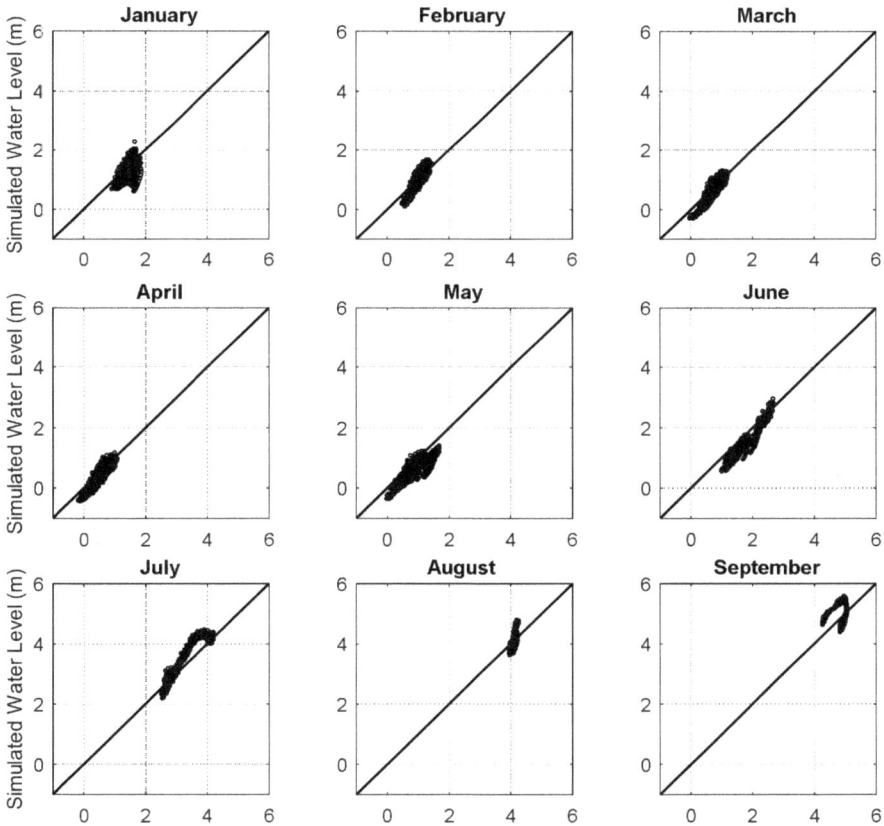

Figure 9. *Cont.*

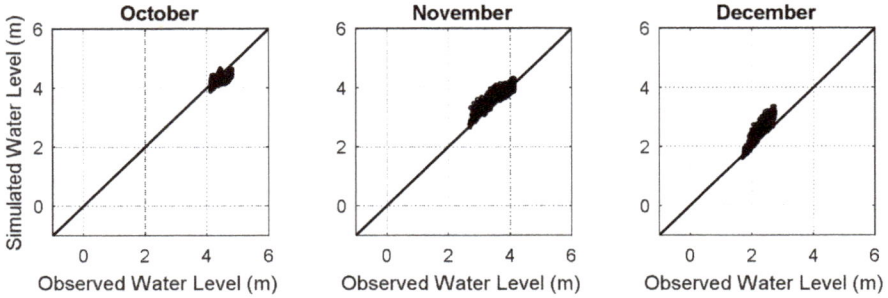

Figure 9. Scatter plots of simulated and observed hourly water level (relative to MSL) of each month in 2000 at Tan Chau station.

5.2. Model Validation

The results of the S-SWMM validation for 2001, 2002 and 2011 are shown in Tables 6–8, respectively.

Table 6. General performance rating for the model for the validation period (year 2001) using the NSE and RMAE indicators (hourly resolution).

| Station | Chau Doc | | Tan Chau | | Can Tho | |
Month	NSE	RMAE	NSE	RMAE	NSE	RMAE
January	−1.08	0.22	−0.05	0.13	0.66	0.21
February	0.85	0.32	0.95	0.18	0.79	0.21
March	0.12	0.06	0.64	0.00	0.75	0.11
April	0.19	0.01	0.62	0.02	0.78	0.23
May	0.14	0.08	0.49	0.12	0.76	5.98
June	0.01	0.20	−0.22	0.28	0.73	0.52
July	0.56	0.00	0.05	0.07	0.68	0.29
August	0.57	0.09	0.88	0.01	0.67	0.06
September	−12.15	0.01	−13.35	0.04	0.79	0.06
October	0.76	0.00	0.31	0.03	0.63	0.06
November	0.37	0.07	0.61	0.03	0.71	0.03
December	−0.52	0.17	−0.05	0.12	0.68	0.05

Table 7. General performance rating for the model for the validation period (year 2002) using the NSE and RMAE indicators (hourly resolution).

| Station | Chau Doc | | Tan Chau | | Can Tho | |
Month	NSE	RMAE	NSE	RMAE	NSE	RMAE
January	−2.00	0.30	−0.33	0.16	0.66	0.16
February	0.83	0.29	0.96	0.14	0.68	0.19
March	0.15	0.17	0.76	0.01	0.75	0.22
April	−0.21	0.08	0.42	0.10	0.67	0.28
May	0.28	0.01	0.54	0.17	0.77	0.24
June	0.46	0.13	−0.02	0.28	0.74	0.15
July	0.70	0.05	0.67	0.09	0.65	0.41
August	0.46	0.12	0.89	0.01	0.74	0.14
September	−3.49	0.03	−0.81	0.02	0.75	0.13
October	0.69	0.03	0.70	0.01	0.75	0.08
November	−1.57	0.17	−0.01	0.09	0.70	0.01
December	−1.83	0.23	−0.53	0.15	0.66	0.04

Table 8. General performance rating for the model for the validation period (2011) using the NSE and RMAE indicators (hourly resolution).

Station Month	Chau Doc		Tan Chau		Can Tho	
	NSE	RMAE	NSE	RMAE	NSE	RMAE
January	−4.76	0.80	−5.34	0.74	0.45	0.54
February	0.47	0.67	0.63	0.63	0.72	0.60
March	−0.50	0.38	0.03	0.37	0.64	0.53
April	−0.10	0.29	0.51	0.27	0.67	0.86
May	0.03	0.29	0.62	0.21	0.68	1.79
June	−0.16	0.19	0.68	0.01	0.69	1.29
July	−4.17	0.33	−0.03	0.09	0.72	0.30
August	−3.61	0.36	0.11	0.11	0.75	0.12
September	−5.96	0.23	−0.06	0.08	0.78	0.10
October	−17.46	0.07	−0.41	0.02	0.77	0.04
November	−1.57	0.20	−0.06	0.11	0.64	0.10
December	−6.38	0.45	−3.80	0.35	0.58	0.23

Based on the results in Tables 6–8 and the classifications for NSE and RMAE, error classification for NSE and RMAE indicators for each station in 2001, 2002 and 2011 are shown in Tables 9 and 10 respectively.

Table 9. Error classification (%) for NSE indicator in 2001, 2002 and 2011.

Station	Classification for NSE			
	Very Good	Good	Satisfactory	Unsatisfactory
Chau Doc	8.33	5.56	5.56	80.55
Tan Chau	13.89	8.33	19.44	58.33
Can Tho	19.44	63.89	13.89	2.78

Table 10. Error classification (%) for RMAE indicator in 2001, 2002 and 2011.

Station	Classification for RMAE				
	Excellent	Good	Reasonable	Poor	Bad
Chau Doc	55.56	36.11	5.56	2.78	0.00
Tan Chau	77.78	16.67	2.78	2.78	0.00
Can Tho	50.00	25.00	13.89	2.78	8.33

Tables 9 and 10 show that a vast majority of RMAE values at Chau Doc, Tan Chau and Can Tho stations are in excellent and good classifications. Specifically, the percentages of data/model comparison that fall into excellent and good categories are 91.67%, 94.45% and 75% at Chau Doc, Tan Chau and Can Tho, respectively. However, it is noted that 2.78% of RMAE values fall into the poor category for all 3 stations, while at Can Tho 8.33% RMAE values also fall into the bad category. With respect the NSE indicator, 83%, 13.89% and 22.22% of NSE values fall in very good and good categories (in combination) at Can Tho, Chau Doc, and Tan Chau stations, respectively. The majority of NSE values at Chau Doc (80.55%) and Tan Chau (58.33%) are unsatisfactory.

For the RMAE indicator, the poor and bad classifications are generally in the dry months (from December to June), while the good and excellent categories are generally in the wet months (from July to November). However, it is noteworthy that at Can Tho, the target area pertaining to this study, the differences between the simulated monthly mean water levels and the measured monthly mean water levels are small even during wet months, the maximum value is 0.15 m. As the main focus of this study is flooding during wet months at Can Tho, the weaker data/model comparison, especially during dry months, at the secondary stations of Chau Doc and Tan Chau do not have major implications on achieving study objectives.

For the NSE indicator, unsatisfactory data/model comparisons are shown mainly at Chau Doc and Tan Chau stations, while at Can Tho NSE is unsatisfactory on very few occasions (2.78%), and that too during dry months only. While it is possible to improve the NSE values for the two secondary stations by adding nodes and conduits to the simplified model, that will invariably increase the runtime for each simulation, which detracts from the main purpose of this study and will not add any great value to our main objective of developing a model that is able to provide probabilistic estimates of flooding at Can Tho.

Thus, for the purpose of obtaining accurate and fast predictions of flood water levels at Can Tho, the simplified SWMM model performance can be considered good enough.

5.3. Performance Comparison between the Previous Models and the Simplified SWMM Model for Entire Mekong Delta

The characteristics and simulation time associated with the process based hydrodynamic models used in previous studies of the Mekong Delta and the simplified SWMM model used here are shown in Table 11.

Table 11. Characteristics and simulation time of different models for the entire Mekong Delta (for a single core on an Intel(R) Core(TM) i5-4210M CPU @ 2.6GHz 2.6GHz processor in a computer with 8.0 GB of memory (RAM)).

No	Model Name	Number of Nodes and Links		Simulation Time for one Year(min)
		Node/Junction	Link/Cross-Section	
1	ISIS model (1D)	572	3036	90
2	VRSAP model	1505	2111	-
3	Hydro-GIS model	-	13,262	-
4	Detailed SWMM model	575	592	10
5	Simplified SWMM model	37	40	1

6. Conclusions

In this study we have shown that it is possible to simulate river water levels, with an acceptable level of accuracy (good-excellent), at a location of interest in a complex, deltaic river system like the lower Mekong with a relatively simple (simplified from thousands to several tens of cross sections) and fast (reduction of simulation time from 1.5 h to around one minute, for a 1 year simulation) numerical model. The simplified model was achieved by iteratively simplifying a complex hydraulic model, with the focus on accuracy of the water levels at a single point of interest (in this case, near Can Tho city).

The overarching objective of developing this simplified and fast model was to subsequently use it for probabilistic flood simulations and for stakeholder based co-design applications. Probabilistic flood modelling involves running thousands of realizations of the model for a given scenario. Even with advanced computer facilities available today, this requirement makes it prohibitive to use some of the 'traditional' models that take hours to run one simulation. Towards this end the simplified SWMM model presented in this article provides a feasible solution. The computing cost is still considerable (e.g., thousand simulations will cost around 17 core-h), but with a modern computer with sixteen cores and adequate amount of memory, such a simulation should be complete within a little more than an hour.

The application of the SWMM, a model developed to simulate drainage/sewerage systems, has been shown to be capable of simulating river systems with complex boundary conditions with the positive results. Additionally, with the advantage of being an open source model and a simple user interface, it would be an appropriate option for multi-stakeholder co-design meeting. Furthermore, the SWMM model also provides a well-documented, clear application programming interface (API) making it possible to embed the model in computer software applications written in a variety of

J. Mar. Sci. Eng. **2018**, *6*, 55

computer languages like C/C++, Fortran and Python. This opens up the possibility of using the resulting model as a basis for innovative applications such as serious-gaming.

The simplification of a model can sometimes lead to degradation of the precision and accuracy of its results. The appropriateness of a simplified model should be looked at in the context of the ultimate intended use of model outcomes. The intended application of the simplified SWMM model for the Mekong delta is two-fold: (a) to derive probabilistic river water level estimates to be used as input for a probabilistic urban flood model; and (b) to be used as a co-design tool in multi-stakeholder environments. In terms of both these utilities, the level of uncertainty associated with the input information is significant. For example, future sea level projections and upstream flow conditions have large uncertainties associated with them (more in latter than in the former). Therefore, striving for high precision modelling output under these realities does not improve the accuracy of the ultimate intended outcomes. In such circumstances, we argue, it is appropriate to sacrifice some accuracy to achieve efficiency and practicality.

Author Contributions: H.N., A.P. and R.R. conceptualized the study. H.N. collated the data, did all the modelling and analysis of results with the guidance of A.P. and R.R. All authors contributed to preparing the manuscript.

Acknowledgments: H.N. is supported by IHE Delft projects OPTIRISK, DURA FR Research fund, and AXA CC&CR. The authors would like to thank the Mekong River Commission for providing 1-D ISIS model for the entire Mekong Delta. The authors would also like to thank CH2M company for providing the Flood Modeller Pro licence. R.R. is supported by the AXA Research fund and the Deltares Strategic Research Programme 'Coastal and Offshore Engineering'.

Conflicts of Interest: The authors declare no conflict of interest. The founding sponsors had no role in the design of the study; in the collection, analyses, or interpretation of data; in the writing of the manuscript, and in the decision to publish the results.

References

1. Small, C.; Nicholls, R.J. A global analysis of human settlement in coastal zones. *J. Coast Res.* **2003**, *19*, 584–599.
2. Valiela, I. *Global Coastal Change*; Blackwell: Oxford, UK, 2006; p. 368.
3. Hallegate, S.; Green, C.; Nicholls, R.J.; Corfee-Morlot, J. Future flood losses in major coastal cities. *Nat. Clim Chang.* **2013**, *3*, 802–806. [CrossRef]
4. Hinkel, J.; Nicholls, R.J.; Tol, R.S.J.; Wang, Z.B.; Hamilton, J.M.; Boot, G.; Vafeidis, A.T.; McFadden, L.; Ganopolski, A.; Klein, R.J.T. A global analysis of erosion of sandy beaches and sea-level rise: An application of DIVA. *Glob. Planet. Chang.* **2013**, *111*, 150–158. [CrossRef]
5. Ranasinghe, R. Assessing climate change impacts on open sandy coasts: A review. *Earth Sci. Rev.* **2016**, *160*, 320–332. [CrossRef]
6. Gorgoglione, A.; Gioia, A.; Iacobellis, V.; Piccinni, A.F.; Ranieri, E. A Rationale for Pollutograph Evaluation in Ungauged Areas, Using Daily Rainfall Patterns: Case Studies of the Apulian Region in Southern Italy. *Appl. Environ. Soil Sci.* **2016**, *2016*, 9327614. [CrossRef]
7. Suriya, S.; Mudgal, B.V. Impact of Urbanization on Flooding: The Thirusoolam Sub Watershed—A Case Study. *J. Hydrol.* **2012**, *412–413*, 210–219. [CrossRef]
8. Zope, P.E.; Eldho, T.I.; Jothiprakash, V. Impacts of land use-land cover change and urbanization on flooding: A case study of Oshiwara River Basin in Mumbai, India. *Catena* **2016**, *145*, 142–154. [CrossRef]
9. Huong, H.T.L.; Pathirana, A. Urbanization and climate change impacts on future urban flooding in Can Tho city, Vietnam. *Hydrol. Earth Syst. Sci.* **2013**, *17*, 379–394. [CrossRef]
10. Pathirana, A.; Denekew, H.B.; Veerbeek, W.; Zevenbergen, C.; Banda, A.T. Impact of urban growth-driven landuse change on microclimate and extreme precipitation—A sensitivity study. *Atmos. Res.* **2014**, *138*, 59–72. [CrossRef]
11. Domeneghetti, A.; Castellarin, A.; Brath, A. Assessing ratingcurve uncertainty and its effects on hydraulic model calibration. *Hydrol. Earth Syst. Sci.* **2012**, *16*, 1191–1202. [CrossRef]
12. Tri, V.P.D.; Trung, N.H.; Tuu, N.T. Flow dynamics in the Long Xuyen Quadrangle under the impacts of full-dyke systems and sea level rise. *VNU J. Sci. Earth Sci.* **2012**, *28*, 205–214.
13. Akpo, E.; Crane, T.A.; Vissoh, P.V.; Tossou, R.C. Co-production of knowledge in multi-stakeholder processes: Analyzing joint experimentation as social learning. *J. Agric. Educ. Ext.* **2015**, *21*, 369–388. [CrossRef]

14. Mekong River Commission (MRC). *Overview of the Hydrology of the Mekong Basin*; Mekong River Commission: Vientiane, Laos, 2005; pp. 73–75, ISSN 1728 3248.

15. Mekong River Commission (MRC). *State of the Basin Report 2010*; Mekong River Commission: Vientiane, Laos, 2010; ISBN 9789932080571.

16. Kuenzer, C.; Gue, H.; Huth, J.; Leinenkugel, P.; Li, X.; Cech, S. Flood mapping and flood dynamic of the Mekong Delta: ENVISAT ASAR-WSM base time series analyses. *Remote Sens.* **2013**, *5*, 687–715. [CrossRef]

17. Nguyen, H.N. *Human Development Report 2007/2008 Flooding in Mekong River Delta, Viet Nam*; Human Development Report; United Nations Development Programme: New York, NY, USA, 2008; Volume 4.

18. Balica, S.F.; Dinh, Q.; Popescu, I.; Vo, T.Q.; Pham, D.Q. Flood impact in the Mekong Delta, Vietnam. *J. Maps* **2014**, *10*, 257–268. [CrossRef]

19. Nicholls, R.J.; Wong, P.P.; Burkett, V.R.; Codignotto, J.O.; Hay, J.E.; McLean, R.F.; Ragoonaden, S.; Woodroffe, C.D. Coastal systems and low-lying areas. In *Climate Change 2007: Impacts, Adaptation and Vulnerability, Contribution of Working Group II to the Fourth Assessment Report of the Intergovernmental Panel on Climate Change*; Cambridge University Press: Cambridge, UK, 2007.

20. Wong, P.-P.; Losada, I.J.; Gattuso, J.P.; Hinkel, J.; Khattabi, A.; McInnes, K.L.; Saito, Y.; Sallenger, A. Coastal Systems and Low-Lying Areas. In *Climate Change 2014: Impacts, Adaptation, and Vulnerability. Part A: Global and Sectoral Aspects*; Contribution of Working Group II to the Fifth Assessment Report of the Intergovernmental Panel on Climate Change; Cambridge University Press: Cambridge, UK; New York, NY, USA, 2014.

21. Hapuarachchi, H.A.P.; Takeuchi, K.; Zhou, M.C.; Kiem, A.S.; Georgievski, M.; Magome, J.; Ishidaira, H. Investigation of the Mekong River basin hydrology for 1980–2000 using the YhyM. *Hydrol. Process.* **2008**, *22*, 1246–1256. [CrossRef]

22. Kingston, D.G.; Thompson, J.R.; Kite, G. Uncertainty in climate change projections of discharge for the Mekong River Basin. *Hydrol. Earth Syst. Sci.* **2011**, *15*, 1459–1471. [CrossRef]

23. Eastham, J.; Mpelasoka, F.; Ticehurst, C.; Dyce, P.; Ali, R.; Kirby, M. Mekong River Basin Water Resources Assessment: Impacts of Climate Change. In *CSIRO Water for a Healthy Country National Research Flagship Report*; CSIRO: Canberra, Australia, 2008; Volume 153, ISSN 1835-095X.

24. Hoanh, C.T.; Jirayoot, K.; Lacomne, G.; Srunetr, V. *Impacts of Climate Change and Development on Mekong Flow Regimes First Assessment—2009*; MRC Management Information Booklet Series No. 4; Mekong River Commission: Vientiane, Laos, 2011; Volume 16.

25. Västilä, K.; Kummu, M.; Sangmanee, C.; Chinvanno, S. Modelling climate change impacts on the flood pulse in the lower Mekong floodplains. *J. Water Clim. Chang.* **2010**, *1*, 67–86. [CrossRef]

26. Räsänen, T.A.; Koponen, J.; Lauri, H.; Kummu, M. Downstream hydrological impacts of hydropower development in the Upper Mekong Basin. *Water Resour. Manag.* **2012**, *26*, 3495–3513. [CrossRef]

27. Räsänen, T.A.; Someth, P.; Lauri, H.; Koponen, J.; Sarkkula, J.; Kummu, M. Observed river discharge changes due to hydropower operations in the Upper Mekong Basin. *J. Hydrol.* **2017**, *545*, 28–41. [CrossRef]

28. Tran, P.; Marincioni, F.; Shaw, R.; Sarti, M.; Van An, L. Flood risk management in Central Viet Nam: Challenges and potentials. *Nat. Hazards* **2008**, *46*, 119–138. [CrossRef]

29. Hoang, L.P.; Biesbroek, R.; Tri, V.P.D.; Kummu, M.; van Vliet, M.T.H.; Leemans, R.; Kabat, P.; Ludwig, F. Managing flood risks in the Mekong Delta: How to address emerging challenges under climate change and socioeconomic developments. *Ambio* **2018**, 1–15. [CrossRef] [PubMed]

30. Wassmann, R.; Hien, N.X.; Hoanh, C.T.; Tuong, T.P. Sea level rise affecting the Vietnamese Mekong Delta: Water elevation in the flood season and implications for rice production. *Clim. Chang.* **2004**, *66*, 89–107. [CrossRef]

31. Royal HaskoningDHV; WUR; Deltares; Rebel. *Mekong Delta Plan: Long-Term Vision and Strategy for a Safe, Prosperous and Sustainable Delta*; Prepared under the Strategic Partnership Arrangement on Climate Change Adaptation and Water Management between the Netherlands and Vietnam: Hanoi, Vietnam; Amersfoort, The Netherlands, 2013; p. 126.

32. National Institute for Urban and Rural Planning (NIURP) under Vietnam Ministry of Construction. *Development Strategies (CDS) for Medium-Size Cities in Vietnam: Can Tho and Ha Long*; Vietnam Ministry of Construction: Hanoi, Vietnam, 2010.

33. CCCO & ISET. *Peri-Urban Development Planning and Flooding Problems: Story of New Urban Areas in Can Tho City, Viet Nam*; ISET: Hanoi, Vietnam, 2015.

34. Lauri, H.; De Moel, H.; Ward, P.J.; Räsänen, T.A.; Keskinen, M.; Kummu, M. Future changes in Mekong River hydrology: Impact of climate change and reservoir operation on discharge. *Hydrol. Earth Syst. Sci.* **2012**, *16*, 4603–4619. [CrossRef]

35. Piman, T.; Lennaerts, T.; Southalack, P. Assessment of hydrological changes in the lower Mekong basin from basin-wide development scenarios. *Hydrol. Process.* **2013**, *27*, 2115–2125. [CrossRef]

36. Pokhrel, Y.; Burbano, M.; Roush, J.; Kang, H.; Sridhar, V.; Hyndman, D. A Review of the Integrated Effects of Changing Climate, Land Use, and Dams on Mekong River Hydrology. *Water* **2018**, *10*, 266. [CrossRef]

37. Le, T.V.H.; Nguyen, H.N.; Wolanski, E.J.; Tran, T.C.; Haruyama, S. The combined impact on the flooding in Vietnam's Mekong River delta of local man-made structures, sea level rise, and dams upstream in the river catchment. Estuarine Coastal. *Estuar. Coast. Shelf Sci.* **2007**, *71*, 110–116. [CrossRef]

38. Le, T.V.H.; Haruyama, S.; Nguyen, H.N.; Tran, T.C. Study of 2001 Flood Using Numerical Model in the Mekong River Delta, Vietnam. In Proceedings of the International Symposium on Floods in Coastal Cities under Climate Change Conditions, Khlong Nueng, Thailand, 23–25 June 2005; pp. 65–73.

39. Van, P.D.T.; Popescu, I.; Van Griensven, A.; Solomatine, D.P.; Trung, N.H.; Green, A. A study of the climate change impacts on fluvial flood propagation in the Vietnamese Mekong Delta. *Hydrol. Earth Syst. Sci.* **2012**, *16*, 4637–4649. [CrossRef]

40. Rossman, L.A. *Storm Water Management Model User's Manual*; EPA: Cincinnati, OH, USA, 2015; pp. 1–353.

41. Pathirana, A. SWMM5-EA-A tool for learning optimization of urban drainage and sewerage systems with genetic algorithms. In Proceedings of the 11th International Conference on Hydroinformatics, New York, NY, USA, 17–21 August 2014; CUNY Academic Works: New York, NY, USA, 2014.

42. Nash, J.E.; Sutcliffe, J.V. River flow forecasting through conceptual models: Part 1. A discussion of principles. *J. Hydrol.* **1970**, *10*, 282–290. [CrossRef]

43. Sutherland, J.; Walstra, D.J.R.; Chesher, T.J.; van Rijn, L.C.; Southgate, H.N. Evaluation of coastal area modelling systems at an estuary mouth. *Coast. Eng.* **2004**, *51*, 119–142. [CrossRef]

Journal of
Marine Science and Engineering

MDPI

Article

Optimized Reliability Based Upgrading of Rubble Mound Breakwaters in a Changing Climate

Panagiota Galiatsatou *, Christos Makris and Panayotis Prinos

Division of Hydraulics and Environmental Engineering, School of Civil Engineering,
Aristotle University of Thessaloniki, 54124 Thessaloniki, Greece; cmakris@civil.auth.gr (C.M.);
prinosp@civil.auth.gr (P.P.)
* Correspondence: pgaliats@civil.auth.gr; Tel.: +30-2310-995-708

Received: 20 June 2018; Accepted: 18 July 2018; Published: 2 August 2018

Abstract: The present work aims at presenting an approach on implementing appropriate mitigation measures for the upgrade of rubble mound breakwaters protecting harbors and/or marinas against increasing future marine hazards and related escalating exposure to downtime risks. This approach is based on the reliability analysis of the studied structure coupled with economic optimization techniques. It includes the construction of probability distribution functions for all the stochastic variables of the marine climate (waves, storm surges, and sea level rise) for present and future conditions, the suggestion of different mitigation options for upgrading, the construction of a fault tree providing a logical succession of all events that lead to port downtime for each alternative mitigation option, and conclusively, the testing of a large number of possible alternative geometries for each option. A single solution is selected from the total sample of acceptable geometries for each upgrading concept that satisfy a probabilistic constraint in order to minimize the total costs of protection. The upgrading options considered in the present work include the construction or enhancement of a crown wall on the breakwater crest, the addition of the third layer of rocks above the primary armor layer of the breakwater (combined with crest elements), the attachment of a berm on the primary armor layer, and the construction of a detached low-crested structure in front of the breakwater. The proposed methodology is applied to an indicative rubble mound breakwater with an existing superstructure. The construction of a berm on the existing primary armor layer of the studied breakwater (port of Deauville, France), seems to be advantageous in terms of optimized total costs compared to other mitigation options.

Keywords: reliability; economic optimization; coastal structure; upgrading; rubble mound breakwater; climate change; extreme value theory

---

1. Introduction

Global climate change is expected to cause significant long-term changes in mean sea level (MSL), characteristics of the wave fields and storm surges, as well as in the trajectory and intensity of storms [1,2]. Significant fluctuations in the frequency and the intensity of storms, and in the wave climate, were observed in the recent past in the North Sea [3–7], without however identifying significant general trends. Studies conducted in larger areas, i.e., in the Northern Atlantic or along the European coastline, proved certain changes in the wind fields, in storm surge levels, as well as in the wave climate [8–13].

The general inception of a changing climate, characterized by extreme marine events of higher intensity and frequency and MSL rise [14], increases vulnerability and exposure of port and harbor structures to different failure modes, resulting in their inability to fulfill their requirements. The increased probability of failure of such structures under higher (than those designed for) hydraulic loading conditions and the limited residual service lifetime of many of them, combined with the fact that economic activity is assembled in trade, cruise home, cargo and fishing ports, and harbors, creates an urgent need for a reliable upgrading solution to maintain or increase the safety level of engineered water basins in the future, while ensuring that construction and performance costs are kept low. Since most port or harbor structures to date have been designed neglecting the effects of climate change, an upgrading procedure is required to ensure new optimal geometries characterized by a low probability of future failure, minimizing the total expected damages.

Port or harbor structures ensure that downtime risks (defined as the stoppage of operations within the basins due to malfunction of the protection system) are kept low. Such risks are a combination of the failure probability of a structure and its relevant impact, usually estimated as the product of the occurrence probability of the event and its consequences [15]. Coastal risks arise due to hazards consisting of a source or an initiator event (i.e., high wave/sea level), a receptor (i.e., protected basins, local communities, and local infrastructure) and a pathway between them (i.e., failure of defenses). The predicted failure probability results from the sources and pathways of risk, while the expected damages, including economic, social, or environmental impacts, are estimated from the receptors. Rubble mound breakwaters, consisting of large heaps of loose elements, primarily aim at protecting coastal areas against flooding and erosion and ports, harbors, or marinas against wave action, significantly reducing the flooding hazard of the protected areas in case of extreme marine storms. However, the predicted sea level rise and a possible increase in extreme marine storms will expose such structures to larger and longer waves, which in turn will lead to greater wave overtopping and transmission, and greater water penetration into the protected areas. Hence, functionality and safety of such structures has to be re-evaluated under climate change conditions. The proposed work aims at contributing to the reinforcement of selected port, harbor or marina rubble mound structures in order to withstand the threat of increased hydraulic loading due to climate change effects.

1.1. Literature Review

Studying the effects of climate change on coastal and harbor structures started receiving considerable attention since the early 21st century [16–18]. Suh et al. [19] developed a methodology that incorporates the effects of climate change in the design of caisson breakwaters. Becker et al. [20], Suh et al. [21], and Sánchez-Arcilla et al. [22] studied the possible effects of climate change on port and harbor operations and related infrastructure. Chini & Stansby [23] and Mase et al. [24] investigated the influence of both sea level rise (SLR) and future wave and storm surge climate on overtopping of a sea wall, and the stability of composite breakwaters with wave-dissipating blocks, respectively. Hoshino et al. [25] applied a methodology to change the intensity of tropical cyclones in Japan under future climate conditions and assessed the performance of selected sea defenses to the new forcing conditions. Sekimoto et al. [26] proposed time-dependent return periods and a modified residual life for coastal structures to make decisions on adaptation strategies under climate change. Isobe [27] used simple formulas and diagrams to provide rough quantitative estimates of global warming impact on coastal defense structures and proposed a roadmap for the coastal zone to respond to SLR and tropical cyclone intensification. Lee et al. [28] developed a risk-based system to evaluate safety of coastal structures subject to SLR, emphasizing reinforcement needs for current standards of protection maintained under climate change conditions. Within THESEUS (Innovative Technologies for Safer European Coasts in a Changing Climate, 2009–2013) research project [29], the vulnerability and resilience of coastal structures to climate change effects were studied and environmentally friendly upgrading solutions were investigated. Burcharth et al. [30] presented methods to upgrade typical embankments under climate change conditions and estimated the associated construction costs to determine the most effective alternative. Within CCSEAWAVS (Estimating the Effects of Climate Change on Sea Level and Wave Climate of the Greek seas, Coastal Vulnerability and Safety of Coastal and Marine Structures, 2012–2015) research project [31], a first assessment of the vulnerability of coastal and harbor structures to climate change in Greece was attempted and indicative upgrading methods were proposed [32,33].

The overall framework of integrated risk management includes analysis and assessment of risks and implementation of risk reduction/mitigation options [34]. Since the 1953 major flood, the Dutch authorities have used risk-based principles in the design and maintenance of flood defenses, i.e., [35,36]. Mai et al. [37,38] and Jonkman et al. [39] applied risk-based models to design flood protection systems, and Lee et al. [28] developed a risk-based system to evaluate safety of coastal structures subject to sea level rise, emphasizing reinforcement needs for current standards of protection maintained under climate change conditions. Within the EU research project FLOODsite (Integrated flood risk analysis and management methodologies, 2004–2009) [40], an inherent part of a risk-based approach to designing flood defenses reliability (corresponding to the probability of failure under given hydraulic loading conditions) analysis has been further developed to support a range of decisions and adopt different levels of complexity. Research on reliability analysis within FLOODsite [40] focused on developing methods and techniques to incorporate up-to-date knowledge on the different failure modes of the structures considered, as well as on the interactions between these failure mechanisms.

Reliability-based optimal structural design has been applied by numerous authors in the past, i.e., [41–44]. Castillo et al. [45], Dai Viet et al. [46], Prasad Kumar [47], Suh et al. [19], and Galiatsatou & Prinos [48] selected optimized geometries of rubble mound or caisson breakwaters combining structural reliability and methods of economic optimization. In this framework possible economic benefits from including Wave Energy Converters as part of the upgrading scheme (e.g., [49]) could be considered, yet only in cases of low energy sea-sates. Van Gelder et al. [50] provided an outline of methods and tools for reliability analysis of flood defenses. Buijs et al. [51] established time-dependent reliability models for flood defense systems, combing stochastic, hierarchical, and parametric processes to represent their main deterioration mechanisms. Kim & Suh [52] and Naulin et al. [53,54] performed reliability analysis of coastal and harbor structures to determine failure probabilities including different mechanisms. Buijs et al. [51] and Nepal et al. [55] established time-dependent reliability models for coastal embankments using stochastic deterioration models.

1.2. Scope of Work

Safety assessments of flood defense assets are increasingly performed with the technique of structural reliability. Reliability-based design utilizes the probability of failure as a measure of the performance of the coastal or harbor structure. A maximum failure probability is defined and the structure should meet the requirements. In order to make a probabilistic optimization, the mechanisms that lead to structural instability or losing of a specific level of functionality should be considered. Therefore, different failure modes and their relation to overall failure of the structure should be represented in fault trees. Quantitative analysis starts at the level of failure modes with the definition of limit state functions and the probabilistic analysis of all random input variables. The total failure probability of the structure can be defined from a quantification of the probability of occurrence of each failure mode. Reliability-based optimization can be performed on the grounds of minimizing the total cost function of the structure, thus finding a cost-effective design that also meets certain probabilistic requirements.

The present work aims at presenting an improved approach for evaluating the performance of rubble mound breakwaters in a changing climate and at suggesting and implementing appropriate mitigation measures for upgrading to face the increasing marine hazards and exposure to downtime risks. Therefore, the study mainly intends to assess the future safety levels of existing structures under climate change conditions and to propose measures for structural upgrading utilizing a reliability-based procedure. The whole domain of upgrading structures to face climate change impacts is an original research field and reliability-based approaches will significantly contribute to handle the problem posed by balancing the compromises and investments made on the side of the defense system/structure, due to increased safety levels, with the accompanying benefits (limiting losses) in the protected areas. The comparative advantages and main innovation of the proposed work with regard to former research studies on reliability-based approaches are mainly focused on this argument. Thus, it combines the quickly progressing framework of reliability-based optimization, with upgrading mitigation measures and engineering solutions to reinforce port and harbor structures against climate change effects. Such approaches are expected to significantly outperform deterministic methods and can significantly contribute to include hazard, as well as vulnerability determination in the decision process, and to realize appropriate and sustainable measures to minimize port or harbor downtime risks.

The proposed approach can fully consider uncertainties in stress and resistance parameters, it can examine different mitigation measures for upgrading port and harbor breakwaters in a changing climate, and compare their effectiveness within the general framework of minimizing the total downtime risks of the protected areas. It can contribute to increasing protection of ports/harbors against climate change with minimum possible impacts (only economic impacts are examined here) during the service lifetime of the upgraded structures and to achieve high benefits (corresponding to limiting losses) for the protected areas reflected to the regional and national economy. It can also contribute to facilitate adaptation of coastal areas to climate change in a quite adverse and unstable financial environment, which hinders the construction of new and modern defenses. The proposed reliability-based approach can provide a basis for reasonable decision-making and can allow the choice of further rationalized safety levels of port/harbor basins if the consequences of downtime and the costs of protection are made explicit.

In Section 2 of the present work the main aspects of the proposed methodology for reliability-based upgrading are presented. The governing loading variables of the studied structures are represented as random variables and appropriate distribution functions are selected to represent uncertainties in their estimates for future climate conditions. Upgrading options are suggested and the main failure modes of the structures are considered. Appropriate limit state functions are then defined in accordance with the selected failure criterion. A maximally acceptable failure probability is defined and reliability analysis is performed to define the space of acceptable geometries of the structures considered. Finally, the cost of all acceptable geometries within the solution space is estimated for each mitigation option and an economical optimization is performed. In Section 3 the proposed methodology is implemented to a selected rubble mound breakwater. Different optimized and upgraded geometries of the structure are proposed and estimated to withstand loads under expected future climate conditions. Section 4 presents a summary of the presented work and methodology together with the main conclusions of the study.

2. Methodology for Optimized Reliability-Based Upgrading of Rubble Mound Breakwaters

2.1. Proposed Upgrading Scheme

As a more holistic and robust approach compared to conventional design methods, a novel reliability-based upgrading method is introduced in the present work to compare different mitigation options on their effectiveness to withstand the adverse consequences of climate change and to select an optimal solution that minimizes port/harbor downtime risks.

The methodology of reliability-based upgrading followed in this work consists of the following actions:

1. Fitting appropriate cumulative distribution functions (CDFs) to stochastic variables of the marine climate, i.e., offshore wave height and sea level due to storm surge for present climate conditions.
2. Transfer extracted CDFs to the selected breakwater site.
3. Definition of CDFs for future sea states considering climate change scenarios or assumptions.
4. Investigation of the ability of selected breakwaters to withstand impacts due to future sea states, given their design characteristics.
5. Definition of different mitigation options for rubble mound breakwaters to limit damages under future climate conditions.
6. Creation of a fault tree that gives a logical succession of all events leading to port/harbor downtime for each alternative mitigation option (or combination of options if applicable).

7. Investigation of both Ultimate Limit States (ULS) and Serviceability Limit States (SLS) of the rubble mound breakwater to define its total probability of failure.
8. Generation of a large number of possible alternative geometries for each upgrading concept.
9. Definition of an acceptable flooding probability.
10. Selection of an optimal geometry from the sample of acceptable geometries for each upgrading concept, minimizing the total costs of protection. The methodology described above is schematically presented in Figure 1.

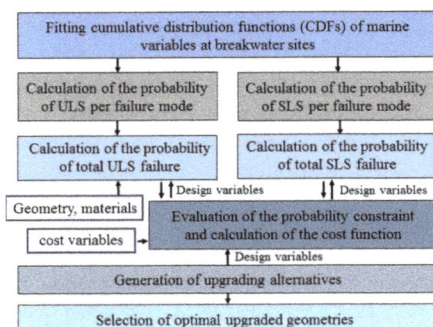

Figure 1. Schematic representation of reliability-based upgrading of a rubble mound structure.

A reliability analysis should start with the definition of an unwanted "top event", which in the present work coincides with failure of the rubble mound breakwater to fulfill its requirements. Such an event can be defined as port/harbor downtime and can be caused under both extreme marine conditions, or under restrictive conditions for the normal prosecution of daily port/harbor operations. The former can cause the collapse of the studied structure (which amounts to large quantities of water entering the protected basin causing severe damage to the structure itself and to port infrastructure), resulting in the loss of its protective properties. The latter are characterized by excessive wave height in the harbor basin, resulting in suspension of port operations. Collapse of the rubble mound breakwater belongs to the Ultimate Limit States (ULS), while cessation of port operations on a daily basis belongs to the Serviceability Limit States (SLS). A fault tree is constructed containing the mechanisms that cause failure of the rubble mound structure (failure modes).

Principal failure mechanisms of conventional rubble mound breakwaters (with or without crest elements) include:

- Failure or instability of the seaside armor layer (e.g., [56]),
- Failure of the rear-side slope (e.g., [57]),
- Scouring of the toe structure (e.g., [58]),
- Excessive overtopping (e.g., [59]),
- The slip cycle (e.g., [60]),
- Sliding and tilting of existing superstructure/crown-wall (e.g., [61]),
- Excessive settlement (e.g., [62]).

The analysis begins with the determination of limit state functions and the description of CDFs of variables in these functions. For port/harbor structures both ULS and SLS are examined. A reliability-based optimized breakwater geometry results from minimizing the total costs (adding upgrading and possible costs of failure) during the service lifetime of each upgraded geometry of the breakwater. Failure costs are estimated considering the failure probability for each type of upgraded geometry of the breakwater within its service lifetime.

Upgrading conceptual options to be examined are presented in Figure 2, and include:

1. Increasing the breakwater crest level by adding a crown wall or existing crest elements being heightened and strengthened.
2. Adding a new protective layer of armor units on the windward slope of the studied rubble mound breakwater combined with new or elevated existing crest elements.
3. Construction of a berm structure in front of the seaside slope.
4. Construction of a low-crested structure in the windward front of the existing breakwater.

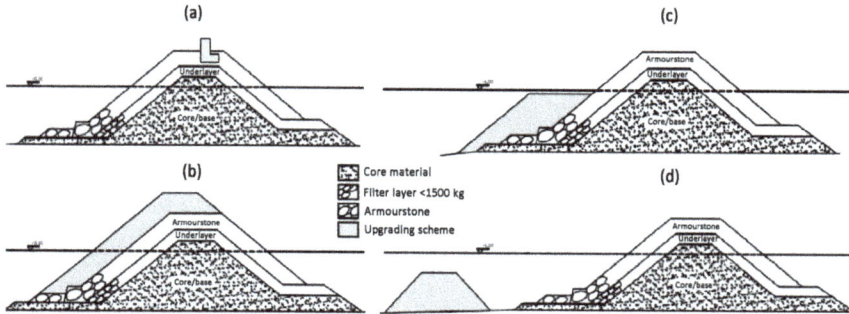

Figure 2. Sketch portrayal of four basic upgrading conceptual options of a rubble mound breakwater used for mitigation against climate change: (**a**) crown wall or heightening of existing crest element; (**b**) additional protective armor layer; (**c**) additional berm on the seaside breakwater slope; (**d**) low-crested structure in front of existing breakwater.

The failure probability which minimizes the total lifetime costs for each of the four mitigation options and corresponds to an optimized geometry will be finally selected. The requirement of minimization of the lifetime costs can be combined with a direct limitation of the solution space in the form of a constraint on the port/harbor downtime probability.

2.2. Probability Distribution Functions for Variables of the Marine Climate

Considering ULS for the studied structure, assumed to happen under extreme marine conditions, extrapolation of the marine variables to high return periods has to be performed. Methods and techniques of the well-established univariate and multivariate Extreme Value Theory (EVT) can be used for such purpose. Considering SLS, maximum daily wave height conditions are analyzed, using a CDF for short term prediction. Extreme value (EV) methods are powerful statistical methods for drawing an inference about the extremes of a process, using only data on relatively extreme values of it. The statistical methodology is motivated by EVT, which relies on the assumption that the limiting models suggested by the asymptotic theory continue to hold at finite but extreme levels [63]. Univariate analysis of extreme marine conditions (i.e., extremes of the significant wave heights and storm surges) can be performed using the Generalized Extreme Value (GEV) distribution function for annual maxima. The GEV distribution function is of the form:

$$G_\theta(x) = \exp\left\{-\left[1 + \xi\left(\frac{x-\mu}{\sigma}\right)\right]^{-1/\xi}_+\right\} \tag{1}$$

where $\sigma > 0$ and $\theta = (\mu, \sigma, \xi)$ the vector of parameters, namely the location (μ), scale (σ), and shape (ξ) parameters, determined by the tail behavior of G_θ. The model of Equation (1) can be fitted to annual maxima wave heights and storm surges and the parameter vector for both variables can be estimated by means of the L-moments estimation procedure.

To estimate the design wave height for the port/harbor structure considered, the extreme wave height CDF at the breakwater's site is approximated using a statistical approach proposed by Kim and Suh [52] and Suh et al. [64]. The latter is based on the assumption that the wave height CDF in coastal water reduces in the mean and in the standard deviation compared with the respective theoretical model for deep water waves, so that the coefficient of variation (CV) remains constant. This assumption seems quite realistic for extreme wave heights, as their CDF becomes narrower and is shifted toward smaller values as they propagate in coastal waters, while its shape does not undergo any significant changes. Therefore it is assumed that extreme wave heights at the breakwater site are also represented by a GEV distribution with shape parameter, ξ, equal to the one estimated for deep water conditions. The CV for the GEV distribution function is:

$$CV = \frac{\frac{\sigma}{\xi}\sqrt{\Gamma(1+2\xi) - \Gamma(1+\xi)^2}}{\mu + \sigma\frac{\Gamma(1+\xi)-1}{\xi}} \tag{2}$$

where $\Gamma(\cdot)$ corresponds to the Gamma CDF. To estimate the GEV parameters for wave height at the breakwater site, the constant CV of Equation (2) is combined with the formula used to calculate the design wave height of the structure i.e., the 100-year nearshore wave height (return period commonly used in the design of coastal and harbor structures):

$$z_{100} = \mu - \frac{\sigma}{\xi}\left\{1 - [-\log(1-p)]^{-\xi}\right\} \tag{3}$$

where $p = 1/T$ (T is the return period of 100 years) and z_{100} is the 100-year design wave height of the studied breakwater.

Considering SLS for the studied structure, daily maxima of significant wave height are analyzed. More specifically, it is assumed that maximum daily offshore random wave heights (both breaking and nonbreaking) follow a Rayleigh CDF, when based on energy flux-balance single-parameter transformation models or daily observations [65]:

$$F_\theta(x) = 1 - \exp\left[-\frac{(x-\mu)^2}{2\sigma^2}\right] \tag{4}$$

with parameter vector $\theta = (\mu, \sigma)$. L-moments estimation procedure is used to calculate the parameter vector of the Rayleigh CDF, whose CV is given by:

$$CV = \frac{\sigma\sqrt{2 - \pi/2}}{\mu + \sigma\sqrt{\pi/2}} \tag{5}$$

Following the methodology used for extreme wave conditions, the CV of wave heights CDF at the breakwater site is assumed to be equal to the one for deep water wave conditions. Goodness of fit for all CDFs of the marine variables has been checked by means of the Kolmogorov–Smirnov test (significance level set to 5%).

2.3. Reliability Functions for Upgrading Options

Reliability-based design hinges on the use of the probability of failure as a measure of the structure performance. To calculate the probability of failure for a certain limit state, it is necessary to know the difference between the resistance of the structure, R_Z, and the load it is exposed to, S_Z. The reliability function, Z, for a certain limit state is therefore defined as:

$$Z = R_Z - S_Z \tag{6}$$

If $Z > 0$ the structure is in a safe state, while for $Z \leq 0$ the failure domain is defined. The probability of failure can generally be expressed as:

$$P_f = P(Z \leq 0) \tag{7}$$

In the present work reliability functions are defined for both ULS and SLS based on formulas describing the main failure modes of rubble mound breakwaters. These reliability functions contain variables of the marine climate at the windward side of the studied breakwater, as well as variables describing geometrical and material properties of the studied structure. Therefore:

$$Z = f(Wave\ Height,\ Sea\ Water\ Level,\ Geometrical\ Parameters,\ Material\ Parameters) \tag{8}$$

For each limit state, a reliability function is therefore defined, CDFs of input variables (Section 2.2) are considered and level II and III reliability methods [66] are used to estimate the failure probabilities corresponding to each failure mode. Level II reliability methods include the linearization of the reliability function at an appropriately defined design point of the failure space. A standard Normal distribution is used to approximate the CDF of each variable present in the reliability function. Level III reliability methods calculate the failure probability by considering the probability density functions (PDFs) of all variables involved, based on Monte Carlo simulation techniques. In this case the probability of failure is approximated as n_f/n where n is the total number of simulations and n_f the number of simulations for which the condition of Equation (7) is satisfied. In the following, the reliability functions Z are provided as Z_{ij}, with the indices i and j referring to the mitigation option ($i = 1$–4) and the failure mechanism ($j = 1$–4), respectively.

A fault tree is constructed giving a logical succession of all events leading to the unwanted "top event", defined as port/harbor downtime, containing both the ULS and SLS of the rubble mound breakwater. For the former, the objective of the structure is the protection of the port area against flooding, while for the latter excessive wave height within the protected basin, causing problems to the standard port operations, defines the failure probability of the breakwater. In this study it is assumed that operations within the port, harbor or marina are only active during normal weather conditions [46]. Extreme marine conditions resulting in ULS failure of the protective structure are assumed to cause partial downtime of the port, harbor, or marina for quite a long period of time, since the basin will then be unprotected until the repair stage is completed.

A rubble mound breakwater usually consists of many parts, such as the core, the armor layers (underlayers and cover layer), the concrete cap and/or crown wall on the crest, and the toe structure. Failure of each one of them under extreme marine conditions can cause malfunction of the entire structure. Only principle failure modes are considered here as the main types of instability under extreme marine climate conditions (ULS), namely instability of the seaside primary armor layer, excessive wave overtopping, and scouring of the breakwater toe. The wave height inside the protected basins in the SLS will be considered as a combination of wave refraction-diffraction via the entrance of the protective system and wave transmission through and overtopping the breakwaters [46]. Figure 3 presents the fault tree corresponding to the port downtime "top event".

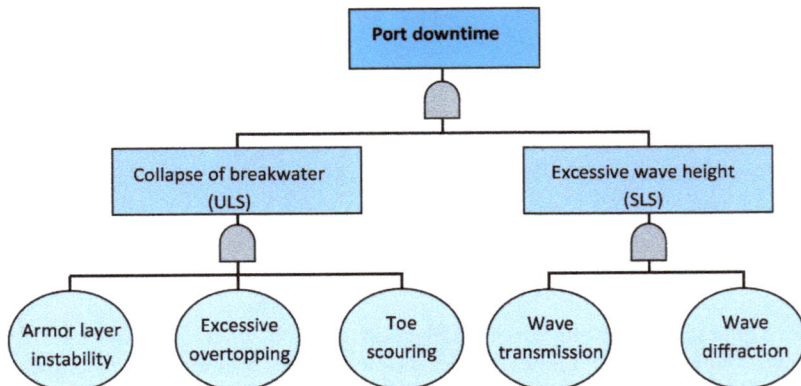

Figure 3. Fault tree for the port downtime "top event".

Future marine climate conditions are generally expected to increase the probability of failure of coastal, port, or harbor structures. For the structures that shall not be able to meet their future requirements, an upgrading method should be suggested and applied. The construction of a crown wall or parapet or strengthening and heightening of an existing superstructure (mitigation option 1) is expected to have no prominent effects on the global stability of the breakwater considered. Nevertheless, it is expected to increase the hydraulic performance of the structure by reducing overtopping and to indirectly contribute to the structure's stability by minimizing hydraulic loads and impact of wave forces on the structure [67]. The addition of a new seaside armor layer (mitigation option 2) is expected to have a mild effect on global stability of the structure [66], while overtopping will decrease due to possible elevation of existing crest elements. A berm structure (mitigation option 3) protects the breakwater toe against scouring and also increases the hydraulic performance of the structure by reducing overtopping [66]. Finally, a detached low-crested structure in the seaward front of the existing breakwater (mitigation option 4) assists in the transformation of the incident wave characteristics that impact the breakwater, inducing lower significant wave heights and peak spectral periods due to breaking, bottom friction, and percolation, and therefore increases both the stability of the main structure and also its defense against overtopping [68]. For each aforementioned mitigation option a large number of possible alternative geometries are implemented and failure probabilities are calculated using the fault tree of Figure 3.

2.3.1. Addition of a Crown Wall or Heightening of an Existing Crest Element

When a new crown wall or parapet is added on the breakwater crest, or when existing crest elements are heightened and strengthened (mitigation option 1; Figure 2a), hydraulic stability of the rock seaside primary armor layer of the structure can be estimated by means of the Hudson stability formula [69]:

$$\frac{H_{su}}{\Delta D_n} = (K_D \cot \theta)^{1/3} \tag{9}$$

where H_{su} [m] is the significant wave height in front of the studied rubble mound breakwater corresponding to its ULS, $\Delta = (\rho_r/\rho_w) - 1$, and ρ_r and ρ_w are the rock and water density [ton/m^3] respectively, D_n [m] is the mean nominal diameter of armor elements (i.e., the equivalent cube length of median rock), K_D is the stability coefficient of the formula, and θ is the seaward slope angle of the breakwater. The reliability function, Z_{11}, used for primary armor layer stability is given by:

$$Z_{11} = \Delta D_n (K_D \cot \theta)^{1/3} - H_{su} \tag{10}$$

Hydraulic stability of the seaside primary armor layer of a rubble mound breakwater can also be assessed by means of the semi-empirical formulas of van der Meer [70] for plunging and surging type breaking conditions for rock structures. The formula corresponding to the former conditions is given below:

$$\frac{H_{su}}{\Delta D_n} = 6.2 P^{0.18} \left(\frac{S}{\sqrt{N}}\right)^{0.2} \zeta_m^{-0.5}, \ \zeta_m = \tan\theta / \sqrt{s_{om}} \tag{11}$$

where P is the permeability coefficient of the structure, N is the number of waves, S is the damage level (ratio of area eroded in a given cross-section), ζ_m is the surf similarity parameter or Irribaren number, $s_{om} = H_{su}/L_p$ is the average wave steepness, and L_p is the wave length corresponding to the peak spectral period T_p. The respective reliability function is given by:

$$Z_{11} = 6.2 \Delta D_n P^{0.18} \left(\frac{S}{\sqrt{N}}\right)^{0.2} \zeta_m^{-0.5} - H_{su} \tag{12}$$

For the ULS of excessive wave overtopping, the formula proposed by Coeveld et al. [71] which considers the effect of relatively small crest elements on top of permeable structures is combined with the formula of EurOtop [72] which calculates overtopping discharge for structures with no crest elements. Coeveld et al. [71] introduced a reduction factor for the influence of crest elements on wave overtopping, emphasizing the fact that the crest element height significantly modifies overtopping discharge. The suggested prediction formula for wave overtopping is:

$$Q' = 1.55 \cdot \exp\left(-4\frac{R_{c2}}{H_{su}} - 0.4\frac{G_c}{H_{su}} - 2\frac{N_L}{H_{su}}\right) \text{ with } Q' = \frac{q_m(\text{with crest elements})}{q_m(\text{no crest elements})} \tag{13}$$

where R_{c2} [m] is the height of the crown element with respect to crest level, G_c [m] is the crest width in front of the crest element, N_L [m] is the nose length of crest element, and q_m [m^2/s] is the overtopping discharge per linear meter. The overtopping discharge for a breakwater with no crest elements is considered as follows [72]:

$$\frac{q_m}{\sqrt{g H_{su}^3}} = 0.2 \cdot \exp\left(-2.6\frac{R_c}{H_{su}\gamma_f}\right) \tag{14}$$

where, R_c [m] is the crest freeboard with no crest elements and γ_f is the influence factor for roughness. The reliability function for excessive wave overtopping used for mitigation option 1, Z_{12}, is:

$$Z_{12} = q_{allow} - 0.31 \cdot \exp\left(-2.6\frac{R_c}{H_{su}\gamma_f}\right)\sqrt{g H_{su}^3} \cdot \exp\left(-4\frac{R_{c2}}{H_{su}} - 0.4\frac{G_c}{H_{su}} - 2\frac{N_L}{H_{su}}\right) \tag{15}$$

where q_{allow} is the maximum allowable overtopping discharge. The formula of van der Meer et al. [58] is used to estimate the damage of the breakwater toe in depth-limited breaking conditions:

$$\frac{H_{su}}{\Delta D_n} = \left(0.24\frac{h_b}{D_n} + 1.6\right) N_{odtoe}^{0.15} \tag{16}$$

where h_b [m] is the (incident breaking wave) water depth at the toe expressed as the sum of the MSL, the tide and the storm surge and N_{odtoe} is the number of displaced units within a strip of width D_n at the breakwater toe. The reliability function for toe stability, Z_{13}, is:

$$Z_{13} = \left(0.24\frac{h_b}{D_n} + 1.6\right) \cdot N_{odtoe}^{0.15} \cdot \Delta D_n - H_{su} \tag{17}$$

2.3.2. Addition of A Third Protective Armor Layer Combined with Heightening Crest Elements

When a third rock layer is added on the seaward side of the primary armor layer of the studied breakwater (mitigation option 2; Figure 2b), a mild effect is expected on the global stability of the armor stone. A third layer is assumed to improve the wave energy dissipation, and therefore, an increase of about 10% can be expected in the stability of the structure [66]. The reliability function, Z_{21}, used for primary armor layer stability is given by:

$$Z_{21} = 1.1 \Delta D_n (K_D \cot \theta)^{1/3} - H_{su} \tag{18}$$

The formula of van der Meer [70] can be also used to assess reliability function Z_{21} for mitigation option 2 (Equation (12) including a 10% increase in the structure stability).

If the addition of a third layer on the breakwater seaside primary armor layer is combined with crest elements, i.e., a crown wall on the breakwater crest, Equation (15) can be used to express the reliability function for excessive wave overtopping, Z_{22}. In case the addition of a third armor layer is not combined with elevated superstructures on breakwater crest, the EurOtop [72] formula is used to construct the reliability function for excessive wave overtopping:

$$Z_{22} = q_{allow} - 0.2 C_r \exp \left(-2.6 \frac{R_c}{H_{su} \gamma_f} \right) \sqrt{g H_{su}^3} \quad C_r = 3.06 \exp \left(-1.5 \frac{G_c}{H_{su}} \right) \tag{19}$$

where C_r is the reduction factor due to the effect of armored crest berm and G_c [m] is the crest berm width. For toe stability the reliability function of Equation (17) can be utilized also for this case (Z_{23}).

2.3.3. Construction of a Berm Structure in Front of the Breakwater Seaside Slope

Constructing a berm on the windward side of the studied breakwater (mitigation option 3; Figure 2c), a significant effect is generally expected on the global stability and the overtopping discharge of the rubble mound breakwater. The primary armor layer of the breakwater should suffer reduced damage due to the transmission of decreased wave height on the berm crest. The reliability function for primary armor layer stability, Z_{31}, becomes:

$$Z_{31} = \Delta D_n (K_D \cot \theta)^{1/3} - K_t H_{su} \tag{20}$$

where K_t is the transmission coefficient which is estimated utilizing the formula of D'Angremond et al. [73] for (berm) low-crested structures [74] with crest widths $B/H_{su} \leq 10$:

$$K_t = -0.4 \frac{R_c}{H_{su}} + 0.64 \left(\frac{B}{H_{su}} \right)^{-0.31} \cdot [1 - \exp(-0.5\xi)] \tag{21a}$$

The previous formula was extended by Briganti et al. (2003) [75] to cover crest widths of $B/H_{su} > 10$:

$$K_t = -0.35 \frac{R_c}{H_{su}} + 0.51 \left(\frac{B}{H_{su}} \right)^{-0.65} [1 - \exp(-0.41\xi)] \tag{21b}$$

where R_c [m] is the freeboard height, which is negative if the low-crested structure is submerged below MSL [76], B [m] is the berm width, ξ is the surf similarity parameter $\xi = \tan\theta / s^{0.5}$, and $s = H_{su}/L_p$ is the wave steepness. Equation (21) (a,b) has a minimum value of 0.075 and a maximum of 0.8. Herein 3-D wave transmission effects are not considered in a holistic approach, yet overtopping, percolation, transmission, and diffraction are treated separately. In future implementations of K_t in the framework of reliability-based upgrading studies, the combined effects of wave diffraction and 2-D transmission (overtopping and filtration) for a 3-D K_t calculation [77] will be examined in order to avoid possible overestimations. It should be noted that the notion of damage does not exist for a berm breakwater, because such a rubble mound breakwater is considered reshaping. Reliability function Z_{31} can be also assessed based on the formula of van der Meer [70] (Equation (12) including the wave transmission coefficient, K_t, to decrease hydraulic loading on the seaside primary armor layer of the breakwater). For excessive wave overtopping the formula of EurOtop [59] for berm breakwaters can be used [66,69]:

$$\frac{q_m}{\sqrt{gH_{su}^3}} = 0.2C_r \cdot \exp\left(-2.6\frac{R_c}{H_{su}\gamma_f\gamma_b}\right) \tag{22}$$

$$\gamma_b = 1 - k_B(1 - k_h), \; k_B = \frac{B_{berm}}{L_{berm}} \; k_h = 0.5 - 0.5\cos\left(\pi\frac{h_{berm}}{x_{berm}}\right) \tag{23}$$

where γ_b is the berm coefficient, B_{berm} is the berm width [m], L_{berm} is a length consisting of B_{berm}, and two horizontal distances corresponding to the projection of one H_{su} [m] above and one below the berm reference level, h_{berm} is the reference level of the berm width with respect to sea water level and x_{berm} is considered to be equal to $2H_{su}$ if the berm is below still water level (SWL) and $R_{u2\%}$ if berm is above SWL [66]. Therefore, reliability function for excessive wave overtopping for mitigation option 3, Z_{32}, is:

$$Z_{32} = q_{allow} - 0.2C_r\exp\left(-2.6\frac{R_c}{H_{su}\gamma_f\gamma_b}\right)\sqrt{gH_{su}^3} \tag{24}$$

For toe stability the reliability function of Equation (17) can be utilized also for this case (Z_{33}).

2.3.4. Construction of A Low-crested Structure in Front of Existing Breakwater

Considering the construction of a low-crested structure in front of the rubble mound breakwater (mitigation option 4; Figure 2d), the fault tree of the ULS for the studied structure also includes the main failure mechanisms for the low-crested structure. Figure 4 presents the ULS fault tree for mitigation option 4 considering the primary armor layer instability for the low-crested structure.

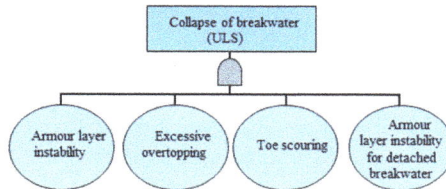

Figure 4. Fault tree for the ULS of a rubble mound breakwater with a detached structure.

For mitigation option 4, the primary armor layer of the studied breakwater suffers reduced damage caused by the wave transmission over the low-crested structure. Equations (20) and (21)(a,b) can be used to estimate the reliability of the structure for the failure mechanism of primary armor layer instability, Z_{41}. Considering excessive wave overtopping, the reliability function, Z_{42}, used is:

$$Z_{42} = q_{allow} - 0.2C_r\exp\left(-2.6\frac{R_c}{K_tH_{su}\gamma_f}\right)\sqrt{g(K_tH_{su})^3} \tag{25}$$

where K_t is the transmission coefficient estimated utilizing Equations (21a) and (21b) [74,76]. Equation (17) can again be used as the reliability function for toe stability, Z_{43}. According to DELOS Project Guidelines [78] for environmental design of low-crested coastal defense structures, it is recommended to select the stone size of the low-crested structure utilizing the formula:

$$\frac{H_{su}}{\Delta D_{n50det}} = 0.06\left(\frac{R_{cdet}}{D_{n50det}}\right)^2 - 0.23\frac{R_{cdet}}{D_{n50det}} + 1.36 \tag{26}$$

where R_{cdet} [m] is the freeboard of the low-crested structure (negative, if the structure is submerged under MSL) and D_{n50det} [m] is the mean nominal diameter of the rock armor of the low-crested structure. Application limits of the aforementioned formula are defined by $-3 < R_{cdet}/D_{n50} < 2$. The corresponding reliability function is:

$$Z_{44} = \left(0.06 \left(\frac{R_{cdet}}{D_{n50det}} \right)^2 - 0.23 \frac{R_{cdet}}{D_{n50det}} + 1.36 \right) \Delta D_{n50det} - H_{su} \tag{27}$$

2.3.5. Serviceability Limit State of a Rubble Mound Breakwater

Excessive wave height in the protected basin is considered as a SLS of the breakwater under study for all four mitigation options considered. The wave height transformation inside the protected basin is considered as a combination of wave refraction and shoaling, wave diffraction due to sub-aerial protection works, wave transmission through and overtopping of the breakwater [46]. The derived reliability function to represent the SLS is:

$$Z_{SLS} = H_{allow} - \left(K_{dif} + K_{trans} \right) H_{ss} \tag{28}$$

where H_{allow} is the maximum allowable wave height inside the protected basin and H_{ss} [m] is the incident significant wave height corresponding to the SLS of the studied breakwater. The transmission coefficient, K_{trans}, is substituted in Equation (28) by means of a simple formula used in "The Rock Manual" [66]:

$$K_{trans} = 0.46 - 0.3 \frac{R_c}{H_{ss}} \tag{29}$$

where R_c [m] is the freeboard of the studied breakwater. Application limits of the aforementioned formula are defined by $-1.13 < R_c/H_{ss} < 1.2$. The diffraction coefficient is estimated using the formula of Kraus [79]:

$$K_{dif} = \sqrt{0.5 \left[\tanh \left(\frac{s_{max} \theta_D}{W_D} \right) + 1 \right]} \tag{30}$$

where s_{max} is the maximum directional concentration parameter due to wave refraction in shallow water, defined in terms of the deepwater parameter $s_{max,o}$, using the water depth at the breakwater tip, deepwater wavelength and the angle of the wave crest with respect to bottom contours where the wave enters the shallow water [80], θ_D (radians) is the wave diffraction angle and W_D is given by:

$$W_D = -0.000103 \cdot s_{max}^2 + 0.270 \cdot s_{max} + 5.31 \tag{31}$$

Equation (30) is considered to be rather simplistic for the estimation of wave diffraction, therefore more sophisticated numerical models could be used instead (e.g., [81–83], etc.).

2.4. Quantification of Total Costs

Breakwater design or upgrading requires a detailed representation of hydraulic boundary conditions (wave height, mean water level, storm surge, and tide), as well as of geotechnical parameters (gradation of bottom material, D_{n50} values of sediment and rocks, and bottom friction coefficients) and topographical conditions (geometries of structures, obliqueness of wave impact, and bathymetry in the vicinity of the structure) and other features. Considering the stochastic variables of the marine climate, the distribution functions derived earlier (Section 2.2) are used for the variables entering the limit state functions. Boundary conditions of both present and future climate conditions are used as input in the present work. A fault tree is constructed including both ULS and SLS for the breakwater under study for every mitigation option (Section 2.3). A large number of possible alternative geometries (cross sections) for each mitigation option can then be tested, according to their performance for the limit state functions. A failure probability is defined using probabilistic and numerical methods for each geometric option. To decide whether a geometry option is applicable in one structural concept, an acceptable value of the failure probability should be prescribed. The combination of the calculated failure probabilities and the probability constraint define the set of acceptable geometries as [35]:

$$\mathbf{D} = \left\{ \mathbf{z} \middle| P_f(\mathbf{z}) \le P_{f,max} \right\} \tag{32}$$

where z is the vector of the design variables, P_f is the failure probability, and $P_{f,max}$ is the maximum acceptable failure probability of the studied rubble mound breakwater.

Equation (32) provides a large number of alternative geometries, which fulfill the probabilistic constraint. To select among these solutions, it would be really useful to estimate the costs of every alternative geometric setup for each mitigation option considered. Therefore, the geometry is selected to satisfy the probabilistic constraint while minimizing the total economic expenses. In the present study, only the costs of upgrading and the expected costs of a possible failure of the upgraded structure in its expected lifetime are taken into account in the optimization process [46]. For each upgrading concept and geometry, a construction cost is estimated, depending on structure length, L, the upgrading methodology, and construction materials as follows:

$$I_{cons} = L \cdot A(\mathbf{z}) \cdot I(\mathbf{z}) \tag{33}$$

where z is the vector of design variables in each upgrading mitigation option, A corresponds to the areas of the cross sections of the breakwater layers, and I represent the respective costs by volume.

The expected costs of failure are calculated for the expected lifetime of the upgraded breakwater, which in the present work is set to 100 years and represent economic damages in case of failure, as well as costs of repair. The formula to extract the costs of failure is [46]:

$$I_{failure} = \sum_{i=1}^{M} \frac{C_{ULS} P_{f,ULS}(\mathbf{z})}{(1+r)^i} + \frac{365 C_{SLS} P_{f,SLS}(\mathbf{z})}{(1+r)^i} \tag{34}$$

where C_{ULS} and C_{SLS} are damage costs and $P_{f,ULS}$ and $P_{f,SLS}$ are probabilities of failure in case of ULS and SLS, respectively, r is the interest rate, and M is the reference period for ULS and SLS failure. The probability of failure for the ULS, $P_{f,ULS}$, is expressed on an annual scale, while the probability of failure for the SLS, $P_{f,SLS}$, is expressed on a daily scale (see Section 2.2). ULS economic costs, C_{ULS}, include: (a) structural and (b) economic damage costs [46]. Structural costs correspond to damages to the upgraded breakwaters and other infrastructure. Economic damage costs include loss of direct income caused by the downtime of port/harbor or marina operations, expenses for alternative transportation of goods and passengers, loss of indirect income caused by bad reputation and competitive activities, and indirect economic damage to the local and national economy. ULS economic damage costs can be approximated considering that the protected basin has to cut its capacity for a long period of time, depending on the upgrading concept and the repair strategy.

3. Upgrading of A Rubble Mound Breakwater in the Port of Deauville

In the present work different upgrading mitigation options (Section 2.3) are applied within a reliability analysis and lifetime cost minimization framework to a selected rubble mound breakwater located in the marine area of the northwestern French city of Deauville, so that the structure and the associated protected basin can withstand future climate conditions. The studied breakwater resembles the one protecting the marina built within the residential complex of port Deauville to compensate for the saturation of the yacht basin at port Morny. The main characteristics of the existing breakwater, i.e., basic layers, layer materials, geometry of layer units, crest width, base and crest level, etc. are preserved. Performed minor changes aid the implementation of the proposed methodological framework. The rubble mound breakwater of the port of Deauville is 770 m long.

The typical cross-section of the breakwater includes a core made of limestones, a secondary (inner) armor layer made of quarry rock (rip-rap) weighting up to 0.5, and a primary (outer) armor layer made of blocks reaching a maximum weight of 3.5 to 5 ton, with W_{50} = 2.5 ton. The windward slope of the primary armor layer is 1/2, while the leeward slope of the breakwater is 2/3. The crest height is +7 m (+9 m) from the seabed level in front of the upstream (downstream) slope, and +12 m from reference level of the submerged basement. Figure 5 shows the (typical) cross-section of the studied rubble mound breakwater of the present work, located in the port of Deauville on the northern coast of France in the English Channel.

The marine conditions that were available to us for the Deauville port area represent daily data for joint offshore waves and sea levels under high-tide conditions, for a period of 10^4 years [84,85]. The aforementioned database has been created combining the method of Hawkes et al. [86] to model the relationship of simultaneous wave and surge observations with Monte Carlo simulation [84]. The method of Hawkes et al. [86] has been applied to in situ measurements (tide gauges, wave data buoys) and the database ANEMOC covering the period from 1 January 1979 to 31 August 2002 [87]. The resulting Monte Carlo simulations include 7.06×10^6 couples of significant wave height, storm surge, tide, and water level corresponding to the peak of the semi-diurnal tide, i.e., one record every 12−13 h. The available datasets are based on stationary simulations and are considered to represent present climate conditions, ignoring the possible effects of climate change on the marine environment. Future marine climate conditions are assessed based on existing literature (i.e., [88,89]). Indicative values of significant wave heights in the broader Area of Le Havre correspond to 4.5–5.5 m, based on a 24-year period analysis (during 1979–2002) by using the Generalized Paretto Distribution with a 2.5 m threshold for 100-years return period.

Annual maxima of offshore significant wave height and storm surge are extracted from the 10000-year simulated time series, and the GEV distribution (Equation (1)) is fitted to the data samples using the method of *L*-moments [90]. The extreme tails of offshore wave heights and storm surges are described by the GEV parameter vectors θ_1 = (4.19, 0.48, −0.078) and θ_2 = (0.69, 0.14, 0.109), respectively. The Normal distribution function $N(0.066, 0.006^2)$ is fitted to offshore wave steepness data during extreme conditions. Maximum daily wave heights in the area are fitted by a Rayleigh distribution (Equation (4)) using *L*-moments and the resulting parameter vector is θ_3 = (−0.26, 1.11).

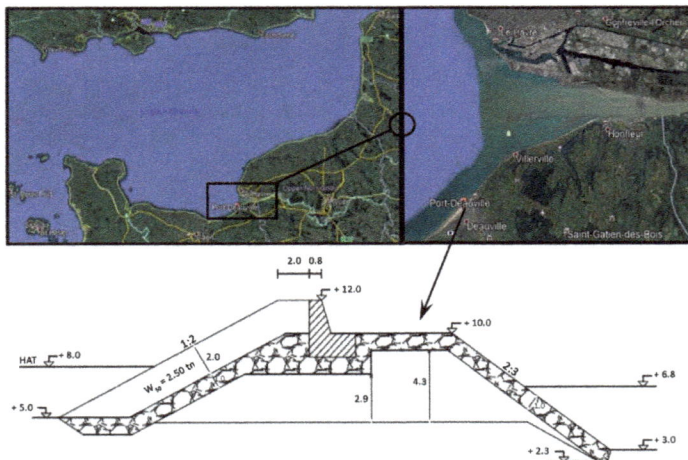

Figure 5. Typical cross-section of the studied rubble mound breakwater in the port of Deauville, Le Havre, northern France.

To estimate the distribution function of extreme significant wave heights at the breakwater site, the approach described in Section 2.2, viz. Equations (2) and (3) can be used for present climate conditions. To implement the approach, the design wave height of the studied breakwater is assumed to correspond to a return period of 100 years, which is considered to be quite a common return period for designing port and harbor protection structures. Applying Equations (2) and (3), the GEV distribution of extreme wave height for present climate conditions at the breakwater site is characterized by a parameter vector $\theta_1' = (1.69, 0.22, -0.078)$. For regular wave conditions (represented by the CDF of daily maximum offshore wave heights), the approach of Section 2.2 (Equation (5)) results in the Rayleigh distribution with parameter vector $\theta_3' = (-0.11, 0.45)$ characterizing the SLS of the studied breakwater. The level of the highest astronomical tide (HAT) at the breakwater location is estimated to reach almost 3 m from seabed. Considering future climate conditions, some general estimations are used in the present study to assess extreme marine conditions, caused by lack of data for future sea states at the site of interest.

Three estimates of sea level rise due to global warming in 2100 are available from [84,89,91]:

1. an optimistic scenario of +40 cm,
2. a pessimistic scenario of +60 cm,
3. an extreme scenario of +1 m, compared to the year 2000.

Because of shortage of available information, the estimated extreme value distribution function for the stochastic component of total sea level, namely the storm surge, is assumed unchanged compared to present climate conditions. However, it has been assumed that the total sea level presents an increase by the end of the 21st century based on the optimistic scenario of ONERC [89]. According to a CETMEF (Centre d'Etudes Techniques Maritimes et Fluviales) study [88] extreme waves could increase by 0 to 30 cm in the North-East Atlantic. Along the French coast, an increase of the offshore wave height between 20 and 40 cm is expected until 2100, according to greenhouse gasses emission scenarios B1 and A2, respectively. In the present study an increase of 40 cm in the 100-years extreme wave height has been assumed. Therefore, the extreme tail of wave heights is described by the parameter vector $\theta_1' = (1.69, 0.33, -0.078)$, retaining the mean and the shape parameters of the present climate distribution function unchanged and changing the scale parameter, so that the 100-years wave height increases by 40 cm. Changes applied to the scale parameter of the GEV distribution function imply that the future marine climate presents an increase in extreme waves variability compared to present climate conditions, while their mean extreme values hardly change during this interval (the location parameter remains unchanged). The shape parameter of the GEV, which governs the tail behavior of extreme values, remains unchanged with respect to present climate conditions to facilitate the extraction of wave extremes future distribution. However, an increase in this parameter could also be assumed, since it describes whether extreme events are likely to occur and it is well known that climate change is associated with changes in both the magnitude and frequency of such events.

For all four mitigation options, the minimization problem can be represented as:

$$
\begin{aligned}
&P_f(\mathbf{z}) \le P_{f,\max} = 0.01 \\
&\min I_{total}(\mathbf{z}) = I_{cons} + I_{failure} = \\
&= I(\mathbf{z}) + \sum_{i=1}^{M} \frac{C_{ULS} P_{f,ULS}(\mathbf{z})}{(1+r)^i} + \frac{365 C_{SLS} P_{f,SLS}(\mathbf{z})}{(1+r)^i} \\
&(H_{ss}, q_m) \le (H_{allow}, q_{allow})
\end{aligned}
\tag{35}
$$

The maximum tolerable wave height in the port basin during regular weather conditions is set at $H_{allow} = 0.5$ m. The maximum allowable overtopping discharge for extreme wave conditions is $q_{allow} = 0.01$ m^3/s/m [72]. The number of displaced units within a strip with width D_n at the toe of the structure is set at $N_{odtoe} = 2$, allowing some flattening out of the toe of the structure. The permeability factor of the structure is $P = 2$, the number of waves $N = 3000$, while the damage level of the van der Meer [70] formula for hydraulic stability is set at $S = 2–3$ (initial damage).

The diffraction coefficient at the breakwater site is estimated using the formula of Kraus [79] as $K_{dif} = 0.217$ (Equations (30) and (31)) assuming an angle of wave crest with respect to bottom contours where the wave enters shallow water equal to 30°. For each upgrading concept and geometry, a construction/upgrading cost is estimated, depending on structure length, the upgrading methodology, and construction materials.

Typical approximated values of upgrading costs (per unit length) used in this work include (actual values are not considered critical, and could be arbitrarily selected out of current prices based on market research):

1.   1500 € for a parapet wall being strengthened and heightened by 50 cm,
2.   2500 € for adding a third rock layer on the primary armor layer of the studied breakwater,
3.   5000 € for constructing a berm at the seaside slope of the existing structure,
4.   25,000 € for constructing a low-crested structure in front of the studied breakwater.

The expected costs of failure of the upgraded structures represent damages or losses in case of the marina downtime and costs of repair. The economic costs of failure are estimated separately for the ULS and the SLS. ULS economic costs include: (a) structural and (b) economic damage costs [46]. Structural costs correspond to damages to the upgraded breakwaters and other infrastructure. Economic damage costs include loss of direct income (port revenue) caused by the downtime of marina operations, expenses for alternative transportation of goods and passengers, loss of indirect income caused by bad reputation and competitive activities, and indirect economic damage to the local and national economy. In the present work, it is assumed that the cost of repairing the upgraded breakwater in case of failure reaches 20% of the initial upgrading expenses. The marina of Deauville, which can accommodate up to 850 yachts with 150 berths reserved for visitors, is utilized in this paper to assess total failure costs (for both ULS and SLS) for the optimization process. To estimate the direct economic damage, it is assumed that when a collapse by an extreme storm occurs, the marina has to cut its capacity by half and this can last for up to half a year. Indirect economic damage is also included in the optimization process by increasing the aforementioned economic damage costs by 50%. SLS costs, C_{SLS}, are estimated per day of downtime of the operations in the protected basin. Daily losses of direct and indirect income are also estimated for each downtime day. Indirect economic damage is also accounted for by increasing the cost of daily suspension of marina operations by 50%. The reference time, M, for the ULS and the SLS in the present work is set to $M = 100$ years and the interest rate is considered 5%.

3.1. Results of Mitigation Option 1: Crest/Crown Elements

Regarding mitigation option 1, with crest elements added to the structure or existing crown structures being heightened and strengthened, the upgrading procedure for the studied breakwater is presented in Figure 6. The height of the existing crest element is in fact raised to face the increased future marine hazards. The crown structure exceeds the breakwater crest by a dimension equal to W, which serves as a variable of the total cost optimization process (see Section 2.4). Figures 7 and 8 present the total costs, estimated using Equations (33) and (34), as functions of the crest element height with respect to the breakwater crest. Figure 7 presents results of the optimization process allowing 0%–5% damage ($K_D = 4$), while Figure 8 displays optimized costs allowing 5%–10% damage to the primary armor layer ($K_D = 4.9$). For almost no damage to the primary armor layer, the optimization procedure results in $W_{opt} = 1.4$ m, while the minimum total cost is estimated to be $I_{tot} \approx 11.868 \times 10^6$ €. The respective values are $W_{opt} = 1.4$ m and $I_{tot} \approx 8.517 \times 10^6$ € for 5%–10% damage to the primary armor layer. Therefore, the amount of allowed damage to the primary armor layer does not seem to influence results of the reliability-based optimization process in terms of the magnitude of the crown wall elevation. The upgrading costs (I_{cons}) cover almost 54% and 75% of the total costs (I_{tot}), respectively. It should be noted that application of van der Meer [70] formula for armor layer stability (Equations (11) and (12)) produced quite similar results with the ones shown in Figure 7.

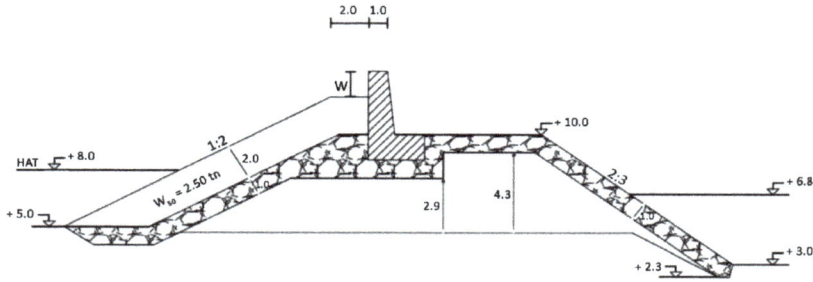

Figure 6. Cross-section of the studied rubble mound breakwater in Deauville, France, for mitigation option 1.

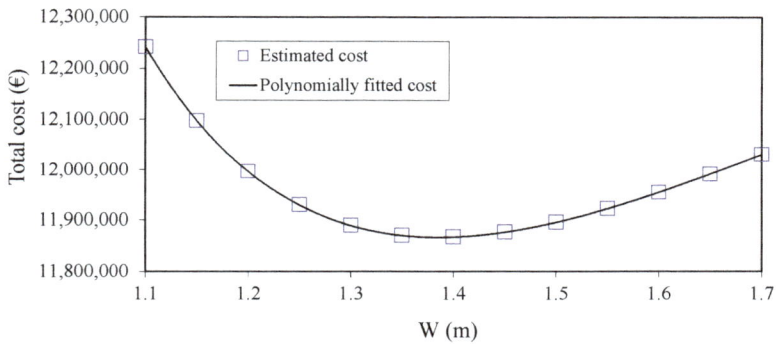

Figure 7. Total costs (€) as function of additional height of the crest element, W (m) for $K_D = 4$.

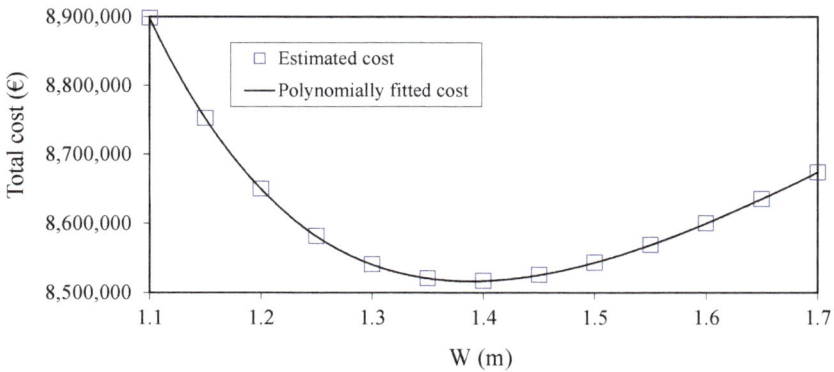

Figure 8. Total costs (€) as function of additional height of the crest element, W (m) for $K_D = 4.9$.

3.2. Results of Mitigation Option 2: Additional Armour Layer with Enahncement of Crest/Crown Elements

In mitigation option 2 a third layer is added to the primary armor layer of the structure, thus the upgrading procedure is presented in Figure 9. In the case of inadequate covering by the third layer of the crest width, it has to be combined with a crest element in the form of mitigation option 1. The third armor layer should be of the same mean nominal diameter D_{n50} (with $W_{50} = 2.5$ ton) as the two underlayers of the studied breakwater, therefore the height, W, of the concrete element exceeding the breakwater crest is the sole variable that has to be estimated during the total cost optimization process.

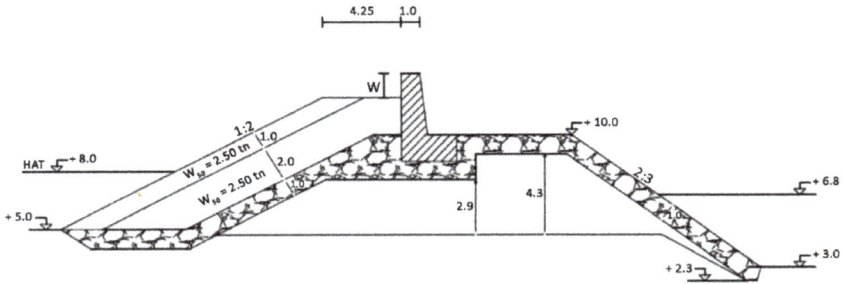

Figure 9. Cross-section of the studied rubble mound breakwater in Deauville France, for mitigation option 2.

Figures 10 and 11 present the total costs as a function of the additional height of the crest element. Figure 10 presents results of the optimization process allowing 0%–5% damage ($K_D = 4$), while Figure 11 allows 5%–10% damage to the primary armor layer ($K_D = 4.9$). For almost no damage to the primary armor layer, the optimization procedure results in $W_{opt} = 1.3$ m, while the minimum total cost is estimated $I_{tot} \approx 10.643 \times 10^6$ €. For 5%–10% damage to the primary armor layer the respective values of optimal crest height and total cost are $W_{opt} = 1.3$ m and $I_{tot} \approx 9.732 \times 10^6$ €. The upgrading costs (I_{cons}) cover almost 87% and 94% of the total costs (I_{tot}), respectively. In regard to mitigation option 2, the amount of damage to the primary armor layer mainly affects the third layer of rocks added above the two existing underlayers. Therefore the primary armor layer suffers reduced damage compared to mitigation option 1 (no protective armor layer) and the maintenance on the armor stone for mitigation option 2 is significantly reduced. Results similar to Figure 10 are also produced using the van der Meer [74] formula to calculate hydraulic stability of the seaside primary armor layer of the studied structure (Equations (11) and (12)).

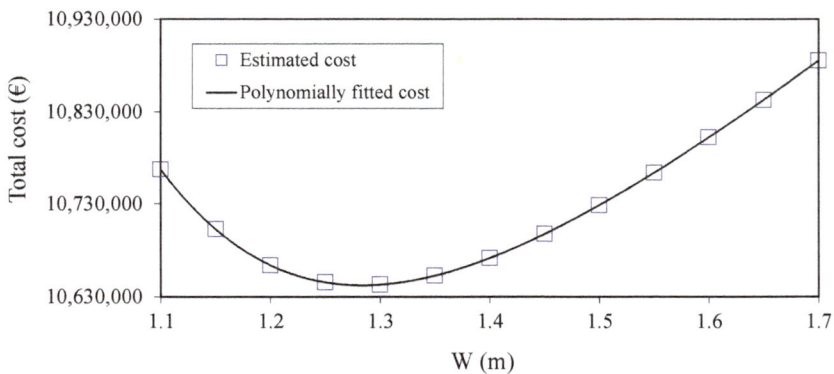

Figure 10. Total costs (€) as a function of the additional height of the crest element, W (m), when a third layer is added above the primary armor layer ($K_D = 4$).

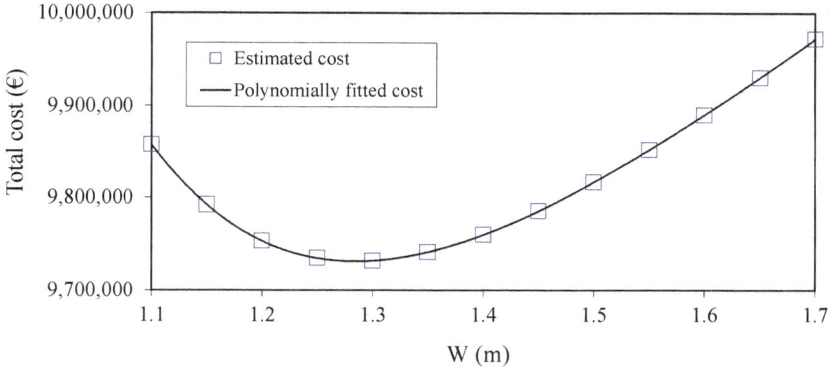

Figure 11. Total costs (€) as a function of the additional height of the crest element, W (m), when a third layer is added above the primary armor layer ($K_D = 4.9$).

3.3. Results of Mitigation Option 3: Additional Berm

The upgrading procedure of mitigation option 3 (berm added to the structure) is presented in Figure 12. The berm width, B_{berm}, and the reference level of the berm width with respect to sea level, h_{berm}, are estimated during the total cost optimization process. The mean nominal weight of the rocks used to construct the berm is 2.5 ton, similar to the one used for the rocks of the primary armor layer. Figure 13 presents the total costs, estimated using Equations (33) and (34), as a function of the berm width, B_{berm}; the lowest total cost has been observed for $h_{berm} = 1$ m. However, Figure 14 presents the total costs, as a function of the B_{berm} for $h_{berm} = 0.8$ m. The differences in optimized amounts of total costs are not significant between the two crest berm depths. More specifically, for $h_{berm} = 1$ m, the optimized berm width is $B_{bermopt} = 9.8$ m, while the minimum total cost is estimated $I_{tot} \approx 7.002 \times 10^6$ €. For $h_{berm} = 0.8$ m, the optimized berm width is $B_{bermopt} = 9.4$ m, while the minimum total cost is estimated $I_{tot} \approx 7.11 \times 10^6$ €. The upgrading costs (I_{cons}) cover almost 76% and 78% of the total costs (I_{tot}), respectively.

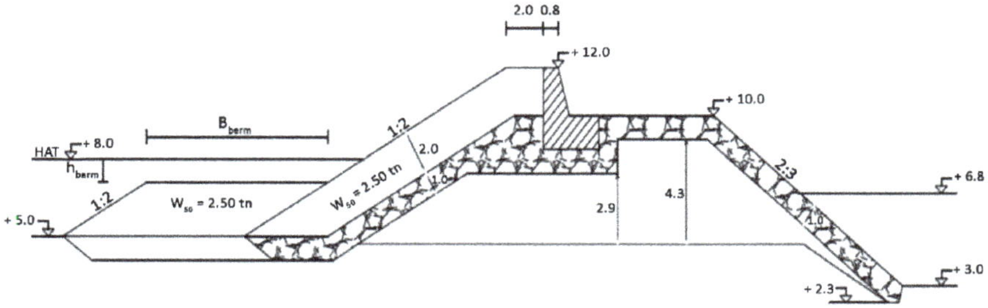

Figure 12. Cross-section of the studied rubble mound breakwater in Deauville, France for mitigation option 3.

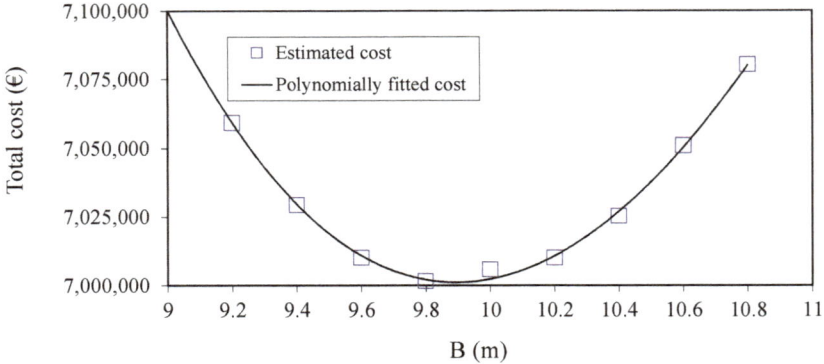

Figure 13. Total costs (€) as a function of the berm width, *B* (m), for h_{berm} = 1 m.

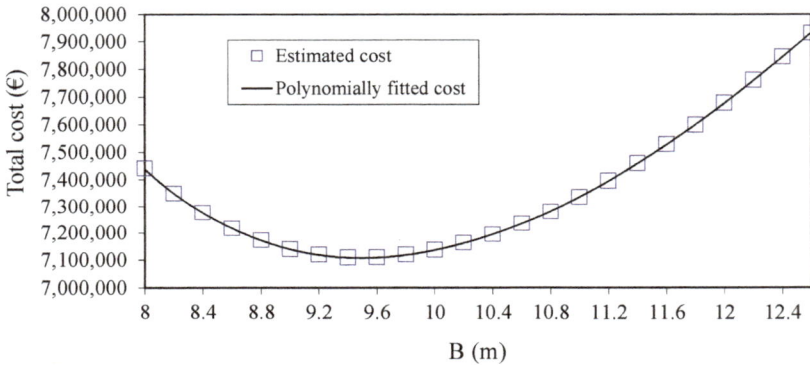

Figure 14. Total costs (€) as a function of the berm width, *B* (m), for h_{berm} = 0.8 m.

3.4. Results of Mitigation Option 4: Detached Low-Crested Structure

Mitigation option 4 includes the construction of a protective structure in the upwind front of the studied breakwater. The detached low-crested structure is designed to have both its slopes equal to 1/2. The crest width, *B*, the mean nominal diameter, D_{n50det}, and the freeboard level, R_{cdet}, of the low-crested structure are the variables of the total cost optimization process. The basic constructional constraints for mitigation option 4 impose a minimum height and crest width of the low-crested structure equal to at least three stones, namely $\geq 3 \times D_{n50det}$. For the armor layers of the low-crested breakwater, the armor units are considered to be imbricated rip-rap type stones with a general size requirement of $D_{n50det} \geq 0.9144$ m, i.e., the armor units are custom order quarry rocks of Class I with D_{n50det} = 1.3–1.65 m corresponding to typical weights of W_{50} = 2.5–6 t. The defending detached structure is a submerged breakwater with R_{cdet} ranging from 0 to −2 m (by 0.25 m; only two representative cases are shown herein). Figure 15 presents the total costs, estimated using Equations (33) and (34), as a function of the crest width, *B*, and the mean nominal diameter, D_{n50det} for R_{cdet} = −0.5 m (the negative sign means 0.5 m below SWL). Figure 16 presents a similar plot for R_{cdet} = −1 m. The optimized low−crested structure geometry is B_{opt} = 4.6 m and D_{n50opt} = 1.53 m for R_{cdet} = −0.5 m, therefore the detached breakwater height is *h* = 4.6 m in waters with depth equal to *d* = 5.1 m for a total cost of $I_{tot} \approx 11.816 \times 10^6$ €. Similarly, for R_{cdet} = −1 m the optimized low-crested structure geometry is B_{opt} = 4.3 m and D_{n50opt} = 1.43 m, therefore, the detached breakwater height is *h* = 4.3 m in water depth of $d = 5.3$ m for a total cost $I_{tot} \approx 10.433 \times 10^6$ €. The upgrading costs (I_{cons}) cover almost 88% and 87% of the total costs (I_{tot}), respectively.

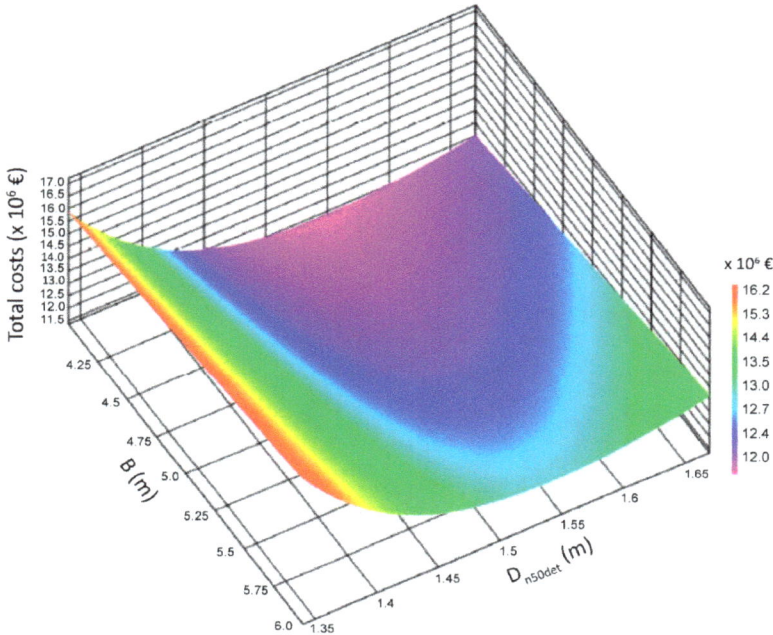

Figure 15. Total costs (€) as a function of the crest width, B (m), and the mean nominal diameter of rocks, D_{n50det} (m), for the detached low-crested structure of $R_{cdet} = -0.5$ m.

4. Summary and Conclusions

Port or harbor downtime risks are considered to be crucial for all countries characterized by high concentration of critical infrastructure and economic activities in coastal areas, and by high significance of seaport industry, maritime transport, communications due to tourism, and sea-related activities. The latter are determining factors of national economic performance, development, and regional growth. However, it should be considered that a lot of such structures and defenses are aging and have already exceeded their service lifetime. The abovementioned challenges coupled with the general inception of a changing climate are expected to increase exposure and vulnerability of coastal areas to future risks. Hence, functionality and safety of such structures have to be re-evaluated under climate change conditions and appropriate upgrading mitigation measures must be considered.

The approach presented in this work is based on reliability analysis of the studied structure coupled with economic optimization techniques. CDFs for the extremes of the stochastic variables of the marine climate, i.e., extreme significant wave height, are first extracted for the area under study and for present climate conditions and then transformed to represent the marine climate in front of the studied breakwater. CDFs for future sea states are then extracted taking account of climate change scenarios or assumptions. If the breakwater cannot withstand the newly defined marine conditions, different mitigation options for upgrading are suggested. The engineering mitigation options considered in the present work focus on: (1) increasing the breakwater crest level by adding a crown wall or existing crest elements being heightened and strengthened; (2) adding a new protective layer of armor units on the front slope of the studied rubble mound breakwater combined with new or elevated existing crest elements; (3) constructing of a berm structure in front of the seaside slope, and (4) constructing a low-crested structure in front of the existing breakwater. A fault tree is constructed giving a logical succession of all events that lead to port/harbor downtime for each alternative mitigation option including both ULS and SLS. For each upgrading concept a large number of possible alternative geometries are tested, according to their performance for the limit state functions. From the total sample of acceptable geometries for each upgrading concept that satisfy a probabilistic constraint, a single solution is selected minimizing the total costs of protection during the service lifetime of the upgraded structure.

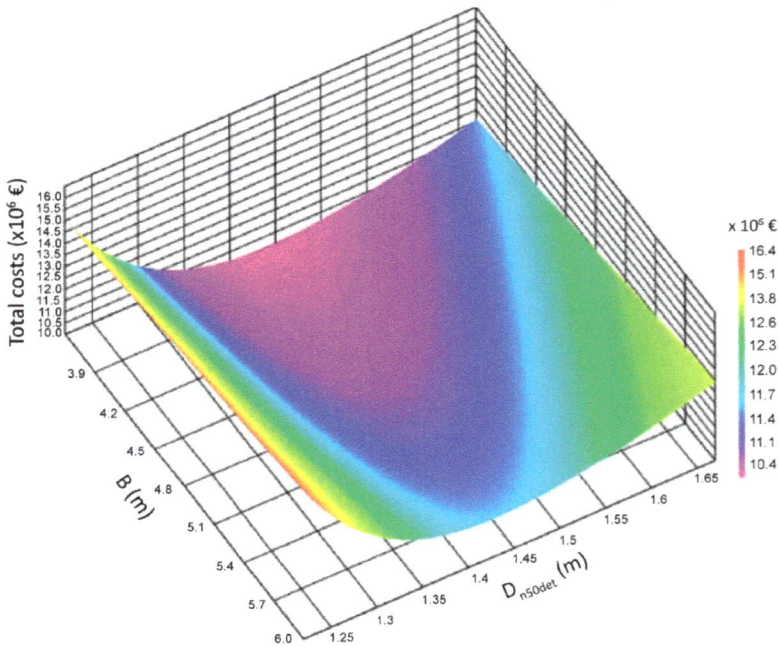

Figure 16. Total costs (€) as a function of the crest width, B (m), and the mean nominal diameter of rocks, D_{n50det} (m), for the detached low-crested structure of $R_{cdet} = -1$ m.

The proposed work aims to propose a reliable and robust framework to upgrade port and harbor rubble mound breakwaters against climate change, combining the conceptual framework of risk with modern optimization techniques and statistical methods. The recommended reliability-based upgrading approach, allows one to determine the safety levels of the respective basins, considering the protection costs and savings of the protected areas. It also allows one to compare different mitigation measures for upgrading port and harbor breakwaters under climate change in terms of their effectiveness within the general framework of minimizing the total downtime risks of the respective basins. The recommended framework introduces risk-based approaches in the field of upgrading coastal structures to face climate change impacts. Such methods are expected to significantly contribute facing increasing marine hazards and minimizing exposure of coastal areas and defenses to more severe future climate conditions.

The presented methodology is applied to a rubble mound breakwater resembling to the existing one in the port of Deauville, France. Three main failure mechanisms are considered here as the main types of instability under extreme marine climate conditions, namely the instability of the primary armor layer, the excessive wave overtopping, and the scouring of the toe of the breakwater. Considering the SLS, the wave height transformation inside the port basin is considered as a combination of wave refraction-diffraction caused by penetration via the entrance of the port, wave transmission through, and overtopping of the breakwater. The construction of a berm in the upwind front of the studied breakwater (mitigation option 3), resulting in optimized values of the parameters of the berm width, $B_{bermopt}$ = 9.8 m, and water depth at the reference level of the berm, $h_{bermopt}$ = 1 m, presented the lowest total costs during the service lifetime of the upgraded structure ($I_{tot} \approx 7.002 \times 10^6$ €) set equal to 100 years. The total costs of mitigation option 3 are lower by 41% and 34% than those estimated for the mitigation option 1 and 2, respectively. Therefore adding a berm on the windward armor slope of the studied breakwater is found to be the most advantageous mitigation option, compared to alternatives comprising increased breakwater crest levels and crown walls, supplement of extra armor layers, and construction of an upwind low-crested defense structure. But it should be emphasized that in the present work the amount of damage to the constructed berm has been ignored, thus our analysis applies to non-reshaping berms. The inclusion of the formulas for limit states of reshaping berm breakwaters (allowing a considerable amount of rearrangements and displacements of rocks to occur) could modify the extracted results under the same hydraulic conditions, as the failure probabilities for global stability of the rubble mound breakwater would decrease, adding to a further reduction of downtime costs for the studied marina.

List of Symbols (in order of appearance)

G_θ	Generalized Extreme Value (GEV) distribution function
θ	Vector of GEV parameters
μ	Location parameter of the GEV
σ	Scale parameter of the GEV
ξ	Shape parameter of the GEV
x	Arbitrary variable of GEV distribution function (e.g., annual maxima of wave heights or storm surge)
CV	Coefficient of variation
$\Gamma(\cdot)$	Gamma Cumulative Distribution Function (CDF)
z_{100}	Design event value for any parameter (e.g., wave height) for a 100-year return period
p	Probability of 1 in T years [1/yrs]
T	Return period [yrs]
F_θ	Rayleigh CDF
H_{su}	Significant wave height (design wave height at the toe of the structure) [m]
Δ	Dimensionless relative buoyant density of rock [= $(\rho_r/\rho_w) - 1$]
ρ_r	Rock density [ton/m^3]
ρ_w	Water density [ton/m^3]
D_n	Mean nominal diameter of armor elements of the rubble mound breakwater [m]
θ	Seaward slope angle of the breakwater [°]
K_D	Dimensionless stability coefficient of the Hudson formula
Z	Reliability function for a limit state
Z_{ij}	Reliability function for mitigation option i and failure mechanism j
i	Index of the mitigation option (i = 1–4)
j	Index of the failure mechanism (j = 1–4)
R_Z	Resistance of the structure
S_Z	Loading of the structure
n	Total number of simulations
n_f	Number of simulations for which the condition Z ≤ 0 is valid
P_f	Failure probability
Z_{11}	Reliability function for breakwater stability under mitigation option 1
P	Permeability coefficient of breakwater structure
N	Number of waves
S	Damage level (ratio of area eroded in a given cross-section)

ξ_m	Surf similarity parameter (Irribaren number)
s_{om}	Average wave steepness
L_p	Wave length corresponding to T_p [m]
T_p	Peak spectral wave period [sec]
Q'	Dimensionless wave overtopping rate
R_{c2}	Height of the crown element with respect to crest level [m]
G_c	Crest width in front of the crest element [m]
N_L	Nose length of crest element [m]
q_m	Overtopping discharge per linear meter [m^2/s]
R_c	Crest freeboard height [m]
γ_f	Influence factor for roughness
z_{12}	Reliability function for wave overtopping of the breakwater under mitigation option 1
q_{allow}	Maximum allowable overtopping discharge per linear meter [m^2/s]
h_b	Water depth at the breakwater toe as sum of MSL, tide and storm surge [m]
N_{odtoe}	Number of displaced units within a strip of width D_n at the breakwater toe
Z_{13}	Reliability function for breakwater toe stability under mitigation option 1
Z_{21}	Reliability function for breakwater's primary armor layer stability under mitigation option 2
C_r	Reduction factor due to the effect of armored crest berm
G_c	Crest berm width [m]
Z_{22}	Reliability function for excessive wave overtopping of the breakwater under mitigation option 2
Z_{23}	Reliability function for breakwater toe stability under mitigation option 2
Z_{31}	Reliability function for breakwater's primary armor layer stability under mitigation option 3
K_t	Wave transmission coefficient (for low-crested structures)
B	Berm width [m]
ξ	Surf similarity parameter or Irribaren number (=$\tan\theta/s^{0.5}$)
s	Irregular wave steepness (=H_{su}/L_p)
Z_{32}	Reliability function for excessive wave overtopping of the breakwater under mitigation option 3
γ_b	Berm coefficient
B_{berm}	Berm width [m]
L_{berm}	Length consisting of B_{berm} and two horizontal distances corresponding to projection of H_{su} above and below the berm reference level [m]
h_{berm}	Reference level of the berm width with respect to sea water level [m]
x_{berm}	Horizontal distance equal to $2H_{su}$ if berm is below still water level (SWL) and $R_{u2\%}$ if berm is above SWL [m]
$R_{u2\%}$	Run-up height (of 2% probability) above SWL [m]
Z_{33}	Reliability function for breakwater toe stability under mitigation option 3
Z_{41}	Reliability function for breakwater's primary armor layer stability under mitigation option 4
Z_{42}	Reliability function for excessive wave overtopping of the breakwater under mitigation option 4
Z_{43}	Reliability function for breakwater toe stability under mitigation option 4
R_{cdet}	Freeboard height of the low-crested structure [m]
D_{n50det}	Mean nominal diameter of the rock armor of the low-crested structure [m]
Z_{44}	Reliability function for selection of D_{n50} of low-crested structures under mitigation option 4
Z_{SLS}	Reliability function for the representation of SLS
H_{allow}	Maximum allowable wave height inside the protected basin [m]
H_{ss}	Incident significant wave height corresponding to SLS of the studied breakwater [m]
K_{trans}	Wave transmission coefficient for sub-aerial breakwaters
K_{dif}	Wave diffraction coefficient for sub-aerial breakwaters
s_{max}	Maximum directional concentration parameter due to wave refraction in shallow water
$s_{max,o}$	Maximum directional concentration parameter in deep water
θ	Wave diffraction angle [rad]
W_D	Diffraction parameter
D	Set of acceptable geometries
z	Vector of design variables in each upgrading mitigation option
P_f	Failure probability of the rubble mound breakwater

$P_{f,max}$	Maximum acceptable failure probability of the rubble mound breakwater
L	Structure length [m]
A	Area of the cross section of the breakwater layers [m^2]
I	Respective costs by volume [€]
I_{cons}	Upgrading costs [€]
$I_{failure}$	Costs of failure [€]
C_{ULS}	Damage costs under ULS [€]
C_{SLS}	Damage costs under SLS [€]
$P_{f,ULS}$	Probabilities of failure in case of ULS
$P_{f,SLS}$	Probabilities of failure in case of SLS
r	Interest rate [%]
M	Reference period for ULS and SLS failure [yrs]
W_{50}	Weight of armor rock corresponding to D_{n50det} [ton]
$N(\cdot)$	Normal distribution function
HAT	Highest astronomical tide [m]
I_{tot}	Minimum total cost [€]
W_{opt}	Optimized weight of armor rock [ton]
$B_{bermopt}$	Optimized berm width [m]
D_{n50opt}	Optimized mean nominal diameter of the rock armor of the low-crested structure [m]
B_{opt}	Optimized crest width of the detached low-crested structure [m]

Author Contributions: P.G. formulated the basic concept of the paper; P.G. and C.M. performed the formal analysis; P.P. conducted the funding acquisition; P.G., C.M. and P.P. contributed on the investigations; P.G. and C.M. implemented the methodology; P.P. was the project administrator; P.G. was in charge of the software; P.P. supervised the research; C.M. validated the presented data; P.G. formulated the visualizations; P.G. wrote the original draft; C.M. and P.P. co-wrote parts of the manuscript, reviewed and edited the paper.

Funding: This research was funded by the European Commission through FP7.2009-1 grant number "Contract 244104-THESEUS Innovative technologies for safer European coasts in a changing climate". Part of the work was funded by NSRF 2007–2013 "CCSEAWAVS project (2012–2015) Estimating the Effects of Climate Change on Sea Level and Wave Climate of the Greek seas, Coastal Vulnerability and Safety of Coastal and Marine Structures".

Acknowledgments: The authors want to thank Philippe Sergent, Scientific Director of CETMEF for providing the datasets used in the paper.

Conflicts of Interest: The authors declare no conflicts of interest.

References

1. IPCC. *Climate change 2007: The Scientific Basis, Contribution of Working Group I to the Fourth Assessment Report of the Intergovernmental Panel on Climate change*; Cambridge University Press: New York, NY, USA, 2007.

2. IPCC. *Managing the Risks of Extreme Events and Disasters to Advance Climate change Adaptation, A Special Report of Working Groups I and II of the Intergovernmental Panel on Climate change*; Cambridge University Press: Cambridge, UK; New York, NY, USA, 2012.

3. WASA-Group. Changing waves and storms in the Northeast Atlantic? *Bull. Am. Meteorol. Soc.* **1998**, *79*, 741–760. [CrossRef]

4. Alexandersson, H.; Tuomenvirta, H.; Schmith, T.; Iden, K. Trends of storms in NW Europe derived from an updated pressure data set. *Clim. Res.* **2000**, *14*, 71–73. [CrossRef]

5. Woth, K.; Weisse, R.; Storch, H. Climate change and North Sea storm surge extremes: An ensemble study of storm surge extremes expected in a changed climate projected by four different regional climate models. *Ocean Dyn.* **2006**, *56*, 3–15. [CrossRef]

6. Weisse, R.; von Storch, H.; Niemeyer, H.D.; Knaack, H. Changing North Sea storm surge climate: An increasing hazard? *Ocean Coast. Manag.* **2012**, *68*, 58–68. [CrossRef]

7. De Winter, R.C.; Sterl, A.; de Vries, J.W.; Weber, S.L.; Ruessink, G. The effect of climate change on extreme waves in front of the Dutch coast. *Ocean Dyn.* **2012**, *62*, 1139–1152. [CrossRef]

8. Wolf, J.; Woolf, D.K. Waves and climate change in the north-east Atlantic. *Geophys. Res. Lett.* **2006**, *33*. [CrossRef]

9. Wang, X.; Swail, V. Climate change signal and uncertainty in projections of ocean wave heights. *Clim. Dyn.* **2006**, *26*, 109–126. [CrossRef]
10. Debernard, J.; Roed, L. Future wind, wave and storm surge climate in the Northern Seas: A revisit. *Tellus* **2008**, *60*, 427–438. [CrossRef]
11. Cid, A.; Castanedo, S.; Abascal, A.J.; Menéndez, M.; Medina, R. A high resolution hindcast of the meteorological sea level component for Southern Europe: The GOS dataset. *Clim. Dyn.* **2014**, *43*, 2167–2184. [CrossRef]
12. Vousdoukas, M.I.; Voukouvalas, E.; Annunziato, A.; Giardino, A.; Feyen, L. Projections of extreme storm surge levels along Europe. *Clim. Dyn.* **2016**, *47*, 3171–3190. [CrossRef]
13. Vousdoukas, M.I.; Mentaschi, L.; Voukouvalas, E.; Verlaan, M.; Feyen, L. Extreme sea levels on the rise along Europe's coasts. *Earths Future* **2017**, *5*, 304–323. [CrossRef]
14. Galiatsatou, P.; Makris, C.; Prinos, P.; Kokkinos, D. Nonstationary joint probability analysis of extreme marine variables to assess design water levels at the shoreline in a changing climate. *Nat. Hazards* **2018**. submitted.
15. FLOODsite. *Language of Risk—Project Definitions*; Report T32-04-01; HR Wallingford Ltd: Wallingford, Oxfordshire, UK, 2009.
16. Townend, I.; Burgess, K. Methodology for Assessing the Impact of Climate Change upon Coastal Defence Structures. In Proceedings of the 29th ICCE, Lisbon, Portugal, 19–24 September 2004; pp. 3953–3965.
17. Okayasu, A.; Sakai, K. Effect of Sea Level Rise on Sliding Distance of a Caisson Breakwater: Optimisation with Probabilistic Design Method. In Proceedings of the 30th ICCE, San Diego, CA, USA, 3–8 September 2006; World Scientific: Singapore, 2006; pp. 4883–4893.
18. Olabarrieta, M.; Medina, R.; Losada, I.J.; Mendez, F.J. Potential Impact of Climate Change on Coastal Structures: Application to the Spanish Littoral. In Proceedings of the Coastal Structures Conference, Venice, Italy, 2–4 July 2007; pp. 1728–1739.
19. Suh, K.D.; Kim, S.W.; Mori, N.; Mase, H. Effect of climate change on performance-based design of caisson breakwaters. *J. Waterw. Port Coast. Ocean Eng.* **2011**, *138*, 215–225. [CrossRef]
20. Becker, A.; Inoue, S.; Fischer, M.; Schwegler, B. Climate change impacts on international seaports: Knowledge, perceptions, and planning efforts among port administrators. *Clim. Chang.* **2012**, *110*, 5–29. [CrossRef]
21. Suh, K.D.; Kim, S.W.; Kim, S.; Cheon, S. Effects of climate change on stability of caisson breakwaters in different water depths. *Ocean Eng.* **2013**, *71*, 103–112. [CrossRef]
22. Sánchez-Arcilla, A.; García-León, M.; Gracia, V.; Devoy, R.; Stanica, A.; Gault, J. Managing coastal environments under climate change: Pathways to adaptation. *Sci. Total Environ.* **2016**, *572*, 1336–1352. [CrossRef] [PubMed]
23. Chini, N.; Stansby, P.K. Extreme values of coastal wave overtopping accounting for climate change and sea level rise. *Coast. Eng.* **2012**, *65*, 27–37. [CrossRef]
24. Mase, H.; Tsujio, D.; Yasuda, T.; Mori, N. Stability analysis of composite breakwater with wave-dissipating blocks considering increase in sea levels, surges and waves due to climate change. *Ocean Eng.* **2013**, *71*, 58–65. [CrossRef]
25. Hoshino, S.; Esteban, M.; Mikami, T.; Takabatake, T.; Shibayama, T. Climate Change and Coastal Defences in Tokyo Bay. In Proceedings of the 33rd ICCE, Santander, Spain, 1–6 July 2012; Volume 1, p. 19.
26. Sekimoto, T.; Isobe, M.; Anno, K.; Nakajima, S. A new criterion and probabilistic approach to the performance assessment of coastal facilities in relation to their adaptation to global climate change. *Ocean Eng.* **2013**, *71*, 113–121. [CrossRef]
27. Isobe, M. Impact of global warming on coastal structures in shallow water. *Ocean Eng.* **2013**, *71*, 51–57. [CrossRef]
28. Lee, C.E.; Kim, S.W.; Park, D.H.; Suh, K.D. Risk assessment of wave run-up height and armor stability of inclined coastal structures subject to long-term sea level rise. *Ocean Eng.* **2013**, *71*, 130–136. [CrossRef]
29. THESEUS. *Innovative Technologies for Safer European Coasts in a Changing Climate*. Contract Number: 244104, 2009–2013. Available online: http://www.theseusproject.eu/ (accessed on 20 July 2018).
30. Burcharth, H.F.; Andersen, T.L.; Lara, J.L. Upgrade of coastal defence structures against increased loadings caused by climate change: A first methodological approach. *Coast. Eng.* **2014**, *87*, 112–121. [CrossRef]
31. CCSEAWAVS (2012–2015). Estimating the Effects of Climate Change on Sea Level and Wave Climate of the Greek Seas, Coastal Vulnerability and Safety of Coastal and Marine Structures. Available online: http://www.thalis-ccseawavs.web.auth.gr/en (accessed on 20 June 2018).

32. Karambas, T.; Koftis, T.; Tsiaras, A.; Spyrou, D. Modelling of Climate Change Impacts on Coastal Structures-Contribution to their Re-Design. In Proceedings of the CEST 2015, Rhodes, Greece, 3–5 September 2015.

33. Koftis, T.; Prinos, P.; Galiatsatou, P.; Karambas, T. An Integrated Methodological Approach for the Upgrading of Coastal Structures due to Climate Change Effects. In Proceedings of the 36th IAHR World Congress, The Hague, The Netherlands, 28 June–3 July 2015.

34. Schanze, J. A conceptual framework for flood risk management research. In *Flood Risk Management Research from Extreme Events to Citizens Involvement, Proceedings of the EFRM Dresden, Germany 6–7 February 2007*; Leibniz Institute of Ecological and Regional Development (IOER): Dresden, Germany, 2007; pp. 1–10.

35. Voortman, H.G. Risk-Based Design of Large Scale Flood Defence System. Ph.D. Thesis, Delft University of Technology, Delft, The Netherlands, 2003.

36. Vrijling, J.K. Probabilistic design of flood defence systems in the Netherlands. *Reliab. Eng. Syst. Saf.* **2001**, *74*, 337–344. [CrossRef]

37. Mai, C.V.; van Gelder, P.H.A.J.M.; Vrijling, J.K. Safety of coastal defences and flood risk analysis. *Saf. Reliab. Manag. Risk* **2006**, *2*, 1355–1366.

38. Mai, C.V.; van Gelder, P.H.A.J.M.; Vrijling, H.; Stive, M. Reliability- and risk-based design of coastal flood defences. *Coast. Eng.* **2008**, *10*, 4276–4288.

39. Jonkman, S.N.; Kok, M.; Van Ledden, M.K.; Vrijling, J.K. Risk-based design of flood defence systems: A preliminary analysis of the optimal protection level for the New Orleans metropolitan area. *J. Flood Risk Manag.* **2009**, *12*, 170–181. [CrossRef]

40. FLOODsite. *Integrated Flood Risk Analysis and Management Methodologies.* Contract Number: GOCE-CT-2004-505420, 2004–2009. Available online: www.floodsite.net (accessed on 20 July 2018).

41. Buijs, F.A.; van Gelder, P.H.A.J.M.; Vrijling, J.K.; Vrouwenvelder, A.C.W.M.; Hall, J.W.; Sayers, P.B.; Wehrung, M.J. Application of Dutch Reliability-Based Flood Defence Design in the UK. In Proceedings of the ESREL Conference, Maastricht, The Netherlands, 15–18 June 2003; Volume 1, pp. 311–319.

42. Burcharth, H.F.; Sørensen, J.D. On Optimum Safety Levels of Breakwaters. In Proceedings of the 31st PIANC Congress, Estoril, Portugal, 14–18 May 2006; pp. 634–648.

43. Stedinger, J.R. Expected probability and annual damage estimators. *J. Water. Resour. Plan. Manag.* **1997**, *132*, 125–135. [CrossRef]

44. Steenbergen, H.M.G.M.; Lassing, B.L.; Vrouwenvelder, A.C.W.M.; Waarts, P.H. Reliability analysis of flood defence systems. *Heron* **2004**, *49*, 51–73.

45. Castillo, C.; Mínguez, R.; Castillo, E.; Losada, M.A. An optimal engineering design method with failure rate constraints and sensitivity analysis. Application to composite breakwaters. *Coast. Eng.* **2006**, *53*, 1–25. [CrossRef]

46. Dai Viet, N.; Verhagen, H.J.; van Gelder, P.H.A.J.M.; Vrijling, J.K. Conceptual Design for the Breakwater System of the South of Doson Naval Base: Optimization versus Deterministic Design. In Proceedings of the PIANC-COPEDEC VII: 7th International Conference on Coastal and Port Engineering in Developing Countries. "Best Practises in the Coastal Environment", Dubai, United Arab Emirates, Paper No. 053. 24–28 February 2008.

47. Prasad Kumar, B. Reliability based design method for coastal structures in shallow seas. *Indian J. Geo-Mar. Sci.* **2010**, *39*, 605–615.

48. Galiatsatou, P.; Prinos, P. Reliability-based design optimization of a rubble mound breakwater in a changing climate. In *Comprehensive Flood Risk Management: Research for Policy and Practice*; Klijn, F., Schweckendiek, T., Eds.; CRC Press: London, UK; Balkema: London, UK, 2013; ISBN 978-0-41-562144-1.

49. Naty, S.; Viviano, A.; Foti, E. Feasibility Study of a WEC Integrated in the Port of Giardini Naxos, Italy. In Proceedings of the 35th International Conference on Coastal Engineering, Antalya, Turkey, 17–20 November 2016.

50. Van Gelder, P.; Buijs, F.; Horst, W.; Kanning, W.; Mai Van, C.; Rajabalinejad, M.; de Boer, E.; Gupta, S.; Shams, R.; van Erp, N.; et al. Reliability analysis of flood defence structures and systems in Europe. In *Flood Risk Management: Research and Practice*; Samuels, P., Huntington, S., Allsop, W., Harrop, J., Eds.; Taylor & Francis Group: London, UK, 2009; ISBN 978-0-415-48507-4.

51. Buijs, F.A.; Hall, J.W.; Sayers, P.B.; van Gelder, P.H.A.J.M. Time-dependent reliability analysis of flood defences. *Reliab. Eng. Syst. Saf.* **2009**, *94*, 1942–1953. [CrossRef]

52. Kim, T.M.; Suh, K.D. Reliability analysis of breakwater armor blocks: Case study in Korea. *Coast. Eng. J.* **2010**, *52*, 331–350. [CrossRef]

53. Naulin, M.; Kortenhaus, A.; Oumeraci, H. Reliability Analysis and Breach Modelling of Coastal and Estuarine Flood Defences. In *Proceedings of the ISGSR 2011, Munich, Germany, 2–3 June 2011*; Vogt, N., Schuppener, B., Straub, D., Bräu, G., Eds.; Bundesanstalt für Wasserbau: Karlsruhe, Germany, 2011; ISBN 978-3-93-923001-4.

54. Naulin, M.; Kortenhaus, A.; Oumeraci, H. Reliability-based flood defense analysis in an integrated risk assessment. *Coast Eng. J.* **2015**, *57*, 1540005. [CrossRef]

55. Nepal, J.; Chen, H.P.; Simm, J.; Gouldby, B. Time-dependent reliability analysis of Flood defence assets using generic fragility curve. In *E3S Web of Conferences (7)*; EDP Sciences: Les Ulis, France, 2016.

56. Van der Meer, J.W. Deterministic and probabilistic design of breakwater armor layers. *J. Waterw. Port Coast. Ocean Eng.* **1988**, *114*, 66–80. [CrossRef]

57. Van Gent, M.R.; Pozueta, B. Rear-side stability of rubble mound structures. In Proceedings of the 29th ICCE, Lisbon, Portugal, 19–24 September 2004; pp. 3481–3493.

58. Van der Meer, J.W.; D'Angremond, K.; Gerding, E. Toe structure stability of rubble mound breakwaters. In Proceedings of the Advances in Coastal Structures and Breakwaters Conference, London, UK, 27–29 April 1995; Institution of Civil Engineers, Thomas Telford Publishing: London, UK, 1995; pp. 308–321.

59. EurOtop. Manual on Wave Overtopping of Sea Defences and Related Structures. An Overtopping Manual Largely Based on European Research, but for Worldwide Application. Van der Meer, J.W., Allsop, N.W.H., Bruce, T., de Rouck, J., Kortenhaus, A., Pullen, T., Zanuttigh, B., Eds.; 2016. Available online: www.overtopping-manual.com (accessed on 25 May 2018).

60. Schiereck, G.J. *Introduction to Bed, Bank and Shore Protection*; CRC Press: Boca Raton, FL, USA, 2003.

61. Nørgaard, J.Q.H.; Andersen, L.V.; Andersen, T.L.; Burcharth, H.F. Displacement of Monolithic Rubble-Mound Breakwater Crown-Walls. In Proceedings of the 33rd ICCE, Santander, Spain, 1–6 July 2012; Volume 1.

62. De Rouck, J.; Van Doorslaer, K.; Goemaere, J.; Verhaeghe, H. Geotechnical design of breakwaters in Ostend on very soft soil. In Proceedings of the 32nd International Conference on Coastal Engineering (ICCE), Shangai, China, 30 June–5 July 2010; Volume 1, p. 67.

63. Coles, S.; Heffernan, J.; Tawn, J. Dependence measures of extreme value analysis. *Extremes* **1999**, *2*, 339–365. [CrossRef]

64. Suh, K.D.; Kwon, H.D.; Lee, D.Y. Some statistical characteristics of large deepwater waves around the Korean Peninsula. *Coast. Eng.* **2010**, *57*, 375–384. [CrossRef]

65. Thornton, E.B.; Guza, R.T. Transformation of wave height distribution. *J. Geophys. Res. Oceans* **1983**, *88*, 5925–5938. [CrossRef]

66. CIRIA, CUR, CETMEF. *The Rock Manual. The Use of Rock in Hydraulic Engineering*, 2nd ed.; CIRIA: London, UK, 2007; Volume 683.

67. Tuan, T.Q.; Vu, M.C.; Le, H.T. Experimental Study on Wave Overtopping at Seadikes with Vertical Crown-Wall. In Proceedings of the 5th International Conference on Asian and Pacific Coasts, Singapore, 22–25 September 2009; World Scientific: Singapore, 2009; pp. 79–85.

68. Gómez Pina, G.; Valdés, J.M. Experiments on Coastal Protection Submerged Breakwaters: A Way to Look at the Results. In Proceedings of the 22th International Conference on Coastal Engineering, Delft, The Netherlands, 2–6 July 1990; pp. 1592–1605.

69. U.S. Army Corps of Engineers (USACE). *Coastal Engineering Manual*; Chapter VI-5—Fundamentals of Design; Burcharth, H.F., Hughes, S.A., Eds.; U.S. Army Corps of Engineers: Washington, DC, USA, 2006.

70. Van der Meer, J.W. Stability of Breakwater Armour Layers-Design Formulae. *Coast. Eng.* **1987**, *11*, 219–239. [CrossRef]

71. Coeveld, E.M.; Busnelli, M.M.; van Gent, M.R.A.; Wolters, G. Wave Overtopping of Rubble Mound Breakwaters with Crest Elements. In Proceedings of the 30th International Conference on Coastal Engineering, San Diego, CA, USA, 3–8 September 2006; Volume 5, pp. 4592–4604.

72. EurOtop. *Wave Overtopping of Sea Defences and Related Structures: Assessment Manual*; Pullen, T., Allsop, N.W.H., Bruce, T., Kortenhaus, A., Schüttrumpf, H., van der Meer, J.W., Eds.; Environment Agency (EA): Bristol, UK; Expertise Netwerk Waterkeren (ENW): Utrecht, The Netherlands; Kuratorium für Forschung im Küsteningenieurwesen (KFKI): Hamburg, Germany, 2007.

73. D'Angremond, K.; van der Meer, J.W.; de Jong, R.J. Wave Transmission at Low-Crested Structures. In Proceedings of the 25th International Conference on Coastal Engineering, Orlando, FL, USA, 2–6 September 1996; pp. 2418–2426.

74. Van der Meer, J.W.; Briganti, R.; Zanuttigh, B.; Wang, B. Wave transmission and reflection at low-crested structures: Design formulae, oblique wave attack and spectral change. *Coast. Eng.* **2005**, *52*, 915–929. [CrossRef]

75. Briganti, R.; van der Meer, J.W.; Buccino, M.; Calabrese, M. Wave Transmission behind Low-Crested Structures. In Proceedings of the Coast Structures Conference, Portland, OR, USA, 26–30 August 2003; ASCE: Reston, VA, USA, 2003; pp. 580–592.

76. Makris, C.V.; Memos, C.D. Wave Transmission over Submerged Breakwaters: Performance of Formulae and Models. In Proceedings of the 17th International Offshore and Polar Engineering Conference (ISOPE), Lisbon, Portugal, 1–6 July 2007; pp. 2613–2620.

77. Vicinanza, D.; Cáceres, I.; Buccino, M.; Gironella, X.; Calabrese, M. Wave disturbance behind low crested structures: Diffraction and overtopping effects. *Coast. Eng.* **2009**, *56*, 1173–1185. [CrossRef]

78. Delos. *D 59 Design Guidelines: Environment Design Guidelines of Low Crested Coastal Defence Structures*; Burcharth, H., Lamberti, A., Eds.; Pitagora Editrice Bologna: Bologna, Italy, 2004.

79. Kraus, N.C. Estimate of breaking wave height behind structures. *J. Waterw. Port Coast. Ocean Eng.* **1984**, *110*, 276–282. [CrossRef]

80. Goda, Y. *Random Seas and Design of Maritime Structures*, 1st ed.; University of Tokyo Press: Tokyo, Japan, 1985.

81. Karambas, T.V.; Memos, C.D. Boussinesq model for weakly nonlinear fully dispersive water waves. *J. Waterw. Port Coast. Ocean Eng.* **2009**, *135*, 187–199. [CrossRef]

82. Memos, C.D.; Karambas, T.V.; Avgeris, I. Irregular wave transformation in the nearshore zone: Experimental investigations and comparison with a higher order Boussinesq model. *Ocean Eng.* **2005**, *32*, 1465–1485. [CrossRef]

83. Viviano, A.; Musumeci, R.E.; Foti, E. A nonlinear rotational, quasi-2DH, numerical model for spilling wave propagation. *Appl. Math. Model.* **2015**, *39*, 1099–1118. [CrossRef]

84. Prevot, G.; Kergadallan, X.; Sergent, P. Overtopping Evolution due to the Climate Change and Consequences on the Structures with an Analytical Method and a Statistical Method. In Proceedings of the AIPCN-France Third Mediterranean Days of Coastal and Port Engineering, Marseilles, France, 22–24 May 2013.

85. Sergent, P.; Prevot, G.; Mattarolo, G.; Brossard, J.; Morel, G.; Mar, F.; Benoît, M.; Ropert, F.; Kergadallan, X.; Trichet, J.-J.; et al. Adaptation of coastal structures to mean sea level rise. *La Houille Blanche* **2014**, *6*, 54–61. [CrossRef]

86. Hawkes, P.J.; Gouldby, B.P.; Tawn, J.A.; Owen, M.W. The joint probability of waves and water levels in coastal engineering design. *J. Hydraul. Res.* **2002**, *40*, 241–251. [CrossRef]

87. Benoît, M.; Lafon, F.; Goasguen, G. Constitution et exploitation d'une base de données d'états de mer le long des côtes françaises par simulation numérique sur 23 ans. Base ANEMOC en Atlantique-Manche-Mer du Nord. *Eur. J. Environ. Civ. Eng.* **2012**, *12*, 35–50. [CrossRef]

88. Morellato, D.; Benoît, M. Vagues et Changement Climatique-Simulation des états de mer dans l'Océan Atlantique de 1960 à 2100 pour trois Scénarios de Changement Climatique. In Proceedings of the Les 8èmes Journées Scientifiques et Techniques, Brest, France, 8–9 December 2010. [CrossRef]

89. ONERC. *Synthèse n°2, Prise en Compte de L'élévation du Niveau de la mer en vue de L'estimation des Impacts du Changement Climatique et des Mesures D'adaptation Possibles*; Direction Générale Energie et Climat Grande Arche: Paroi nord 92 055 La Défense Cedex, France, 2010; 6p.

90. Hosking, J.R.M. L-moments: Analysis and estimation of distribution using linear combination of order statistics. *J. R. Stat. Soc. B* **1990**, *52*, 105–124.

91. LiCCo. *Two Countries, One Sea—A Cross Channel Perspective on Climate Change and the Coast*. Report of the LiCCo Project, Interreg IVA France (Channel)—England Programme. 2014. Available online: http://www.licco.eu (accessed on 20 July 2018).

Journal of
*Marine Science
and Engineering*

MDPI

Article

Regional Scale Risk-Informed Land-Use Planning Using Probabilistic Coastline Recession Modelling and Economical Optimisation: East Coast of Sri Lanka

Ali Dastgheib [1,*], Ruben Jongejan [2], Mangala Wickramanayake [3] and Roshanka Ranasinghe [1,4,5]

[1] Department of Water Science and Engineering, IHE Delft Institute for Water Education, P.O. Box 3015, 2601 DA Delft, The Netherlands; r.ranasinghe@un-ihe.org
[2] Jongejan Risk Management Consulting, Schoolstraat 4, 2611 HS Delft, The Netherlands; ruben.jongejan@jongejanrmc.com
[3] Coast Conservation and Coastal Resource Management Department, 4th Floor, New Secretariat Building, Maligawatte, Colombo 01000, Sri Lanka; mangalawk@gmail.com
[4] Harbour, Coastal and Offshore Engineering, Deltares, P.O. Box 177, 2600 MH Delft, The Netherlands
[5] Department of Water Engineering and Management, University of Twente, P.O. Box 217, 7500 AE Enschede, The Netherlands
* Correspondence: a.dastgheib@un-ihe.org; Tel.: +31-15-2151845

Received: 17 July 2018; Accepted: 12 October 2018; Published: 15 October 2018

Abstract: One of the measures that has been implemented widely to adapt to the effect of climate change in coastal zones is the implementation of set-back lines. The traditional approach of determining set-back lines is likely to be conservative, and thus pose unnecessary constraints on coastal zone development and fully utilising the potential of these high-return areas. In this study, we apply a newly developed risk-informed approach to determine the coastal set-back line at regional scale in a poor data environment. This approach aims to find the economic optimum by balancing the (potential) economic gain from investing in coastal zones and the risk of coastal retreat due to sea level rise and storm erosion. This application focusses on the east coast of Sri Lanka, which is experiencing rapid economic growth on one hand and severe beach erosion on the other hand. This area of Sri Lanka is a highly data-poor environment, and the data is mostly available from global databases and very limited measurement campaigns. Probabilistic estimates of coastline retreat are obtained from the application of Probabilistic Coastline Recession (PCR) framework. Economic data, such as the discount rate, rate of return of investment, cost of damage, etc., are collated from existing estimates/reports for the area. The main outcome of this study is a series of maps indicating the economically optimal set-back line (EOSL) for the ~200-km-long coastal region. The EOSL is established for the year 2025 to provide a stable basis for land-use planning decisions over the next two decades or so. The EOSLs thus determined range between 12 m and 175 m from the coastline. Sensitivity analyses show that strong variations in key economic parameters such as the discount rate have a disproportionately small impact on the EOSL.

Keywords: coastline retreat; coastal risk; economical optimisation; coastal zone management; climate change adaptation

---

1. Introduction

The effects of climate change on hydrodynamic forcing, such as sea level rise, changes in wave conditions, changes in the sediment supply from rivers, etc., is already resulting in changes in the rate of coastal erosion/accretion along the sandy coastlines of the world [1–5].

In light of climate change and especially coastline retreat due to sea level rise, many countries and coastal zone authorities are developing adaptation plans to protect coastal assets and future investments along their coastlines. These measures span a very wide range, from traditional hard structures, such as sea walls, and soft solutions such as (periodic) nourishment campaigns, to more modern strategies such as pre-storm interventions (coastal dune enforcements, transient breakwaters, etc.). One of the measures that has been widely implemented as part of the "accommodate" or "retreat" strategies in coastal zone management is the implementation of coastal set-back lines. A set-back line is a line along the coast, seaward of which certain development activities are prohibited or restricted. Traditionally, this line is determined based on a linear summation of long-term recession due to long-shore sediment transport gradients (estimated based on aerial photos or sediment budget modelling), the impact of storm erosion for a given return period (estimated from coastal profile modelling or historical data) and long-term recession due to sea level rise (estimated by the Bruun rule [6]). This method produces a set-back line that is likely to be conservative, resulting in rather severe constraints on coastal zone development. A risk-informed approach to coastal zone management was first proposed in 2002 [7], which was later refined [8,9]. The latter introduced the concept of the economically optimal set-back line (EOSL). This method, which balances the (potential) economic gain from coastal zone investments and the risk due to coastal erosion was applied to Narrabeen Beach, Sydney, Australia [9,10].

Two types of data are needed for determining the EOSL: economic data and probabilistic estimates of the development of coastline retreat in the future. Economic data, such as the discount rate, rate of return of investment (RoI), cost of damage, etc., may be collected from existing estimates/reports for the area of interest or be based on expert judgement. Probabilistic estimates of coastline retreat can be obtained from the Probabilistic Coastline Recession (PCR) framework [11]. The PCR framework is designed to perform Monte Carlo simulations, that is, to calculate a large number of long, realistic sequences of beach erosion and recovery, and to then statistically analyse the results instead of relying only on single storm effects with specific return period (i.e., deterministic estimates). Model inputs are statistical parameters representing the wave climate, water levels, and gaps between storms to randomly generate model boundary conditions. These parameters are derived using the Joint Probability Method (JPM), which considers not only a single event storm, but series of storm events [12,13]. This method includes allowances for joint probability between all basic erosion variates including: wave height, period and direction, event duration, and the time interval between events.

In this study, the above approach was applied along the east of Sri Lanka (Trincomalee and Batticaloa—Figure 1). The total study area spans over 200 km and in total 83 coastal profiles along the coastline were analysed. For each of the 83 profiles, the cumulative distribution function of probability of coastal retreat in the future was derived, which fed into the determination of the associated EOSL.

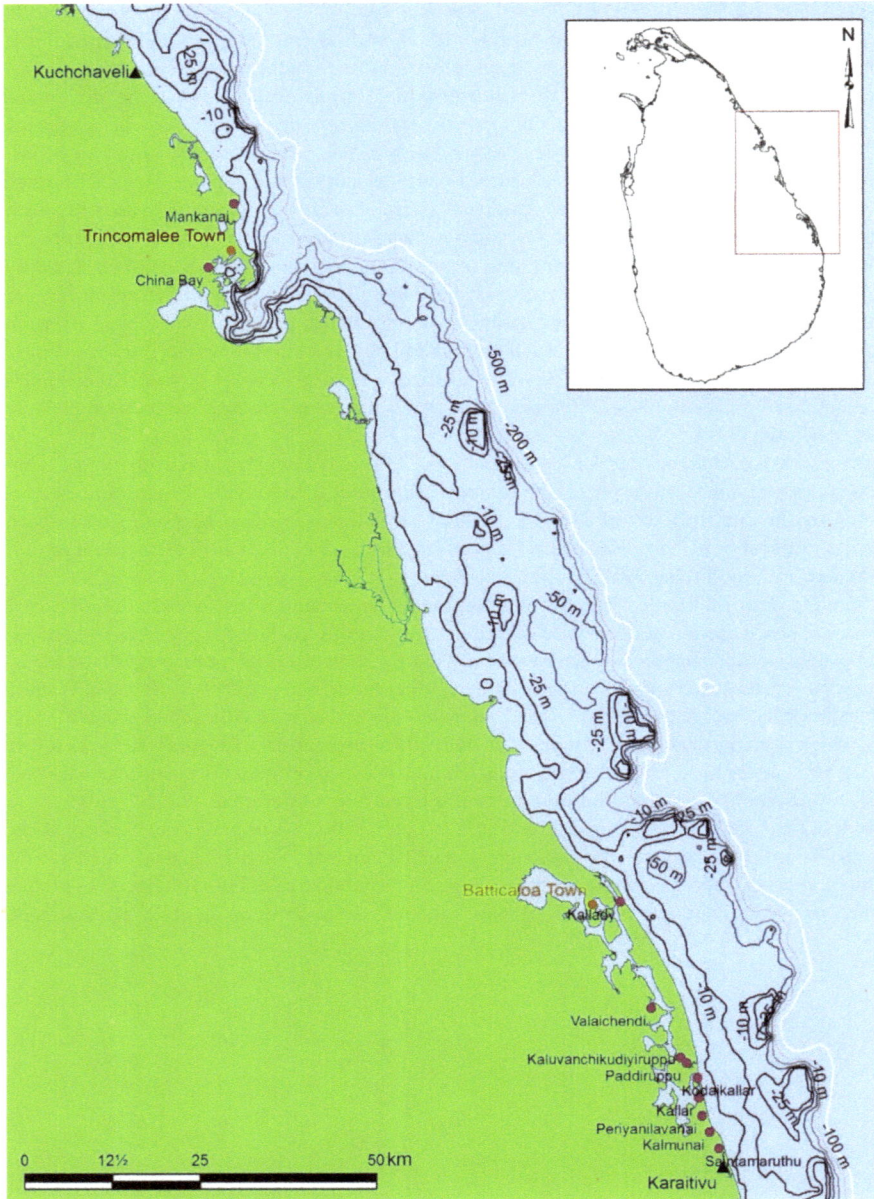

Figure 1. Study area: Trincomalee and Batticaloa district coastlines, along the east coast of Sri Lanka from Kuchchaveli to Karaitivu (~200 km).

2. Study Site

Sri Lanka's coastline is 1680 km long, and the country's geographical location leaves it exposed to storms from several directions that differ in their characteristics according to coastline orientation and season. Furthermore, the coastal zone is extremely low lying (less than 1 m above mean sea level (MSL) up to a distance of 1–2 km inland in most places), leaving it highly exposed to coastal inundation and erosion. At present, large parts of the nation's coastline suffer from storm erosion regularly, as well as chronic coastline recession [14]. With a projected global sea level rise of up to 1 m above present-day MSL (by 2100) and increased intensity and frequency of storms and storm surges, coastal communities and developments around Sri Lanka will likely be severely threatened by permanent/episodic inundation and/or coastal erosion [15].

The conflict situation that prevailed for more than 30 years in the northern and the eastern districts of Sri Lanka has disrupted proper management of the coastal resources and developments in those regions. Furthermore, the socio-economic status of the coastal communities in the entire country has degraded rapidly due to the conflict situation. However, since the end of the civil conflict in 2009, ample opportunities for economic development, particularly along the coast, have emerged due to the many natural resources of the coastal region. Accordingly, the Government of Sri Lanka (GOSL) has given high priority to developing major economic hubs, maritime facilities, tourism and the energy sector within particularly the severely under-developed and under-utilised northern and eastern coast regions. The rapid development of the tourism industry, especially in the eastern coastal region, is a priority for the GOSL from an economic point of view. However, in order to safeguard the unique environment in the region and environment-related livelihoods of the eastern communities, balancing conservation requirements with development activities is needed. With the marked post-war increase of population in coastal towns and accelerated economic activities in the eastern coastal region, the requirement of an integrated management approach of conserving, developing and sustainable utilization of resources has also been recognised. These plans also include a coastal zone and coastal resource management plan for the country [14].

The eastern part of the country is vulnerable to many natural disasters such as cyclones, storms, chronic coastal recession, and also tsunamis (as evidenced by the December 2004 Tsunami). The majority of the eastern communities and developments are located in Trincomalee, Batticaloa, and Amparai districts, all of which are densely populated, mainly within the coastal belt. The main livelihoods in the communities of these areas are fishery, agriculture or lagoon fishery [16,17].

For this study, two different sites in the east coast of Sri Lanka were selected: Trincomalee and Batticaloa. In this section, these two sites are described. At both sites, a long stretch of sandy coastline is interrupted by headlands, river mouths and (seasonal) tidal inlets.

3. Data Collection and Analyses

This study required several types of data. These data were either measured in the field or sourced from global databases. This section describes the data used and their analysis.

3.1. Beach Profiles and Sediment Size

The most important data that was needed for this study were cross-shore beach profiles along the coastline of the study areas. These data were provided by the Coast Conservation Department (CCD) of Sri Lanka. The beach profiling was done for approximately 100 profiles in Trincomalee and Batticaloa districts. The profiles were measured from the berm or dune at the landward end and continued to a depth of 10–15 m, or an offshore distance or 2.0 km from the beach, with 0.5 to 2.0 km spacing between them. For each profile, the beach/dune height and beach slope was calculated. Figure 2 shows the areas in which the profiles used in this study are located. At each profile location, sand samples were also collected for grain size determination, and sediment grading curves were established for each profile.

Based on the geo-morphological conditions of the coastline along the study areas, we defined 12 coastal cells in total (Figure 2; 8 cells in Trincomalee and 4 cells in Batticaloa). A cell-averaged beach profile was subsequently computed based on the measured beach profiles in each cell. These cell-averaged profiles were then used to calibrate the erosion model. (See Section 4.2). As an example, Figure 3 shows the average beach profile for coastal cell B-IV. Table 1 shows the average d50 (intercepts for 50% of the cumulative mass of soil sample) measured in each coastal cell.

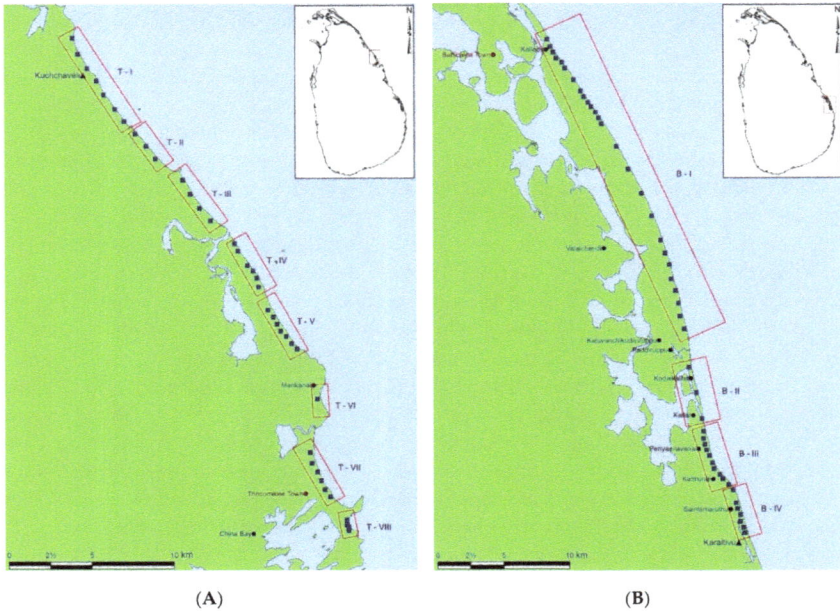

Figure 2. Location of measured beach profiles and adopted zonation in (**A**) Trincomalee and (**B**) Batticaloa.

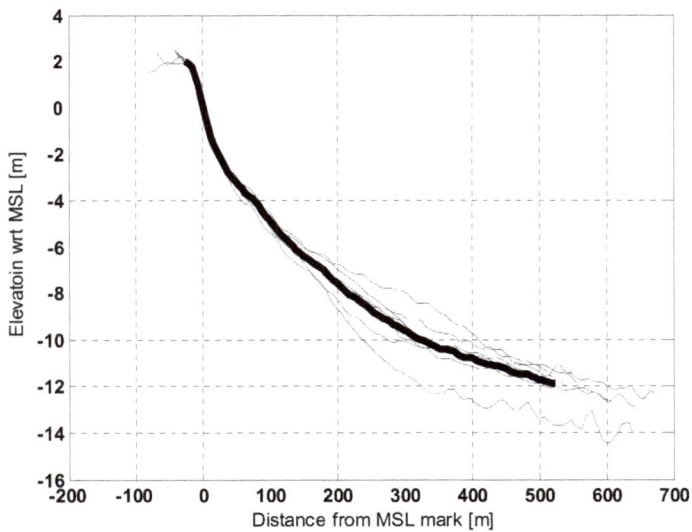

Figure 3. Measured beach Profiles (thin) and computed coastal cell-averaged beach profile (thick) for Coastal cell B-IV (Kalmunai to Karaitivu coastline stretch).

Table 1. Average d50 measured in each coastal cell.

Trincomalee		Batticaloa	
Coastal Cell	Ave. d50 (μm)	Coastal Cell	Ave. d50 (μm)
T-I	213	B-I	553
T-II	285	B-II	377
T-III	268	B-V	441
T-IV	323	B-IV	456
T-V	200		
T-VI	200		
T-VII	240		
T-VIII	270		

3.2. Wave Data

3.2.1. Off-Shore Wave Data

Since existing local wave data in the study area are of very limited duration, here we used wave data from ERA-Interim (ERAi), which is a global reanalysis providing wave characteristics every 6 h from 1979 till 2016 [18] on a 0.5 × 0.5 degree global grid.

3.2.2. Wave Model

The above-described ERAi wave data is for offshore of the study sites, and thus cannot be directly used in the coastal recession models. Therefore, the deep-water ERAi wave data were first transformed to the nearshore (of the coastal cells defined in Section 3.1) using a 2 dimensional spectral wave model. This section provides a description of the wave model, the model setup adopted for this study, and the results obtained.

The spectral wave model used in this study was SWAN (Simulating WAves Nearshore-model), which is a spectral third-generation wave model (e.g., [19,20]). The SWAN model is based on the discrete spectral action balance equation and is fully spectral (in all directions and frequencies). This implies that short-crested random wave fields propagating simultaneously from widely different directions can be accommodated. SWAN computes the evolution of random, short-crested waves in coastal regions with deep, intermediate and shallow water and ambient currents. The SWAN model accounts for (refractive) propagation and represents the processes of wave generation by wind, dissipation due to white-capping, bottom friction and depth-induced wave breaking and non-linear wave–wave interactions (both quadruplets and triads) explicitly with state-of-the-art formulations. The model has been successfully validated in laboratory and (complex) field cases (e.g., [20,21]). It is noted that the SWAN model does not account for diffraction effects.

In this study, all SWAN simulations were carried out in the stationary mode, with the depth-induced breaking model [22]. To account for bed friction in SWAN, the so called 'JONSWAP' model [23] was chosen.

Grid and Bathymetry

To reduce the computational burden and to be able to simulate the phenomena desired in this study, two curvilinear grids were set up: one large-scale regional grid with a resolution of 0.01 degree for the entire east coast of Sri Lanka, and one local grid with minimum resolution of 70 m along the coastline of Trincomalee and Batticaloa. Figure 4 shows the local grid nested within the regional grid. The computational grid in this model covers areas with elevations lower than 2 m above mean sea level.

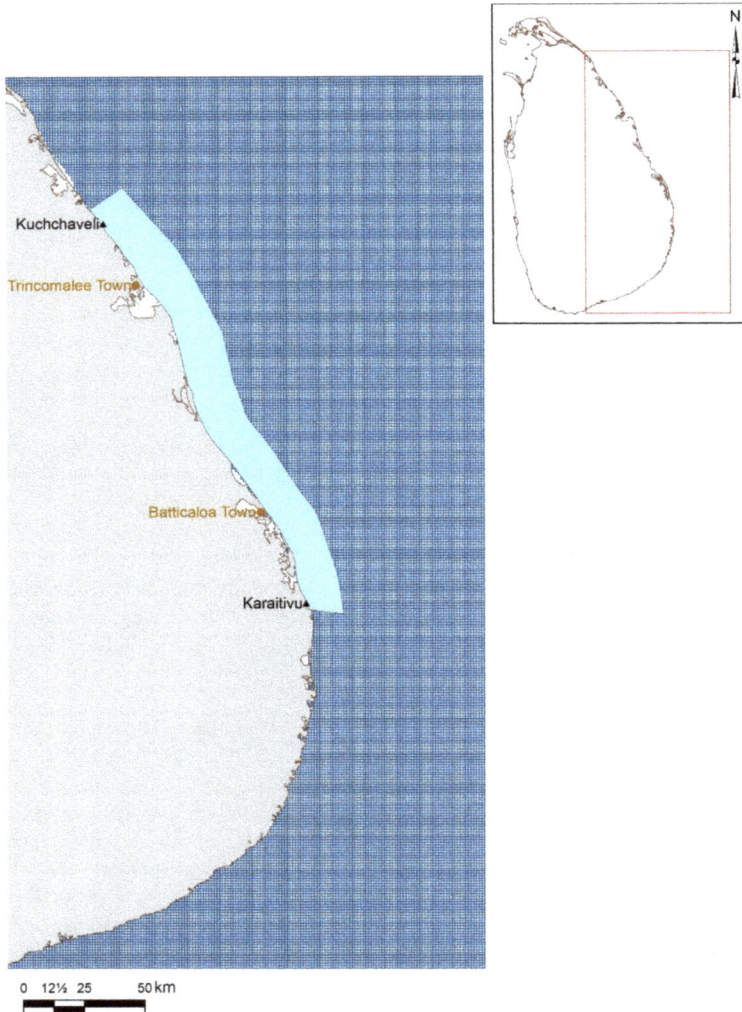

Figure 4. Regional and local wave model grids for the wave model.

For the bathymetry of the wave model, data from General Bathymetric Chart of the Oceans [24], which is a publicly available bathymetry of the oceans, were used. GEBCO data (http://www.gebco.net—last visited 30 August 2017) can be projected on grids at different resolution. These data are not very accurate in the nearshore; therefore, GEBCO data were used for the local grid, with the measured profile data (see Section 3.1).

Boundary Conditions and Forcing

Wave forcing was applied at the open boundaries of the model using nearshore transformed ERA-interim wave data, such as significant wave height, peak period and direction, every 6 h. Figure 5 shows the wave-roses of time series of wave conditions applied at the wave model boundary.

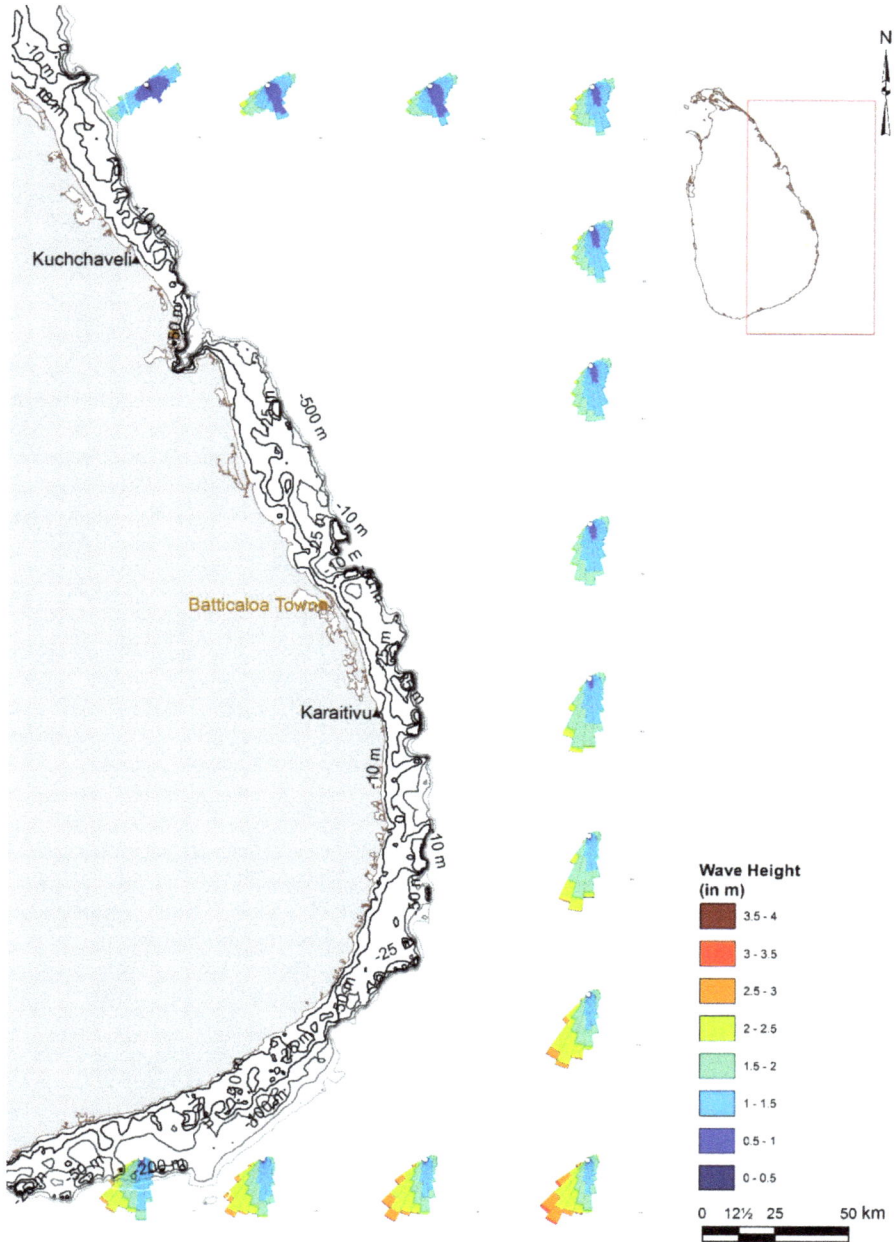

Figure 5. Wave boundary conditions applied to the model as time series of wave conditions.

Simulation and Results

A full simulation of 30 years (1979–2009) was carried out and 6 h wave conditions in front of each coastal cell were recorded. As an example, Figure 6 shows a time series of wave conditions in front of Dutch Bay, Trincomalee (Cell T-VIII).

As can be seen in Figure 6, the dominant angle of wave incidence is SE, but the height of these waves is very small. These SE waves are the waves generated by the SW monsoon (May–September) and are diffracted and refracted around the island of Sri Lanka. On the other hand, E-NE waves with larger heights and E-NE storms are experienced during the NE monsoon (November–February). In the lower panel of Figure 6 this seasonal signal in the wave height is very pronounced. This trend exists for all the coastal cells in the study area.

Storm Detection

For the purpose of this study, it was necessary to define and detect the storms and their characteristics in the 30 years of simulated wave time series for each coastal cell. To estimate storm conditions the following workflow was applied.

For the time series of wave conditions extracted in front of each coastal cell, the 95th percentile significant wave height value for the period of 1979–2009 was set as the cut-off wave height for a POT (peak over threshold) approach of determining storm events. An event was defined as the interval between an adjacent up-crossing and down-crossing of the threshold wave height. The event characteristics were extracted from the wave time series, being the maximum significant wave height, the average direction, the average period, and duration of the event. The duration was defined as the time difference between moments of up-crossing and down-crossing of the event. If the duration of the event was more than 12 h then the event was flagged as a storm. Figures 6 and 7 show this process for the wave extraction point in front of Dutch Bay (Cell T-VIII). In the case of Dutch Bay, 221 storms were detected in the 30 years of data. The overall number of storms detected over the 30-year analysis period based on this workflow in the different coastal cells is around 225 storms, or an average of around 8 storms per year, and as expected, all occurred during the NE monsoon season.

3.2.3. Storm Data Analyses

After detecting storms in all of the coastal cells, statistical analyses needed to be performed on the data. In this section, the methodology which was used for statistical analyses is described using the example of 221 storms detected for Dutch Bay (Cell T-VIII). Figure 8 shows the main storm characteristics for this cell, the maximum significant wave height (Hs) during the storms, and the average peak period of waves (Tp) during the storms, the average wave direction during the storm, and the duration of storms. As shown in Figure 8, storm waves are incident from E-NE (25°–90° Nautical) and occur only during the NE monsoon.

To use these data in coastline recession simulations, a generalised extreme value distribution function (GEV) was fitted to the maximum significant wave height during storms and to the storm durations. Figure 9 shows the cumulative distribution function of GEVs fitted to these variables. Since the wave direction does not have extreme values and for detected storms is limited to a narrow sector, here we used an empirical cumulative distribution function (CDF) for average wave direction during the storms (Figure 9).

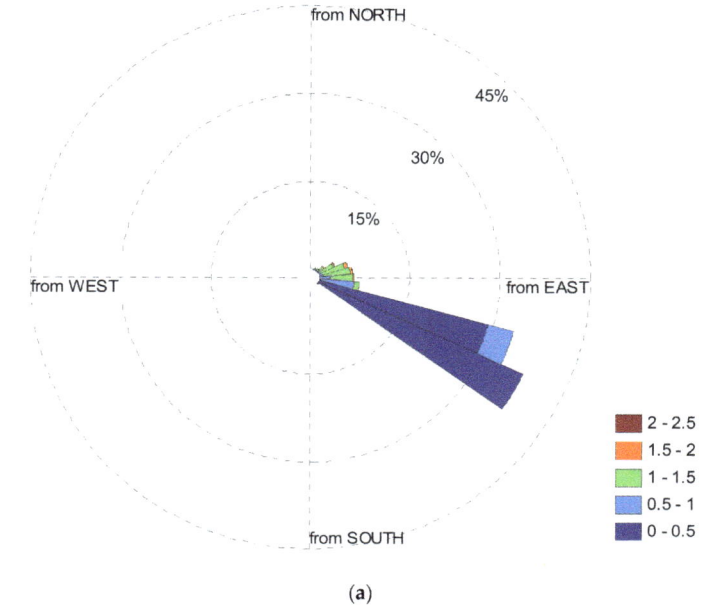

(a)

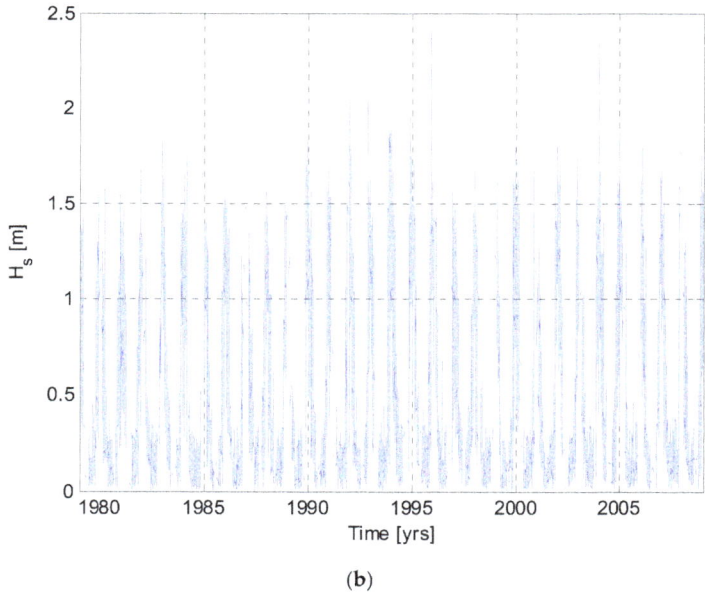

(b)

Figure 6. Nearshore transformed wave rose (**a**), and 30 years of extracted nearshore wave heights (**b**) in front of the Dutch Bay in Trincomalee at the location with 38 m depth.

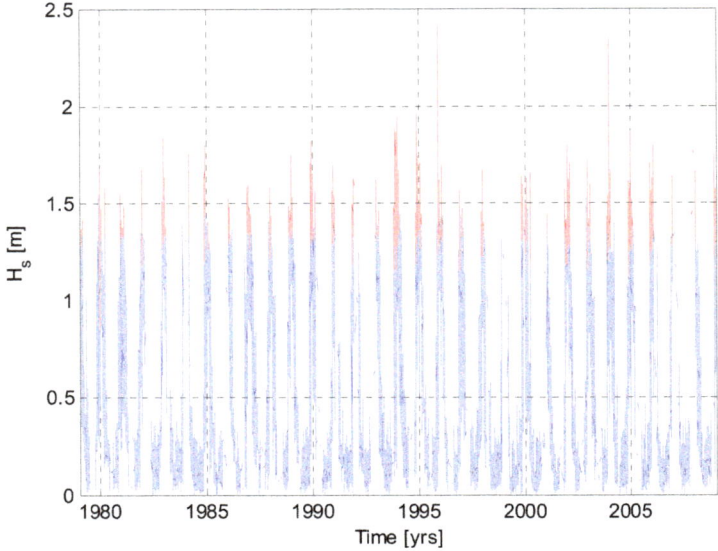

(a)

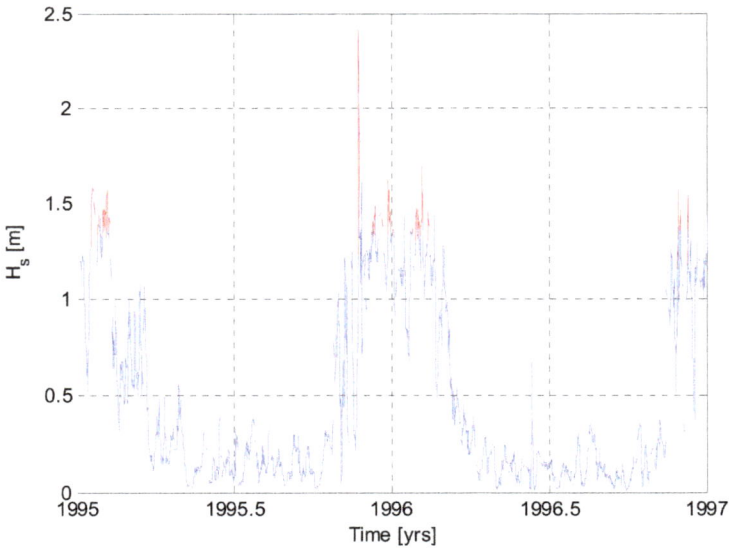

(b)

Figure 7. The process of storm detection for the data point in front of Dutch Bay. Storms identified over the 30-year period from 1979–2009 (**a**), zoom-in plot of storms identified over the 2-year period 1995–1997 (**b**); Blue: All data, Red: identified storm events.

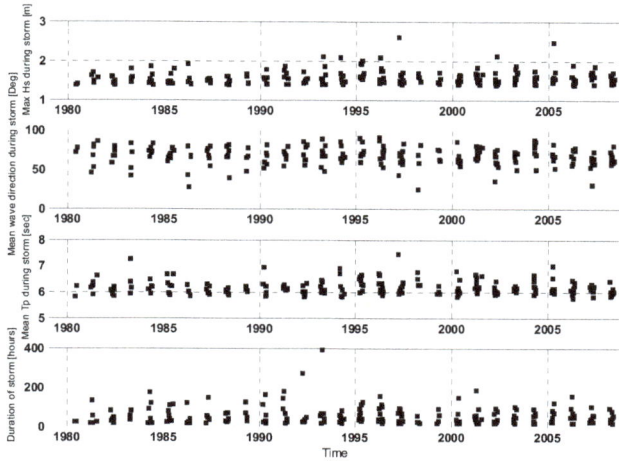

Figure 8. Storms characteristics at Dutch Bay, Trincomalee (Cell T-VIII).

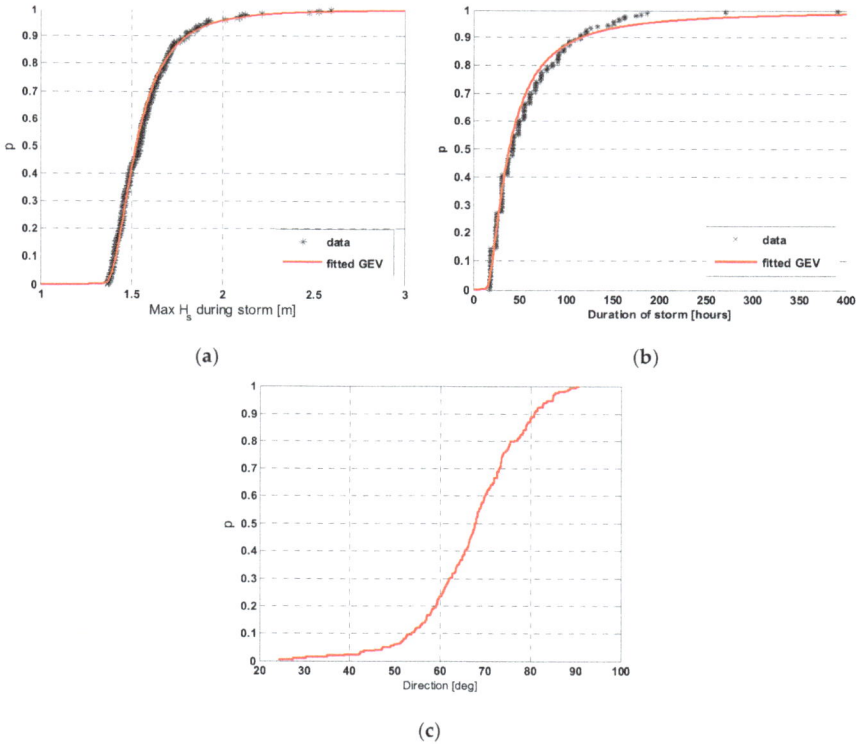

(a)

(b)

(c)

Figure 9. Cumulative distribution function and fitted GEV of maximum significant wave height (H$_s$) (**a**), storm duration (**b**), and Empirical cumulative distribution function of average wave direction (**c**) during 30 years of storms at Dutch Bay (Cell T-VIII).

To complete the statistical analyses, it is also necessary to check the dependencies between different storm characteristics. Similar to previous studies (e.g., [12]), we chose the maximum significant wave height during the storm as the main parameter and checked its dependencies against the other parameters. Figure 10 shows the dependency between maximum significant wave height during the storm and storm duration and the dependency between maximum significant wave height and average wave direction during the storms. The maximum significant wave height and storm duration are (inter)dependent with higher significant wave heights occurring during the longer storms. On the other hand, the maximum significant wave height and the average wave direction are more or less independent of each other.

To generate a dependency distribution that describes the observed dependency between maximum significant wave height and storm duration, we made use of a copula. A copula is essentially a multivariate probability distribution for which the marginal probability distribution of each variable is uniform; in this study, the 'Clayton' copula has been used. Figure 11 shows the joint probability model (JPM) capturing the dependency between maximum significant wave height and storm duration.

For relating the maximum significant wave height and average peak wave period, we used a linear fit of the data, as shown in Figure 12.

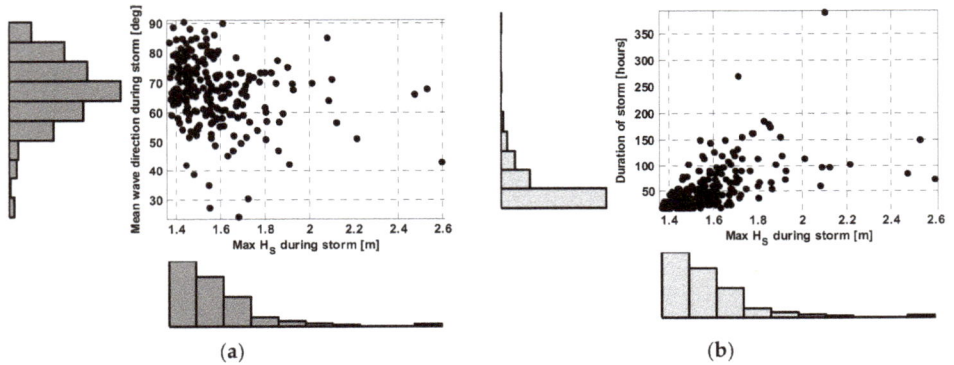

Figure 10. Dependency between maximum significant wave height and average wave direction—(**a** panel), and maximum significant wave height and storm duration—(**b** panel), for Dutch Bay (Cell T-VIII).

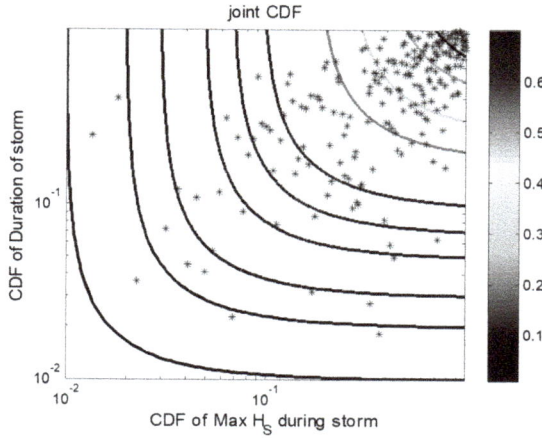

Figure 11. Joint probability model (JPM) of maximum significant wave height (m) and storm duration (Hour) for Dutch Bay (Cell T-VIII).

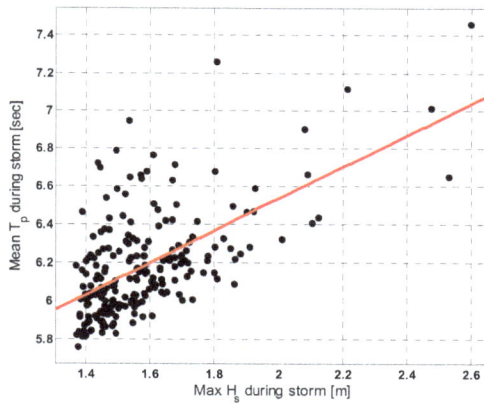

Figure 12. Linear fit between maximum significant wave height and average peak wave period during storms.

3.3. Relative Sea Level Rise

For the purposes of this study, it was also necessary to estimate the time series of projected sea level rise in the study area. The synthesis report of the IPCC reported that over the period 1901–2010, global mean sea level rose by 0.19 (0.17 to 0.21) m, with an average rate of 1.7 mm/year in the 20th Century [25]. The fifth assessment report of the IPCC [26] indicated a transition in the late 19th century to the early 20th century from relatively low mean rates of rise over the previous two millennia to higher rates of rise (high confidence). From 1961 to 2003, the average rate of SLR was 1.8 ± 0.5 mm/year, while between 1993 and 2010 the rate was very likely higher at 3.2 (2.8 to 3.6) mm/year; similarly, high rates likely occurred between 1920 and 1950. IPCC also reported that ocean thermal expansion and glacier melting have been the dominant contributors to 20th century global mean sea level rise [26]. Relative sea level rise (RSLR) over the next 30–100-year period is the sum of three major components: global-mean sea-level change; regional (local) spatial variations in sea-level change; and vertical land movement (subsidence or uplift). The inclusion of regional components of relative sea-level rise is important when developing scenarios for impact and adaptation assessment as they provide a critical link between global climate change and regional/local coastal management strategies [27,28].

IPCC guidelines to determine local RSLR by 2100 have been prescribed [29], where three different options are given based on data availability. Here, we used the 'intermediate' option, together with IPCC AR5 projections of global mean SLR and regional variations in SLR. The suggested 'intermediate' assessment methodology [29] was adopted to derive RSLR scenarios linked to the RCP8.5 IPCC scenario by deriving the coefficients the method as follows:

$$SLR = a_1 t + a_2 t^2 \tag{1}$$

where:

- SLR is global mean sea level rise (m)
- t is number of years starting from 2000 (year)
- a_1 is rate of sea level rise at year 2000 (m/year) (in this case 0.003)
- a_2 is factor of the change in the rate of sea level rise (m/year^2) (in this case 4.5×10^{-5})

To account for regional (local) spatial variations in sea-level change, the difference between global mean sea level rise and ensemble mean regional relative sea level change between 1986–2005 and 2081–2100 for RCP 8.5 [26], is added linearly to the time-varying SLR estimated using Equation (1). This variation includes effects of atmospheric loading, land ice, glacial isostatic adjustment (GIA) and terrestrial water sources. In these calculations, the land subsidence is ignored due to lack of data.

Figure 13 shows the RSLR values taking into account the global mean SLR and regional variations of SLR for the east coast of Sri Lanka.

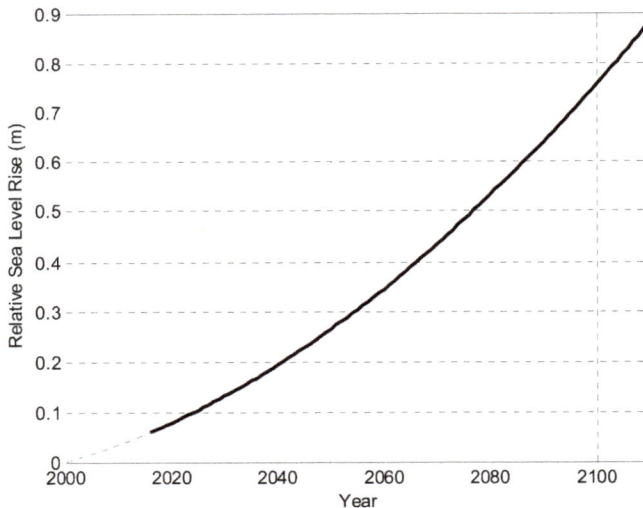

Figure 13. Regional RSLR for RCP 8.5 excluding land subsidence by 2100 (relative to 2000).

4. Application of Probabilistic Coastal Recession (PCR) Simulations

The primary functions of the PCR framework are: randomly generate long (~100 years) time series of storm conditions and the gap between them from pre-determined Joint Probability distributions of storm conditions, calculate erosion during the storm using a profile model, calculate subsequent recovery during the inter-storm periods, and calculate time series of coastline location which is stored as output (Figure 14). This process is repeated ~100,000 times, resulting in the data set necessary for robust statistical analysis [11].

In this study, the (joint) distributions of storm characteristics developed in Section 3 are used to generate the time series of storm conditions and gap between storms (See Section 4.2), and the erosion model of Mendoza and Jimenez [30] (See Section 4.3) is used as the profile model to calculate erosion and coastline retreat.

According to the coast conservation department of Sri Lanka, significant storm surges are very uncommon in the study area. Furthermore, no long-term water level data are available to derive storm surge information in the study area. Recent global storm surge modelling studies [31,32] indicate a 1:100-year return period extreme sea level (combining surge, tide and wave run up) of less than 0.5 m along the east coast of Sri Lanka), and hence storm surges were not considered in this study. Also, since there aren't any studies that have provided reliable projections of climate change driven variations in extreme waves in the study area, such changes were not considered in the simulations undertaken in this study.

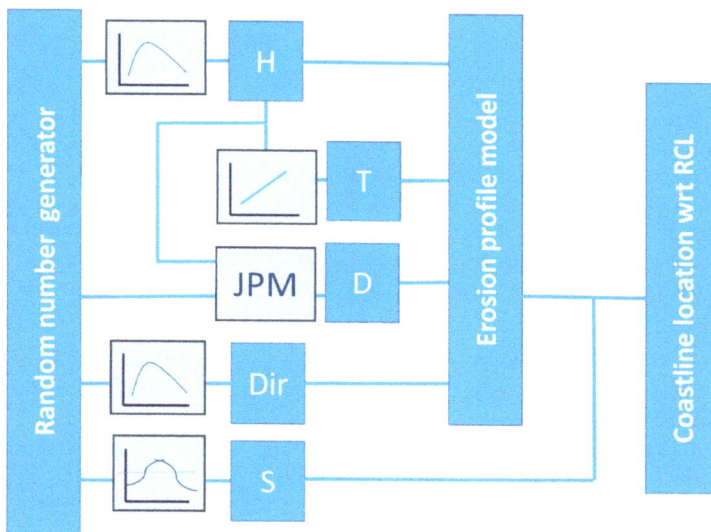

Figure 14. The scheme of the P CR model used in this study. H: maximum Hs in one storm, T: peak period associated with H, D: duration of storm, Dir: Mean direction of storm, S: Gap between two storms.

4.1. Event Generation

PCR simulations require a large number of long records of storm conditions and gaps between storms. This means that first, it is necessary to sample single storms, and then construct a time series of storms.

4.1.1. Sampling Single Storm Conditions (H,T,D,Dir)

In summary, to sample a single storm, the following information, which was determined in Section 3.2.2, is required:

- GEV distributions for $H_{s,max}$ (H) and Storm Duration (D); (e.g., Figure 10)
- Joint probability model, characterised by single values of dependency factor between H & D; (e.g., Figure 11)

- A linear fit between storm wave peak period (T), which is dependent on H; (e.g., Figure 12)
- An empirical distribution for wave direction (Dir), based on measured data. (e.g., Figure 9)

Using this information, a single storm can be randomly generated as follows:

- Sample a random uniform deviation from [0;1] 'a' and use this with the GEV distribution for H to generate a maximum significant for the storm;
- Sample a random uniform deviation from [0;1] 'b' as the dependency parameter as dependency value for H and D;
- Use a and b to determine the deviation 'c' for storm duration (D) from the joint probability model;
- Use 'c' with the GEV distribution for storm duration (D) to generate a storm duration for the storm;
- Use H and the linear fit between H and T to determine storm wave peak period (T);
- Sample a random uniform deviation from [0;1] and use this with the empirical distribution for direction to generate a wave direction for the storm (Dir).

Following this approach, the main storm conditions can be randomly generated from data-fitted distributions.

4.1.2. Constructing a Record of Storms Including Seasonality

A typical PCR model simulation requires a simulation length to be selected and a sequence of storms and subsequent gaps during that period to be generated. For example, a typical situation may involve considering the 100-year period (here taken as the period between 2016 and 2116) to assess the impact of climate change over the next 100 years. The series of storms and the gaps between them during this period are then sampled. Storms are sampled using the methods outlined in Section 4.1. The sampling of the gaps between storms is described below.

Due to the strong seasonality of storms in this case, to be able to generate realistic long-term records of storm waves, we considered the following 3 parameters:

- Gap between storms during the storm season;
- Duration of the 'years', which is the time from start of the first storm in the storm season until the start of the first storm in the next stormy season;
- Duration of 'storm season', which is the time from start of the first storm in the storm season until the end of the last storm in the same storm season;

First, the derived storm time series was analysed to obtain the data set for each of the above-mentioned parameters. Subsequently an empirical distribution function was fitted to the gaps between storms during the storm season and Poisson distribution functions were fitted to the duration of the 'years' and duration of 'storm season'. Then the sequence of storms in a typical simulation length was determined using the following algorithm. Figure 15 shows a schematic of this algorithm.

- Sample a random uniform deviation from [0;1] and use this with the Poisson distribution for Duration of the 'years' to generate the duration of one 'year';
- Sample a random uniform deviation from [0;1] and use this with the Poisson distribution for Duration of the 'storm season' to generate the duration of the 'stormy season' in that 'year';
- Sample a single storm based on the algorithm described in Section 4.1.1;
- Sample a random uniform deviation from [0;1] and use this with the empirical distribution for gap between storms during the storm season to generate one gap corresponding to the storm generated in pervious step;
- Repeat the previous steps until the duration of the 'storm season' is filled with storms;
- Repeat all steps to cover the whole simulation length (e.g., 100 years).

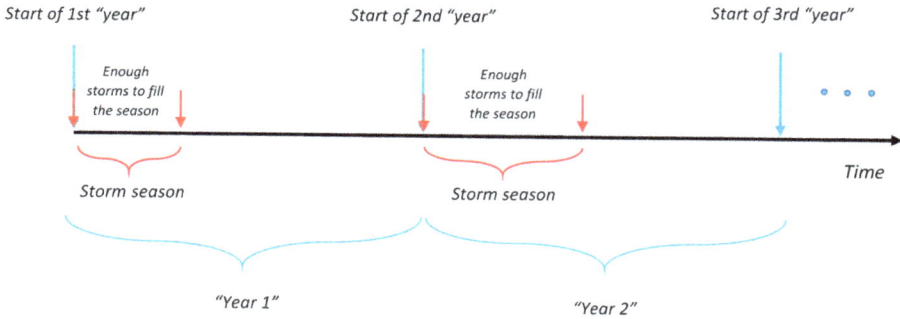

Figure 15. Schematic illustrates the algorithm used for constructing one long-term record of storms.

The storm record generated in this way contains the exact time of each storm and, therefore, based on estimated relative sea level rise (Figure 13), it is now possible to determine the MSL at the time of occurrence of each storm.

4.2. Erosion Model

To calculate the coastal erosion and coastline recession in the PCR model, here we used the structural erosion function (model) proposed by Mendoza and Jimenez [30]. This structural erosion function permits the calculation of the magnitude of eroded volume and beach retreat, by means of an aggregated model which depends only on storm properties (Hs, Tp and duration) and beach morphology (sediment grain size and beach slope).

The structural function is derived by relating the storm-induced eroded volumes simulated with the SBEACH model with a coastal morphodynamic parameter. The selected parameter (JA) is a predictor which comprises the Dean parameter and the beach slope [33],

$$JA = |D_{0,e} - D_0|^{0.5} \times m \tag{2}$$

where D_0 is the Dean parameter ($H/T\,w_s$), $D_{0,e}$ corresponds to its value at equilibrium (2.5 when using deep water waves [34]), w_s is the fall velocity and m is the mean profile slope. This JA parameter was successfully used to predict the magnitude of eroded volumes in beach profile experiments obtained in large wave flumes and, therefore, it is considered a good predictor of storm-induced cross-shore beach profile changes. A linear model is given by Equation (3) [30],

$$\Delta V = C1.JA.dt + C2 \tag{3}$$

This approach can also be used to estimate the shoreline retreat during the storm, by relating the simulated shoreline retreat against corresponding values of the parameter $JA.dt$. To obtain a representative retreat for the subaerial beach, here we calculate a representative beach retreat ΔX_r as

$$\Delta X_r = \Delta V / (B + d^*) \tag{4}$$

where B is the berm height and d^* is the depth down to which erosion of the inner (landward) part of the beach profile occurs.

One of the main points to be considered when applying this approach is that the erosion structural function needs to be calibrated for the study site to properly fit the coefficients in Equation (3).

Due to the absence of pre- and post-storm erosion data at the study site, here we used the Transact (1D) version of the XBeach model [35,36] to calibrate the structural erosion function of Mendoza and Jimenez [30]. In this study, an XBeach 1D model was used in Surfbeat mode for each coastal cell defined in Section 3.1 and used with the average profile and average D50 of that cell in the model. In XBeach simulations, the Soulsby-Van Rijn sediment transport formulation [37] and avalanching process were used. Following other studies (e.g., [30]), we carried out a range of simulations for different possible predictor (*JA.dt*) values for each cell. Using a linear fit between resulting ΔV and JAdt, the values for C1 and C2 were determined for each coastal cell. For the profiles in different coastal cells, different calibration factors were applied. Here it was assumed that, between storms, the beach would recover in a such way that, in the absence of climate change, the coastline would remain in place with an exceedance probability of 50%, and after recovery, the profile would reshape to its original shape, and therefore the beach slope (m) and beach height (B) would remain the same. D50 was taken from the available data, and based on XBeach simulations, the *d** term was set equal to a water depth of 1 m.

4.3. Definition of Reference Coast Line (RCL)

The value of coastal retreat is only useful in planning if it is benchmarked to a reference coastline. Since the results of PCR modelling are intended for subsequent use in economic risk assessments, all the calculated coastal recession estimates were converted to the distance of the coastline after storm erosion from a reference coastline (RCL). The RCL was defined as the line where the exceedance probability of the run-up, estimated using Stockdon's runup formulation [38], over the entire 1979–2009 period was 1%. Practically no development is present seaward of this line.

4.4. PCR Simulations

For every coastal cell along the study area, the methodology shown in Figure 2 was implemented to analyse the wave data from the corresponding observation point. PCR simulations were then carried out for every profile in the respective coastal cell (85 profiles in total). The process of constructing a time series of storms and gaps, and then estimating erosion extents from that time series (see Sections 4.2 and 4.3). The duration of the constructed time series was 96 years (2015–2110) and for each simulation, the most landward location of coastline in each calendar year was recorded.

This enables the construction of an empirical distribution of the maxima of coastline recession for every future year. The simulations needed to be repeated many times before stable results were obtained, especially at lower exceedance probabilities. In practice, for each profile, 100,000 simulations of 96 years were undertaken.

4.5. Results and Discussion

For illustration purposes, here the results of 4 different profiles (T18, T37, B4 and B43) are discussed in detail. Figure 16 shows the exceedance probability curves of coastal recession for these profiles in different years—2025, 2050, 2080 and 2110—and Table 2 gives the values of coastal recession for different exceedance probabilities in 2050 and 2110. For example, Table 2 shows that the probability that coastal recession exceeds 48 m in year 2050 in profile B-4 is 50%. Figure 17 presents the same results, but in a format that is more useful for the public and decision-makers; this shows the increase in the annual probability of coastal recession reaching a fixed location with defined distance from RCL (50 m and 70 m). These types of results for every profile along the study sites, for every meter from the RCL, are used as the input for the economical optimization model. Examples of maps showing the 1% and 50% exceedance probability coastal recession contours are shown in Figure 18.

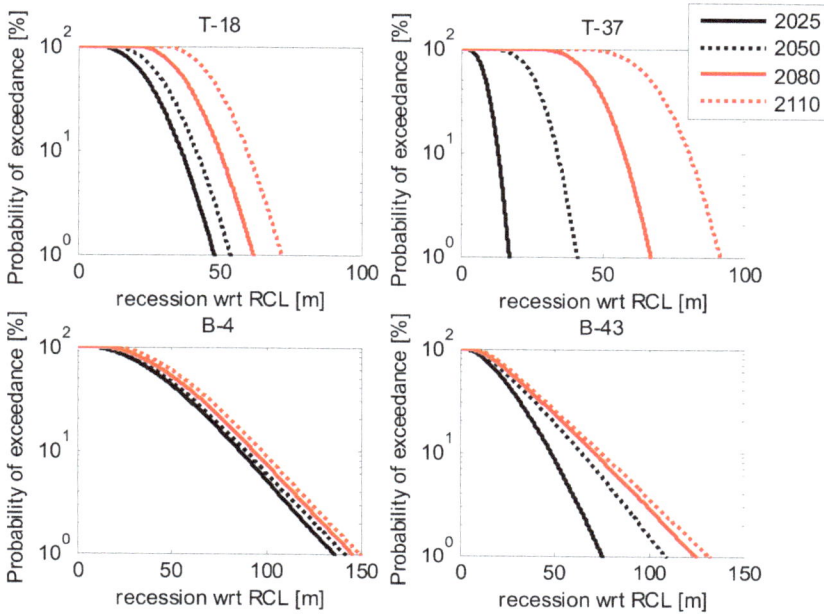

Figure 16. Exceedance probability curves of coastal recession for selected profiles in Trincomalee and Batticaloa districts for different years.

Table 2. Coastal recession magnitudes (m) associated with different exceedance probabilities in 2050 and 2110 for selected profiles.

Profile	2050			2110		
	Probability of Exceedance			Probability of Exceedance		
	1%	10%	50%	1%	10%	50%
B-4	142	89	48	149	98	56
B-43	109	64	28	132	75	32
T-18	54	42	29	72	60	47
T-37	41	34	26	91	80	67

Based on the similar results obtained for all of the 85 modelled profiles, maps of 1% and 50% exceedance probability coastline recession contours were computed for the Trincomalee and Batticaloa study areas. For Trincomalee, the 1% exceedance probability coastline recession in 2110 varies between 37 and 192 m, and for Batticaloa, between 69 and 262 m.

In this study, we used a JPM function between the wave height at the peak of the storm and the storm duration. There are other more complicated methods for the random generation of storms that link more parameters of the storm, e.g., wave height, peak period, total storm energy and wave direction, using multidimensional copulas and mixture Von Mises Fisher distributions. This leads to the generation of wave storm components with a common dependence structure (e.g., [39]). These methods can indeed be applied in the random storm generation of a PCR framework. Also, the choice of erosion model may have an effect on the results. However, in the absence of any storm erosion data, quantifying such effects is not feasible.

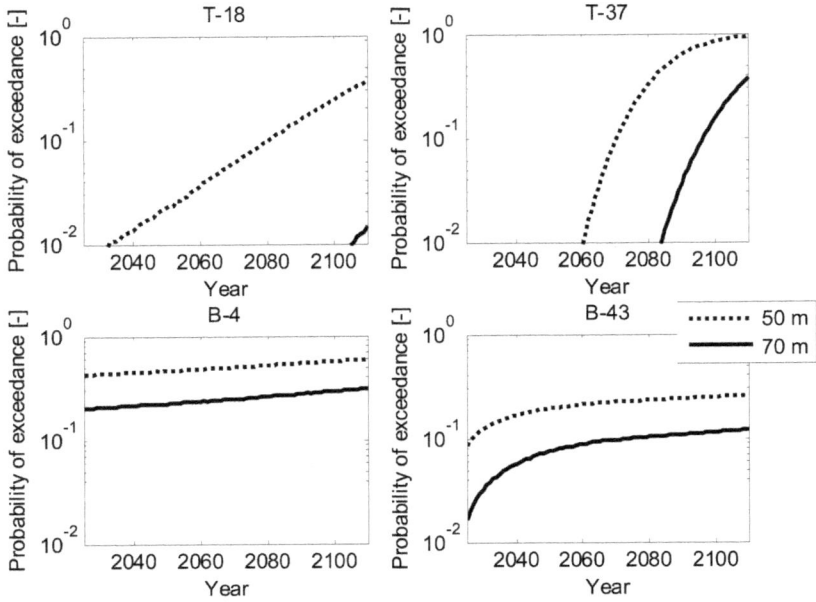

Figure 17. Increase in the annual probability of coastal recession reaching fixed location at 50 or 70 m distances from the RCL.

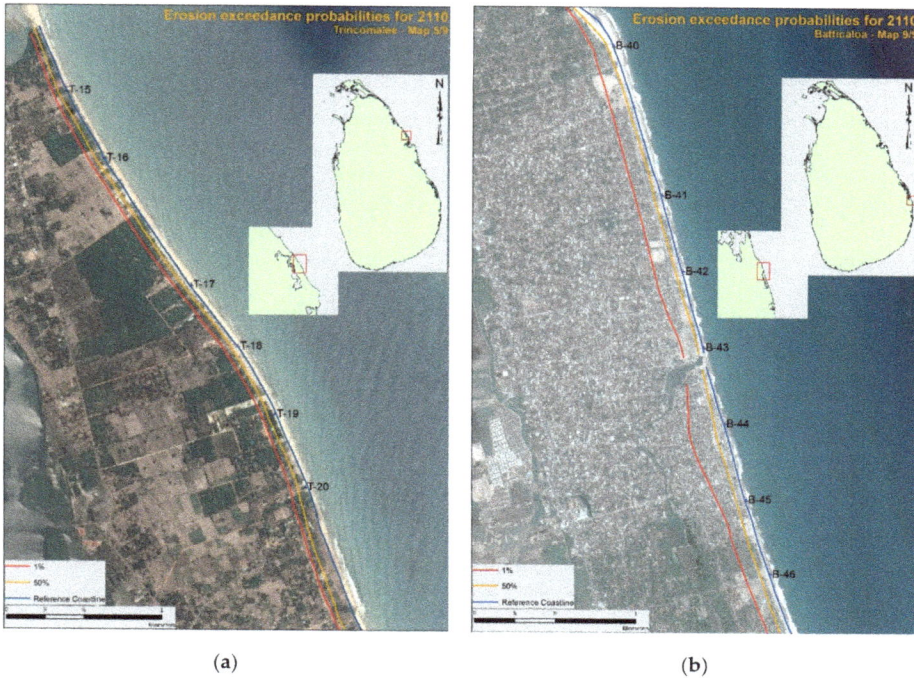

Figure 18. Examples of computed 1% (red) and 50% (orange) exceedance probability coastal recession contours in Trincomalee (**a**) and Batticaloa (**b**).

5. Economically Optimal Setback Lines

5.1. Economic Model

The risk of coastline erosion is mostly an economic issue. Forgoing land-use opportunities in the coastal zone is costly, but so is damage caused by erosion. This makes the establishment of setback lines a balancing act. Positioning a setback line farther inland reduces the economic risk, but increases the associated opportunity cost.

In a world without market imperfections, no individual would invest in a place where the risk of coastline recession would be too high from an economic perspective. In such a world, there would be little reason for governments to intervene and implement setback lines. In practice, however, incorrect risk perceptions [40], irrational behaviour [41], and/or explicit or implicit government guarantees [42,43] may lead to excessive risk-taking, i.e., over-investment in the coastal zone. Governments could stem such over-investment through financial incentives, such as the introduction of a compulsory insurance scheme. The cost of risk bearing would then be factored into the investment decisions of individuals and firms. Alternatively, or additionally, governments could establish setback lines as a means to avoid development close to the coastline. This is, indeed, what is widely done to prevent excessive risk-taking in coastal zones.

An estimate of the optimal position of a setback line (or, the economically optimal balance between risk and reward) can be obtained from the would-be behaviour of rational, well-informed, profit-seeking individuals in a world without market imperfections. Such an economically optimal setback line (EOSL) can serve as a reference for coastal zone managers/planners, who have to make on-the-ground decisions on implementing appropriate setback lines.

Hereafter, we assume the following:

1. Property owners value the risk of coastline recession at expected loss or a multiple thereof. In the presence of efficient insurance markets, the insurance premium would be equal to expected loss, i.e., the product of the probability of damage and potential loss. In that case, every risk-averse individual would purchase insurance. CCD has indicated that the Sri Lankan government is under no circumstances liable for damage caused by coastal erosion. Still, according to the CCD, insurance penetration is low. While this could be due to people's risk preferences, it could also be due to the unavailability of efficiently priced insurance coverage, mistrust or incorrect risk perceptions. Since people are typically risk-averse, here we value the economic risk as a multiple of expected loss.

2. The impact of coastline recession on a property is assumed to be proportional to the footprint of the property being impacted by coastline recession.

3. Each property that is impacted by coastline recession is damaged completely.

4. Damages are restored to their initial condition following a recession impact. This implies that insurance pay-outs cannot be put to alternative use.

5. Restoration takes place in the year in which damage occurs, in the period following the storm season.

The importance of the abovementioned assumptions can easily be investigated via sensitivity analysis, e.g., by increasing or reducing the cost of risk bearing or by varying the discount rate. The optimization procedure itself could also be elaborated to accommodate site-specific conditions and a wider range of decision alternatives, such as protecting sites or nourishing beaches, rather than establishing buffer zones/limiting development. The fact that alternative risk reduction measures (other than set back lines) have been ignored is conservative in the present context, since it implies that future risks may have been overestimated (in the event that protection works were to be implemented).

Under the assumptions above, the net present value $NPV(x)$ of investing at a distance x from today's coastline equals the sum of the investment cost, the present value of the returns on investment and the present value of the cost of risk bearing:

$$NPV(x) = -c(x) + \sum_{i=1}^{n} \frac{c(x) \times r(x)}{(1+\gamma)^i} - \sum_{i=1}^{n} \frac{a \times c(x) \times p_i(x)}{(1+\gamma)^i} \tag{5}$$

where,

$NPV(x)$—The net present value of an investment at a distance x from today's coastline.

γ—Discount rate.

$p_i(x)$—The probability of damage at distance x in year i. This is the probability in year i that the coastline recedes up to a point that is at least a distance x from the reference coastline.

$c(x)$—The investment that is made at a distance x from the initial coastline.

a—The ratio of the certainty equivalent to expected loss. The certainty equivalent is the certain loss that is valued the same as the probability of suffering a loss. For a risk-neutral agent, the certainty equivalent is equal to the expected loss. The factor a could thus be perceived as a risk aversion coefficient.

$r(x)$—Rate of return on the initial investment without accounting for coastline recession risk. The product $r(x) \cdot c(x)$ equals an annual return on investment measured in money (e.g., dollar) terms.

n—The time horizon being considered in years. The value of n equals the number of years in which the return on investment exceeds the certainty equivalent of the risk of coastline recession. This is consistent with the assumption that investors do not willingly incur avoidable losses, but walk away when risks become too high. For practical reasons, we have ignored all cash flows beyond the year 2110, which is the last year for which recession estimates are here computed from the probabilistic coastline recession model.

When the rate of return on investment does not decrease towards the coastline, the economically optimal exceedance probability of today's economically optimal setback line, i.e., $p_1(x)$, can be found at a position x, where the net present value $NPV(x)$, as defined by Equation (5), drops below zero. This exceedance probability follows from:

$$-c(x) + \sum_{i=1}^{n} \frac{c(x) \times r(x)}{(1+\gamma)^i} - \sum_{i=1}^{n} \frac{a \times c(x) \times p_i(x)}{(1+\gamma)^i} = 0 \tag{6}$$

or (after dividing by $c(x)$):

$$-1 + \sum_{i=1}^{n} \frac{r(x)}{(1+\gamma)^i} - \sum_{i=1}^{n} \frac{a \times p_i(x)}{(1+\gamma)^i} = 0 \tag{7}$$

When the rate of return and the probability of damage are both the same from year to year, $n \to \infty$ and Equation (7) reduces to (note that $\lim_{n \to \infty} \sum_{i=1}^{n} \frac{1}{(1+\gamma)^i} = \frac{1}{\gamma}$):

$$p_i(x) = \frac{r(x) - \gamma}{a} \tag{8}$$

Equation (8) follows the intuitive logic that investors would not willingly expose themselves to the risk of coastline recession when the cost of risk bearing is greater than the reward. For example, if risk were valued at expected loss ($a = 1$) and an investment in a coastal zone were to yield a 0.05 annual rate of return compared to 0.03 for an identical investment at an inland location, the efficient setback line would have an exceedance probability of $0.05 - 0.03 = 0.02$ per year. The probability of damage thus essentially acts as a hurdle rate.

The above is illustrated in real-life by fishermen that live as far from the coast as they can without losing sight of their boats. When closeness to the sea offers few benefits and mostly poses risks, it is rational to stay as far away as possible. The same logic explains why luxury hotels can be found along the seafront: closeness to the sea attracts tourists (beach amenity); for luxury hotels, closeness to the sea brings rewards that outweigh a relatively high risk of coastline recession.

When the risk of coastline recession is likely to increase over time due to, e.g., climate change, it becomes economically optimal to introduce a safety margin by lowering the exceedance probability of the setback line that is used for guiding today's land-use planning decisions.

The position of the economically optimal setback line for future investment decisions differs from the position of the EOSL for today's investment decisions. This is because of sea level rise and other trends. This implies that the position of the EOSL should be re-evaluated periodically. Here, we consider the position of the EOSL for the year 2025, taking into account the PCR-output for the period 2025–2110. The EOSL for the year 2025 is the optimal setback line for investments made in the year 2025. While this setback line is somewhat conservative for planning decisions prior to 2025, it provides a stable basis for land-use planning decisions over the next decade or so. The EOSLs thus determined range between 12 m and 175 m from the coastline. It is stressed here, however, that the EOSL position should be re-evaluated periodically (e.g., every 10–20 years).

5.2. Economic Constants

Future rates of return on investment, discount rates and risk preferences are uncertain, especially over periods of 50 years or more. These variables could be treated as stochastic, characterised by probability distributions. Since such a fully probabilistic treatment may trigger discussions about the probabilities assigned to alternative future economic developments, we established base case parameter values and investigated the sensitivity of the EOSL changes in these values. The base case parameter values are the values that were used for determining the EOSL for the different sites. Each site was classified as a high-value zone (e.g., urban area) or a lower-value zone (e.g., rural area). The base case parameters yield somewhat conservative results if all combinations of parameter values are considered equally likely, as discussed in Section 5.4. The parameter values that have been considered are shown in Table 3.

Table 3. The inputs of the sensitivity analyses related to the economic constants.

Variable	Trincomalee		Batticaloa	
	High-Value Zone	**Lower-Value Zone**	**High-Value Zone**	**Lower-Value Zone**
Rate of return on investment relative to the discount rate (Δr) (per year)	0.09 **0.12** (base case) 0.15	0.05 **0.07** (base case) 0.09	0.03 **0.05** (base case) 0.07	0.01 **0.02** (base case) 0.03
Discount rate (γ) (per year)	0.02 **0.03** (base case) 0.04 0.05			
Ratio of the certainty equivalent to expected loss (a) (-)	1.0 (risk-neutral) 1.5 (moderate degree of risk-aversion) **2.0** (base case)			

The parameter values shown in Table 3 are based on information provided by the CCD, historical records of the Sri Lanka core inflation rate and the prime lending rate reported by the Central Bank of Sri Lanka, and anticipated future economic developments. To avoid rates of return on investment that are smaller than the discount rate, the rate of return was defined as $r = \Delta r + \gamma$, with $\Delta r > 0$.

5.3. Results: The Position of the EOSL and Optimal Damage Probabilities

The calculated position of the 2025-EOSL relative to the reference coastline is shown Table 3 for the profiles in Trincomalee and Batticaloa, assuming base case parameter values. The reference coastline (RCL) is the coastline at the start of our simulations in the year 2016. The position of the EOSL varies considerably from profile to profile.

The development of the probability of damage over time at the EOSL is shown in Figure 18 (Trincomalee) and Figure 19 (Batticaloa) for different profiles by continuous lines. For all profiles, the probability of damage at the EOSL increases over time because of, among other things, the effect of sea level rise.

The increasing annual probabilities of damage limit the economic lifetime of a development in the coastal zone. After several years, retreat becomes optimal (note that we conservatively ignore protection as an alternative risk management strategy). This happens when the cost of risk bearing starts to exceed the return on investment, i.e., when $a \cdot p(x) \cdot c(x) \geq r \cdot c(x)$ or $p(x) \geq r/a$, see also Equation (5). For instance, at high-value locations in Batticaloa, the cost of risk bearing starts to exceed the return on investment when the probability of damage becomes greater than $0.08/2 = 0.04$ per year. At moderate/lower-value locations in Batticaloa, retreat becomes optimal when the probability of damage becomes greater than $0.05/2 = 0.025$ per year. In Trincomalee, these limiting probabilities are $0.15/2 = 0.075$ per year and $0.1/2 = 0.05$ per year, respectively. These optimal values have been indicated by dashed lines in Figures 18 and 19. Retreat becomes optimal when the probability of damage at the EOSL starts to exceed the probability of damage at which retreat becomes optimal, i.e., when the continuous and dashed lines cross.

As indicated by Figure 19, the optimal investment horizon for a development at or close to the EOSL is on the order of 20 years (starting from 2025) in Trincomalee. In Batticaloa, it is on the order of 50–80 years, see Figure 20. This difference arises from the relatively rapid changes in the probability distributions of coastline recession in Trincomalee, in combination with relatively high rates of return on investment. High rates of return of investment make short-term investments profitable. Maps of

EOSLs were thus determined for Trincomalee and Batticaloa study areas. Examples of these maps are displayed in Figure 21.

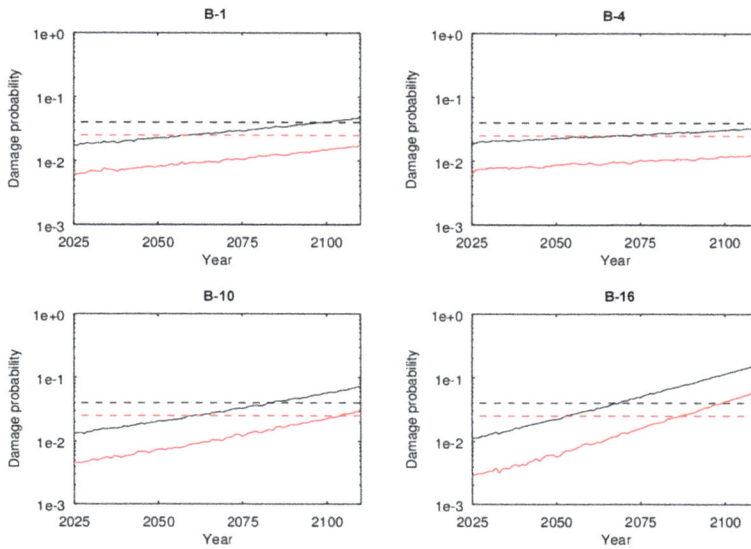

Figure 19. The annual probability of damage at the EOSL for selected profiles in Batticaloa (continuous lines) and the probability of damage above which retreat is optimal (dashed lines). The optimal investment horizon at the EOSL is found where these lines cross. Results for r = 8% per year in black; results for r = 5% per year in red.

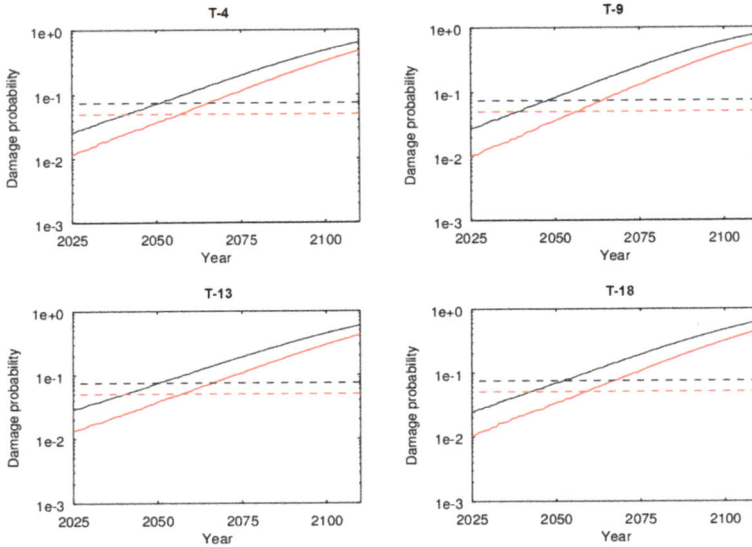

Figure 20. The annual probability of damage at the EOSL for selected profiles in Trincomalee (continuous lines) and the probability of damage above which retreat is optimal (dashed lines). The optimal investment horizon at the EOSL is found where these lines cross. Results for r = 15% per year in black; results for r = 10% per year in red.

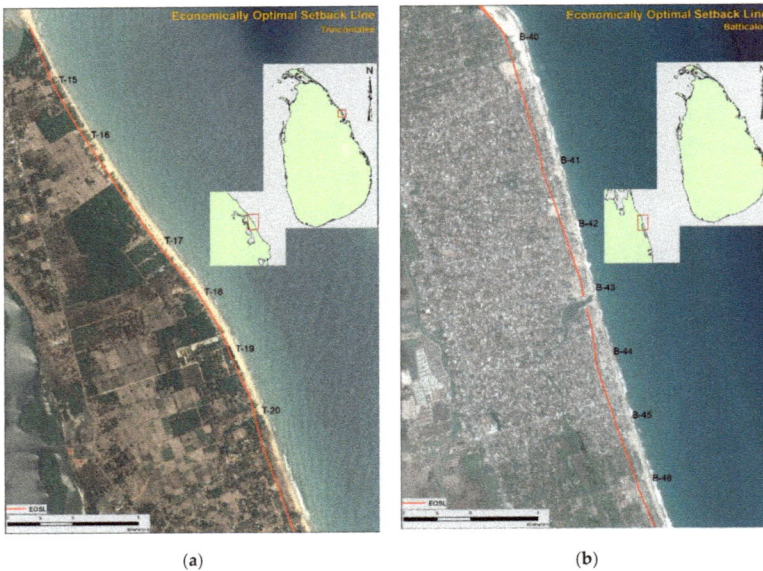

Figure 21. Examples of estimated economically optimal coastal setback lines in Trincomalee (**a**) and Batticaloa (**b**).

5.4. Sensitivity to Variations in Economic Constants

The results of the sensitivity analyses are shown in Figures 22 and 23 for Trincomalee and in Figures 24 and 25 for Batticaloa. In each figure, the first three panels show the effect of varying a single parameter on the position of the EOSL. The last panel of each figure shows the combined effect of varying the economic constants simultaneously. These figures illustrate the following:

1. Decreasing the ratio of the certainty equivalent to expected loss (a) gives the risk of coastline recession less weight and shifts the EOSL seaward. In the sensitivity analyses, we only considered values of a that were smaller than the base case value. This is why the plots only show differences of less than zero. A negative value on the horizontal axis implies the EOSL of the sensitivity analysis lies seaward of the EOSL based on the base case parameter values.

2. Increasing the rate of return on investment relative to the discount rate (Δr) increases the economically optimal probability of damage and shifts the EOSL seaward. Conversely, decreasing Δr reduces the economically optimal probability of damage and shifts the EOSL inland.

3. Increasing the discount rate (γ) while keeping Δr constant decreases the present value of climate change impacts. At the same time, it increases the present value of future returns on investment. This largely explains why variations in the discount rate influence the position of the EOSL to a lower extent than variations in the rate of return on investment.

In Trincomalee, the EOSLs from the sensitivity analyses lie within 5–10 m of the base case EOSL. In Batticaloa, most of the EOSLs are less than 15 m away from the base case EOSL for the single-parameter variations. This is remarkable, considering the relatively large variations that were considered. When all parameters are varied simultaneously, the variance increases. Still, most EOSLs still lie within 20 m of the base case EOSL. It should be noted that some combinations of parameter values imply a state of the world that differs considerably from the state of the world that corresponds to the base case conditions.

In Batticaloa, the EOSL lies, on average, two times farther from the reference coastline than in Trincomalee. Note also that the parameter variations considered in the sensitivity analyses for Trincomalee and Batticaloa are broadly similar in an absolute sense, but rather different in a relative sense. This largely explains the differences in the outcomes of the sensitivity analyses for both sites.

The economic constants could also be treated as stochastic. The distance between the EOSL and the reference coastline then becomes equal to the probability weighted sum of the distances for the various combinations of parameter values. If all combinations from Table 3 were considered equally likely, this EOSL would lie, on average, less than 2 m seaward of the base case EOSL in Trincomalee and about 5–10 m seaward of the base case EOSL in Batticaloa.

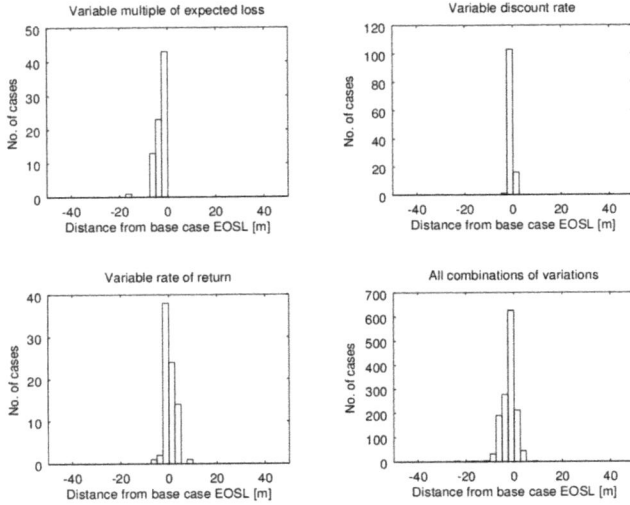

Figure 22. The distance between the EOSL based on the base case parameter values and the EOSL based on different economic constants. Results for profiles in Trincomalee, assuming they all lie in high-value zones. A negative distance implies a position seaward of the base case EOSL.

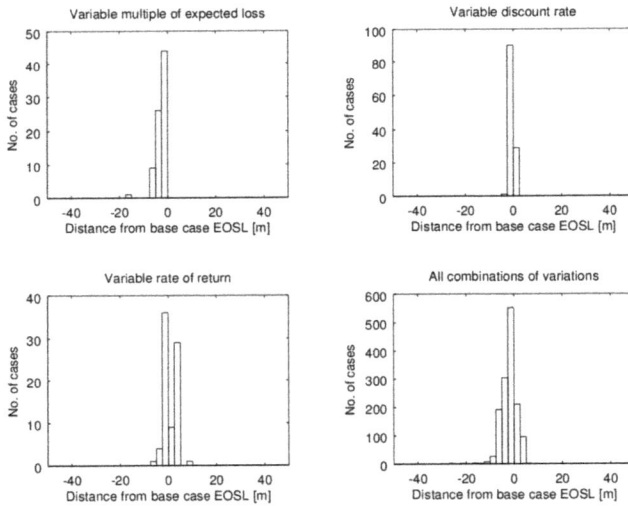

Figure 23. The distance between the EOSL based on the base case parameter values and the EOSL based on different economic constants. Results for profiles in Trincomalee, assuming they all lie in lower-value zones. A negative distance implies a position seaward of the base case EOSL.

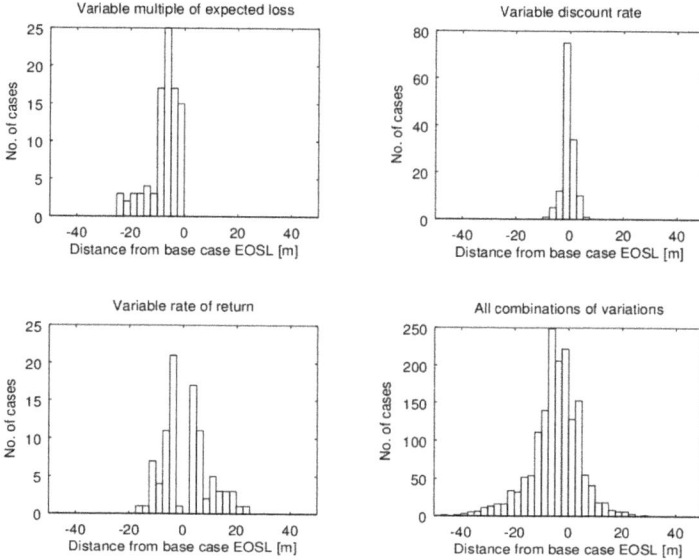

Figure 24. The distance between the EOSL based on the base case parameter values and the EOSL based on different economic constants. Results for profiles in Batticaloa, assuming they all lie in high-value zones. A negative distance implies a position seaward of the base case EOSL.

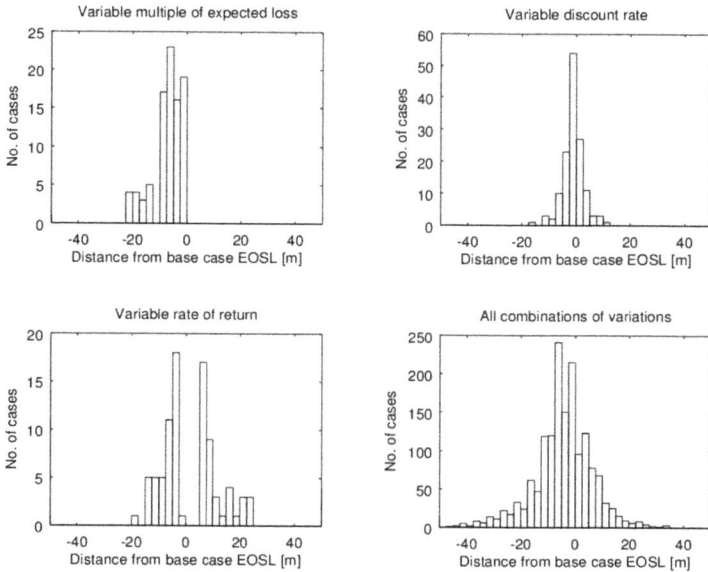

Figure 25. The distance between the EOSL based on the base case parameter values and the EOSL based on different economic constants. Results for profiles in Batticaloa, assuming they all lie in lower-value zones. A negative distance implies a position seaward of the base case EOSL.

6. Conclusions

The Probabilistic Coastline Recession (PCR) model was applied along a ~200 km stretch of the east coast of Sri Lanka, with the ultimate goal of developing economically optimal setback lines for the region, while taking into account the effect of projected sea level rise and storm-induced coastal recession for the period 2015–2110, together with economic considerations.

By combining the output of the PCR model with an economic model that balances risk and return, economically optimal setback lines (EOSL) were established for the study sites. The EOSL shows where the risk of coastline recession starts to outweigh the gains from property development. The EOSL was here established for the year 2025 to provide a stable basis for land-use planning decisions over the next two decades or so. It is advised to re-evaluate the position of the EOSL every 10–20 years, or sooner if observations or new insights give reason for reassessment.

The 1% exceedance probability coastline recession in 2110 varies between 37 and 262 m along the whole study area, and the respective EOSLs range between 12 m and 175 m from the coastline.

Sensitivity analyses show that strong variations in key economic parameters, such as the discount rate, have a disproportionately small impact on the distance between the EOSL and the reference coastline. This can be explained by the strong sensitivity of the exceedance probability of a coastline recession contour to its distance from the coastline. A considerable change in the exceedance probability of the EOSL therefore has relatively little impact on its position.

The optimal investment horizon for a development at or close to the EOSL is on the order of 20 years in Trincomalee and 50–80 years in Batticaloa (starting from 2025). This is because of the relatively high rates of return on investment in Trincomalee, as per estimates provided by the CCD. Developers may not realise that the optimal investment horizon at or close to the EOSL could be considerably shorter than, e.g., 50 years in Trincomalee, and as little as 20 years. As a remedy, the EOSL could be supplemented with a setback line that is specifically designed for decisions that involve a planning or investment horizon greater than 20 years.

Types of risk management actions other than land-use planning have not been considered in the EOSL calculations. This implies that the calculated positions of the EOSLs may be conservative, as future risks may have been overestimated. Protecting sites and/or re-nourishing beaches could be more efficient than retreating when risks increase. The economic model could be extended to include such adaptive pathways. The model could also be improved by modelling the consequences of coastline recession at both the individual and community level in greater detail. This would require seamless integration of the Probabilistic Coastal Recession (PCR) model and the economic model.

When considering model refinements, the availability and accuracy of the required input data should not be overlooked. A model's level of sophistication should be matched by the availability of reliable input data. Otherwise, model improvements merely change the balance between model and parameter uncertainties, without producing significantly greater accuracy.

Author Contributions: The paper and research project was designed by R.R. and M.W. was responsible for collecting and analysing the locally acquired physical and economic data. The coastal hazard identification, modelling and probabilistic analyses were carried out by A.D. and the economical optimization and sensitivity analyses were carried out by R.J. both under the supervision of R.R.

Funding: This study is mainly funded by ADB-IHE Delft knowledge partnership. The counterpart funding required for this project was accommodated through the research budget of the CC&CRM R&D section. These funds enabled the beach profile and bathymetric surveys required for the modelling work. CC&CRM R&D section also provided a substantial amount of in-kind support to this study, for which we are very grateful. The authors are also thankful for the financial support from the AXA Research Fund. RR is supported by the AXA Research fund and the Deltares Strategic Research Programme 'Coastal and Offshore Engineering'.

Acknowledgments: We would like to thank Yasmin Siddiqi, Ellen Pascua and the South Asia Regional Division of ADB (Manila), Palitha Bandara of ADB (Sri Lanka mission), the Coast Conservation and Coastal Resource Management Department of Sri Lanka (CC&CRM Dept.) for making this study a reality. Eng. Sakuntha Padmasiri and Eng. Damith Rupasinghe of CC&CRM (R&D section) are gratefully acknowledged for their continuous and prompt support throughout the two-year study period. A special "Thank you" to the supporting staff from

Trincomalee and Batticaloa districts, especially Eng. Thulsidasan, for the excellent services provided during the study period. Also, the authors want to thank Janaka Bamunawala for his help in preparing the final maps.

Conflicts of Interest: The authors declare no conflicts of interest.

References

1.   Coelho, C.; Silva, R.; Veloso-Gomes, F.; Taveira-Pinto, F. Potential effects of climate change on northwest Portuguese coastal zones. *ICES J. Mar. Sci.* **2009**, *66*, 1497–1507. [CrossRef]

2.   Casas-Prat, M.; Sierra, J.P. Trend analysis of wave direction and associated impacts on the Catalan coast. *Clim. Chang.* **2012**, *115*, 667–691. [CrossRef]

3.   Bonaldo, D.; Benetazzo, A.; Sclavo, M.; Carniel, S. Modelling wave-driven sediment transport in a changing climate: A case study for northern Adriatic Sea (Italy). *Reg. Environ. Chang.* **2015**, *15*, 45–55. [CrossRef]

4.   Ranasinghe, R. Assessing Climate change impacts on Coasts: A Review. *Earth Sci. Rev.* **2016**, *160*, 320–332. [CrossRef]

5.   Dastgheib, A.; Reyns, J.; Thammasittirong, S.; Weesakul, S.; Thatcher, M.; Ranasinghe, R. Variations in the wave climate and sediment transport due to climate change along the coast of Vietnam. *J. Mar. Sci. Eng.* **2016**, *4*, 86. [CrossRef]

6.   Bruun, P. Sea-level rise as a cause of shore erosion. *J. Waterw. Harb. Div.* **1962**, *88*, 117–132.

7.   Vrijling, J.K.; Van Hengel, W.; Houben, R.J. A framework for risk evaluation. *J. Hazard. Mater.* **1995**, *43*, 245–261. [CrossRef]

8.   Jongejan, R.B.; Ranasinghe, R.; Vrijling, J.K.; Callaghan, D.P. A risk-informed approach to coastal zone management. *Aust. J. Civ. Eng.* **2011**, *9*, 47–60. [CrossRef]

9.   Jongejan, R.; Ranasinghe, R.; Wainwright, D.; Callaghan, D.; Reyns, J. Drawing the line on coastline recession risk. *Ocean Coast. Manag.* **2016**, *122*, 87–94. [CrossRef]

10.   Wainwright, D.J.; Ranasinghe, R.; Callaghan, D.P.; Woodroffe, C.D.; Cowell, P.J.; Rogers, K. An argument for probabilistic coastal hazard assessment: Retrospective examination of practice in New South Wales, Australia. *Ocean Coast. Manag.* **2014**, *95*, 147–155. [CrossRef]

11.   Ranasinghe, R.; Callaghan, D.P.; Stive, M. Estimating coastal recession due to sea level rise: Beyond the Bruun rule. *Clim. Chang.* **2012**, *110*, 561–574. [CrossRef]

12.   Callaghan, D.P.; Nielsen, P.; Short, A.; Ranasinghe, R. Statistical simulation of wave climate and extreme beach erosion. *Coast. Eng.* **2008**, *55*, 375–390. [CrossRef]

13.   Callaghan, D.P.; Ranasinghe, R.; Short, A. Quantifying the storm erosion hazard for coastal planning. *Coast. Eng.* **2009**, *56*, 90–93. [CrossRef]

14.   Ministry of Mahaweli Development and Environment. Sri Lanka Coastal Zone and Coastal Resource Management Plan—2018. Government of Sri Lanka. 2018. Available online: http://www.coastal.gov.lk/images/stories/pdf_upload/acts_gazettes_czmp/czcrmp_2018_gazette_2072_58_e.pdf (accessed on 5 December 2018).

15.   Wong, P.-P.; Losada, I.J.; Gattuso, J.-P.; Hinkel, J.; Khattabi, A.; McInnes, K.L.; Saito, Y.; Sallenger, A. *Coastal Systems and Low-Lying Areas. Climate Change 2014: Impacts, Adaptation, and Vulnerability. Part A: Global and Sectoral Aspects*; Contribution of Working Group II to the Fifth Assessment Report of the Intergovernmental Panel on Climate Change; Cambridge University Press: Cambridge, UK; New York, NY, USA, 2014.

16.   Duong, T.M.; Ranasinghe, R.; Luijendijk, A.; Waltsra, D.J.R.; Roelvink, D. Assessing climate change impacts on the stability of small tidal inlets—Part 1: Data poor environments. *Mar. Geol.* **2017**, *390*, 331–346. [CrossRef]

17.   Duong, T.M.; Ranasinghe, R.; Thatcher, M.; Mahanama, S.; Zheng, B.W.; Dissanayake, P.K.; Hemer, M.; Luijendijk, A.; Bamunawala, J.; Roelvink, D. Assessing climate change impacts on the stability of small tidal inlets: Part 2-Data rich environments. *Mar. Geol.* **2018**, *395*, 65–81. [CrossRef] [PubMed]

18.   Berrisford, P.; Dee, D.; Poli, P.; Brugge, R.; Fielding, K.; Fuentes, M.; Kallberg, P.; Kobayashi, S.; Uppala, S.; Simmons, A. *The ERA-Interim archive Version 2.0*; ERA Report Series 1; ECMWF: Reading, UK, 2011; Volume 13177.

19.   Holthuijsen, L.; Booij, N.; Ris, R. A spectral wave model for the coastal zone. In Proceedings of the 2nd International Symposium on Ocean Wave Measurement and Analysis, New Orleans, LA, USA, 25–28 July 1993; pp. 630–641.

20. Ris, R.C. Spectral Modelling of Wind Waves in Coastal Areas. Communications on Hydraulic and Geotechnical Engineering, Report 97-4. Ph.D. Thesis, Delft University of Technology, Delft, The Netherlands, 1997.

21. Ris, R.; Booij, N.; Holthuijsen, L.A. Third-generation wave model for coastal regions, Part II: Verification. *J. Geophys. Res.* **1999**, *104*, 7649–7666. [CrossRef]

22. Battjes, J.; Janssen, J. Energy loss and set-up due to breaking of random waves. In Proceedings of the 16th International Conference Coastal Engineering, Hamburg, Germany, 27 August–3 September 1978; pp. 569–587.

23. Hasselmann, K.; Barnett, T.P.; Bouws, E.; Carlson, H.; Cartwright, D.E.; Enke, K.; Ewing, J.; Gienapp, H.; Hasselmann, D.E.; Kruseman, P.; et al. Measurements of wind wave growth and swell decay during the Joint North Sea Wave Project (JONSWAP). *Dtsch. Hydrogr. Z.* **1973**, *8*, 12.

24. BODC. The GEBCO Digital Atlas" Published by the British Oceanographic Data Centre on Behalf of IOC and IHO. 2003. Available online: http://www.gebco.net (accessed on 15 October 2018).

25. Bindoff, N.; Willebrand, J.; Artale, V.; Cazenave, A.; Gregory, J.; Gulev, S.; Nojiri, Y. Observations: Oceanic climate and sea level. In *Climate Change 2007: The Physical Science Basis*; Contribution of Working Group I to the Fourth Assessment Report of the Intergouvernmental Panel on Climate Change; Cambridge University Press: Cambridge, UK; New York, NY, USA, 2007; pp. 385–432.

26. Church, J.A.; Clark, P.U.; Cazenave, A.; Gregory, J.M.; Jevrejeva, S.; Levermann, A.; Nunn, P.D. *Sea Level Change*; PM Cambridge University Press: Cambridge, UK, 2013.

27. Christensen, J.H.; Hewitson, B.; Busuioc, A.; Chen, A.; Gao, X.; Held, R.; Jones, R.; Kolli, R.K.; Kwon, W.K.; Laprise, R.; et al. Regional Climate Projections. In *Climate Change 2007: The Physical Science Basis*; Contribution of Working Group I to the Fourth Assessment Report of the Intergovernmental Panel on Climate Change; Cambridge University Press: Cambridge, UK; New York, NY, USA, 2007.

28. Parry, M.; Parry, M.L.; Canziani, O.; Palutikof, J.; Van der Linden, P.; Hanson, C. *Coastal Systems and Low-Lying Areas. Climate Change 2007: Impacts, Adaptation and Vulnerability*; Contribution of Working Group II to the Fourth Assessment Report of the Intergovernmental Panel on Climate Change; Cambridge University Press: Cambridge, UK, 2007; pp. 315–356.

29. Nicholls, R.; Hanson, S.; Lowe, J.; Warrick, R.; Lu, X.; Long, A.; Carter, T. *Constructing Sea-Level Scenarios for Impact and Adaptation Assessment of Coastal Area: A Guidance Document*; Supporting Material, Intergovernmental Panel on Climate Change Task Group on Data and Scenario Support for Impact and Climate Analysis; TGICA: Geneva, Switzerland, 2011.

30. Mendoza, E.T.; Jimenez, J.A. Storm-Induced Beach Erosion Potential on the Catalonian Coast. *J. Coast. Res.* **2006**, *48*, 81–88.

31. Muis, S.; Verlaan, M.; Winsemius, H.C.; Aerts, J.C.; Ward, P.J. A global reanalysis of storm surges and extreme sea levels. *Nat. Commun.* **2016**, *7*, 11969. [CrossRef] [PubMed]

32. Vousdoukas, M.I.; Mentaschi, L.; Voukouvalas, E.; Verlaan, M.; Jevrejeva, S.; Jackson, L.P.; Feyen, L. Global probabilistic projections of extreme sea levels show intensification of coastal flood hazard. *Nat. Commun.* **2018**, *9*, 2360. [CrossRef] [PubMed]

33. Jiménez, J.A.; Sánchez Arcilla, A.; Stive, M.J.F. Discussion on prediction of storm/normal beach profiles. *J. Waterw. Port Coast. Ocean Eng.* **1993**, *19*, 466–468. [CrossRef]

34. Viavattene, C.; Jimenez, J.A.; Owen, D.; Priest, S.; Parker, D.; Micou, A.P.; Ly, S. Coastal Risk Assessment Framework Guidance Document. Deliverable No: D.2.3—Coastal Risk Assessment Framework Tool, Risc-Kit Project" (G.A. No. 603458). 2015. Available online: http://www.risckit.eu/np4/file/23/RISC_KIT_D2.3_CRAF_Guidance.pdf (accessed on 15 October 2018).

35. Roelvink, D.; Reniers, A.; van Dongeren, A.P.; de Vries, J.V.T.; McCall, R.; Lescinski, J. Modelling storm impacts on beaches, dunes and barrier islands. *Coast. Eng.* **2009**, *56*, 1133–1152. [CrossRef]

36. Roelvink, D.; McCall, R.; Mehvar, S.; Nederhoff, K.; Dastgheib, A. Improving predictions of swash dynamics in XBeach: The role of groupiness and incident-band runup. *Coast. Eng.* **2017**. [CrossRef]

37. Soulsby, R. *Dynamics of Marine Sands, a Manual for Practical Applications*; Thomas Telford: London, UK, 1997.

38. Stockdon, H.F.; Holman, R.A.; Howd, P.A.; Sallenger, A.H., Jr. Empirical parameterization of setup, swash and run-up. *Coast. Eng.* **2006**, *56*, 573–588. [CrossRef]

39. Lin-Ye, J.; Garcia-Leon, M.; Gracia, V.; Sanchez-Arcilla, A. A multivariate statistical model of extreme events: An application to the Catalan coast. *Coast. Eng.* **2016**, *117*, 138–156. [CrossRef]

40. Kunreuther, H.; Pauly, M. Neglecting Disaster: Why Don't People Insure Against Large Losses? *J. Risk Uncertain.* **2004**, *28*, 5–21. [CrossRef]

41. Slovic, P.; Fischhoff, B.; Lichtenstein, S.; Corrigan, B.; Combs, B. Preference for Insuring against Probable Small Losses: Insurance Implications. *J. Risk Insur.* **1977**, *44*, 237. [CrossRef]

42. Harrington, S.E. Rethinking Disaster Policy; Breaking the cycle of "free" disaster assistance, subsidized insurance, and risky behavior. *Regulation* **2000**, *23*, 40–46.

43. Mishan, E.J. The postwar literature on externalities: An interpretative essay. *J. Econ. Lit.* **1971**, *9*, 1–28.

MDPI

St. Alban-Anlage 66

4052 Basel

Switzerland

Tel. +41 61 683 77 34

Fax +41 61 302 89 18

www.mdpi.com

Journal of Marine Science and Engineering Editorial Office

E-mail: jmse@mdpi.com

www.mdpi.com/journal/jmse